Grace Wahba

ISNM 61:
International Series of Numerical Mathematics
Internationale Schriftenreihe zur Numerischen Mathematik
Série internationale d'Analyse numérique
Vol. 61

Edited by
Ch. Blanc, Lausanne; A. Ghizzetti, Roma;
R. Glowinski, Paris; G. Golub, Stanford;
P. Henrici, Zürich; H.O. Kreiss, Pasadena;
A. Ostrowski, Montagnola; J. Todd, Pasadena

Birkhäuser Verlag
Basel · Boston · Stuttgart

Multivariate Approximation Theory II

Proceedings of the Conference held at the Mathematical Research Institute at Oberwolfach, Black Forest, February 8–12, 1982

Edited by
Walter Schempp, Siegen
Karl Zeller, Tübingen

1982

Birkhäuser Verlag
Basel · Boston · Stuttgart

Editors' addresses:

Prof. Dr. Walter Schempp
Lehrstuhl für Mathematik I
Universität Siegen
Hölderlinstrasse 3
D-5900 Siegen (FRG)

Prof. Dr. Karl Zeller
Mathematisches Institut
Universität Tübingen
Auf der Morgenstelle 10
D-7400 Tübingen (FRG)

Library of Congress Cataloging in Publication Data

Main entry under title:
Multivariate approximation theory II.
(International series of numerical mathematics ; v. 61)
English and German.
Proceedings of the Third International Symposium on Multivariate Approximation Theory.
Includes bibliographies.
1. Approximation theory--Congresses. 2. Functions of several real variables--Congresses. 3. Sard, Arthur. I. Schempp, W. (Walter), 1938– . II. Zeller, Karl. III. International Symposium on Multivariate Approximation Theory (3rd : 1982 : Oberwolfach Mathematical Research Institute) IV. Title: Multivariate approximation theory 2. V. Title: Multivariate approximation theory two. VI. Series.
QA297.5.M845 1982 511'.4 82-14744
ISBN 3-7643-1373-0

CIP-Kurztitelaufnahme der Deutschen Bibliothek

Multivariate approximation theory : proceedings of the conference held at the Math. Research Inst. at Oberwolfach, Black Forest. - Basel ; Boston ; Stuttgart : Birkhäuser
2. February 8–12, 1982. - 1982.
(International series of numerical mathematics ; Vol. 61)
ISBN 3-7643-1373-0
NE: Mathematisches Forschungsinstitut ‹Oberwolfach›; GT

Printed in Switzerland by Birkhäuser AG, Graphisches Unternehmen, Basel
ISBN 3-7643-1373-0

In Memory of

ARTHUR SARD

Born on July 28, 1909 in New York City

Died on August 31, 1980 in Basel

PREFACE

The Third International Symposium on Multivariate Approximation Theory was held at the Oberwolfach Mathematical Research Institute, Black Forest, February 8-12, 1982. The preceding conferences on this topic were held in 1976* and 1979**.

The conference brought together 50 mathematicians from 14 countries. These Proceedings form a record of most of the papers presented at the Symposium.

The topics treated cover different problems on multivariate approximation theory such as new results concerning approximation by polynomials in Sobolev spaces, biorthogonal systems and orthogonal series of functions in several variables, multivariate spline functions, group theoretic and functional analytic methods, positive linear operators, error estimates for approximation procedures and cubature formulae, Boolean methods in multivariate interpolation and the numerical application of summation procedures. Special emphasis was posed on the application of multivariate approximation in various fields of science.

One mathematician was sorely missed at the Symposium. Professor Arthur Sard who had actively taken part in the earlier conferences passed away in August of 1980. Since he was a friend of many of the participants, the editors wish to dedicate these Proceedings to the memory of this distinguished mathematician. A brief appreciation of his life and mathematical work appears as well

---

*"Constructive Theory of Functions of Several Variables". Edited by W. Schempp and Karl Zeller. Lecture Notes in Mathematics, Vol. 571. Berlin-Heidelberg-New York: Springer 1977

**"Multivariate Approximation Theory". Edited by W. Schempp and Karl Zeller. ISNM, Vol. 51. Basel-Boston-Stuttgart: Birkhäuser 1979

as his last paper which includes his personal view of mathematics.

The editors warm thanks are due to the participants of the Symposium and to the contributors who helped to make the conference a success; to Privat-Dozent Dr. F.J. Delvos (Siegen) for his support during the preparation of the Symposium and for his valuable advice; to Dipl.-Math. G. Baszenski (Bochum), Dr. H.-M. Hebsaker (Siegen) and Prof. Dr. H. Posdorf (Dortmund) for their organizational work, to the staff of the Oberwolfach Mathematical Research Institute for the facilities provided, to Mrs. Valerie Meyer (Binningen) for her gracious assistance and to Carl Einsele of Birkhäuser Publishers for his valuable cooperation over the years.

May 1982

Walter Schempp — Siegen

Karl Zeller — Tübingen

Editors

CONTENTS

Zur Tagung.. 13
List of Participants.. 15
Program of the Sessions..................................... 19
Arthur Sard - In Memoriam................................... 23

J. ANGELOS, D. SCHMIDT: Some remarks on product approximations.............................. 27

M.F. BARNSLEY, J.S. GERONIMO, A.N. HARRINGTON, L.D. DAGER: Approximation theory on a snowflake.................... 37

M.F. BARNSLEY, W. RADDATZ: Estimates and bounds for the doubly perturbed eigenvalue problem.................... 43

G. BASZENSKI, F.J. DELVOS, K. HACKENBERG: Remarks on reduced Hermite interpolation.......................... 47

H. BRASS: Ein Beispiel zur Theorie der besten Approximation 59

C.K. CHUI, L.L. SCHUMAKER: On spaces of piecewise polynomials with boundary conditions. I. Rectangles.... 69

W. DAHMEN, C.A. MICCHELLI: Some remarks on multivariate B-splines.. 81

F.J. DELVOS: On discrete trivariate blending interpolation.. 89

F.J. DELVOS, W. SCHEMPP: On precision sets of interpolation projectors.................................. 107

F. DEUTSCH: Which closed convex subsets of an inner product space are Chebyshev? 125

M. EIERMANN, W. NIETHAMMER: Interpolation methods for numerical analytic continuation 131

H. ENGELS, A. MERSCHEN: Blending-Splines auf Dreiecksnetzen 143

W. FREEDEN, R. REUTER: Remainder terms in numerical integration formulas of the sphere 151

M. GASCA, A. LÓPEZ-CARMONA, V. RAMÍREZ: A generalized Sylvester's identity on determinants and its applications to interpolation problems 171

W. HAUSSMANN, E. LUIK, K. ZELLER: Biorthogonality in approximation 185

W. HAUSSMANN, E. LUIK, K. ZELLER: Cubature remainder and biorthogonal systems 191

M.S. HENRY: Product approximation: Error estimates 201

K. JÓNASSON, G.A. WATSON: A Lagrangian method for multivariate continuous Chebyshev approximation problems 211

A. LE MEHAUTE: Construction of surfaces of class $\mathcal{C}^k$ on a domain $\Omega \subset \mathbb{R}^2$, after triangulation 223

J.C. MASON: Minimal projections and near-best approximations by multivariate polynomial expansion and interpolation 241

J. MEINGUET: Sharp "a priori" error bounds for polynomial approximation in Sobolev spaces 255

H.M. MÖLLER: An immediate construction of numerical integration and differentiation formulae.............. 275

F. MÓRICZ: On the approximation by multiple orthogonal series.. 285

T. NISHISHIRAHO: Quantitative theorems on approximation processes of positive linear operators................ 297

K. SALKAUSKAS: Some relationships between surface splines and Kriging.. 313

A. SARD: A view of mathematics............................ 327

W. SCHEMPP: Drei statt einer reellen Variablen?............ 331

R. SCHMIDT: Eine Methode zur Konstruktion von C^1-Flächen zur Interpolation unregelmässig verteilter Daten....... 343

H.S. SHAPIRO: Approximation theory and "domain of dependence" for P.D.E. of hyperbolic type.......................... 363

B. SHEKHTMAN: Properties of spline projections.............. 375

SHEN XIE-CHANG: A survey of recent results on approximation theory in China..................................... 385

G. WAHBA: Vector splines on the sphere, with application to the estimation of vorticity and divergence from discrete, noisy data.................................... 407

ZUR TAGUNG

Vom 8. bis 12. Februar 1982 fand im Mathematischen Forschungsinstitut Oberwolfach eine Tagung über "Mehrdimensionale konstruktive Funktionentheorie" statt. Sie wurde, wie schon die Oberwolfach-Tagungen gleichen Themas der Jahre 1976 und 1979, von den Herausgebern geleitet. Es nahmen insgesamt 50 Mathematiker an der Tagung teil, die aus Belgien, England, Frankreich, Japan, Kanada, den Niederlanden, Österreich, Schottland, Schweden, Spanien, Ungarn, den Vereinigten Staaten von Amerika, der Volksrepublik China und der Bundesrepublik Deutschland kamen.

Das Vortragsprogramm bestand aus 33 Vorträgen, die sich mit Fragen der Darstellung, Approximation und Behandlung reeller Funktionen mehrerer Variablen befaßten. Da Problemstellungen dieser Art zunehmend an theoretischer und praktischer Bedeutung gewonnen haben, wurde bei der Planung großer Wert auf ein ausgewogenes Verhältnis zwischen Themen aus der Theorie der multivariaten Funktionen und ihren numerischen und praktischen Anwendungen gelegt.

Im Bereich der multivariaten Approximationstheorie wurden u.a. H - Mengen, die Polynomapproximation in Sobolev-Räumen, Biorthogonalsysteme und Orthogonalreihen in mehreren Variablen, mehrdimensionale Spline-Funktionen, gruppentheoretische und funktionalanalytische Methoden, rationale Approximation in mehreren Variablen, positive lineare Operatoren und funktionalanalytische Aspekte der Radon-Transformation diskutiert. Schwerpunkte der numerischen Anwendungen waren vor allem neue Fehlerabschätzungen für Approximation und Kubatur, Boolesche Methoden in der mehrdimensionalen Interpolation und der Einsatz von Summationsverfahren. Die praktischen Anwendungen waren außerordentlich breit gestreut. Zu nennen sind hier vor allem Anwendungen in der Geodäsie, Geologie, Limnologie, Meteorologie, Radarortung und medizinischen Tomographie.

Die Tagung verlief in einer überaus freundschaftlichen Atmosphäre, zu der die Gastfreundschaft und zuvorkommende Hilfe der Mitarbeiter des Oberwolfacher Instituts wesentlich beigetragen haben. Ihnen sei an dieser Stelle sehr herzlich gedankt, ebenso wie dem Direktor des Instituts, Herrn Professor Dr. M. Barner, den Vortragenden und den Sitzungsleitern.

Walter Schempp Karl Zeller
Siegen Tübingen

Tagungsleiter

LIST OF PARTICIPANTS

Michael F. Barnsley, School of Mathematics, Georgia Institute of Technology, Atlanta, Georgia 30332, U.S.A.

Günter Baszenski, Rechenzentrum der Ruhr-Universität Bochum, Universitätsstraße 150-NA, D-4630 Bochum 1, Fed. Rep. Germany

Jan Boman, Matematiska Institutionen, Stockholms Universitet, Box 6701, S-11385 Stockholm, Sweden

Helmut Brakhage, Fachbereich Mathematik der Universität Kaiserslautern, Erwin-Schrödinger-Straße, D-6750 Kaiserslautern, Fed. Rep. Germany

Helmut Braß, Lehrstuhl E für Mathematik der Technischen Universität Braunschweig, Pockelsstraße 14 (Forum), D-3300 Braunschweig, Fed. Rep. Germany

Wolfgang Dahmen, Fakultät für Mathematik der Universität Bielefeld, Universitätstraße, D-4800 Bielefeld, Fed. Rep. Germany

Phillipe Defert, Département de Mathématique, Facultés Universitaires de Namur, Rempart de la Vierge, 8, B-5000 Namur, Belgium

Franz Jürgen Delvos, Lehrstuhl für Mathematik I der Universität Siegen, Hölderlinstraße 3, D-5900 Siegen, Fed. Rep. Germany

Frank Deutsch, Department of Mathematics, The Pennsylvania State University, 215 McAllister Building, University Park, Pennsylvania 16802, U.S.A.

Jean Duchon, Mathématique Appliquée, IMAG, B.P. 53X, F-38041 Grenoble Cedex, France

Hermann Engels, Institut für Geometrie und Praktische Mathematik der Rhein.- Westf. Technischen Hochschule Aachen, Abteilung für Numerische Mathematik, Templergraben 55, D-5100 Aachen, Fed. Rep. Germany

Hans G. Feichtinger, Institut für Mathematik der Universität Wien, Strudlhofgasse 4, A-1090 Wien, Austria

Willi Freeden, Institut für Reine und Angewandte Mathematik der Rhein.- Westf. Technischen Hochschule Aachen, Templergraben 55, D-5100 Aachen, Fed. Rep. Germany

Mariano Gasca, Departamento de Ecuaciones Funcionales, Facultad de Ciencias, Universidad de Granada, Granada, Spain

Manfred v. Golitschek, Institut für Angewandte Mathematik und Statistik der Universität Würzburg, Am Hubland, D-8700 Würzburg, Fed. Rep. Germany

Günter Hämmerlin, Mathematisches Institut der Universität München, Theresienstraße 39, D-8000 München 2, Fed. Rep. Germany

David C. Handscomb, Oxford University Computing Laboratory, 19 Parks Road, Oxford OX1 3P2, England

Hans-Martin Hebsaker, Lehrstuhl für Mathematik I der Universität Siegen, Hölderlinstraße 3, D-5900 Siegen, Fed. Rep. Germany

Myron S. Henry, Department of Mathematics, Central Michigan University, Mount Pleasant, Michigan 48859, U.S.A.

Alain le Mehauté, Laboratoire d'Analyse Numérique, Institut National des Sciences Appliquées, 20 avenue des Buttes de Coësmes, F-35043 Rennes Cedex, France

John C. Mason, Department of Mathematics and Ballistics, The Royal Military College of Science, Shrivenham, Swindon, Wilts, SN6 8LA, England

Günter Meinardus, Lehreinheit Mathematik IV der Universität Mannheim, Seminargebäude A5, B 123, D-6800 Mannheim 1, Fed. Rep. Germany

Jean Meinguet, Institut de Mathématique Pure et Appliquée, Université Catholique de Louvain, Chemin du Cyclotron 2, B-1348 Louvain-la-Neuve, Belgium

Charles A. Micchelli, IBM Thomas J. Watson Research Center, P.O. Box 218, Yorktown Heights, New York 10598, U.S.A.

Hans Michael Möller, Fachbereich Mathematik und Informatik der Fernuniversität Hagen, Postfach 940, D-5800 Hagen 1, Fed. Rep. Germany

Ferenc Móricz, Bolyai Institute, University of Szeged, 6720 Szeged, Aradi vértanúk tere 1, Hungary

Manfred W. Müller, Lehrstuhl Mathematik VIII der Universität Dortmund, Postfach 500500, D-4600 Dortmund 50, Fed. Rep. Germany

Gregory M. Nielson, Department of Mathematics, Arizona State University, Tempe, Arizona 85287, U.S.A.

Wilhelm Niethammer, Institut für Praktische Mathematik der Universität Karlsruhe, Englerstraße 2, D-7500 Karlsruhe 1, Fed. Rep. Germany

Toshihiko Nishishiraho, Department of Mathematics, Ryukyu University, Nishihara-Cho, Okinawa 903-01, Japan

Horst Posdorf, Fachhochschule Dortmund, Sonnenstraße 96, D-4600 Dortmund, Fed. Rep. Germany

Peter Pottinger, Erziehungswissenschaftliche Hochschule Rheinland-Pfalz, Abteilung Koblenz, D-5400 Koblenz 1, Fed. Rep. Germany

Manfred Reimer, Lehrstuhl Mathematik III der Universität Dortmund, Postfach 500500, D-4600 Dortmund 50, Fed. Rep. Germany

Dennis C. Russell, Department of Mathematics, York University, Downsview (Toronto), Ontario M3J 1P3, Canada

Kestutis Salkauskas, Department of Mathematics and Statistics, The University of Calgary, 2500 University Drive N.W., Calgary, Alberta T2N 1N4, Canada

Walter Schempp, Lehrstuhl für Mathematik I der Universität Siegen, Hölderlinstraße 3, D-5900 Siegen, Fed. Rep. Germany

Rudolf Scherer, Institut für Praktische Mathematik der Universität Karlsruhe, Englerstraße 2, D-7500 Karlsruhe 1, Fed. Rep. Germany

Hans Joachim Schmid, Mathematisches Institut der Universität Erlangen-Nürnberg, Bismarckstraße 1 1/2, D-8520 Erlangen, Fed. Rep. Germany

Darrell Schmidt, Department of Mathematical Sciences, Oakland University, Rochester, Michigan 48063, U.S.A.

Rita Schmidt, Bereich Datenverarbeitung und Elektronik, Hahn-Meitner-Institut für Kernforschung GmbH, Glienicker Straße 100, D-1000 Berlin 39, Fed. Rep. Germany

Larry L. Schumaker, Center for Approximation Theory, Department of Mathematics, Texas A & M University, College Station, Texas 77843-3368, U.S.A.

Harold S. Shapiro, Department of Mathematics, The Royal Institute of Technology, S-10044 Stockholm 70, Sweden

Boris Shekhtman, Department of Mathematics, University of Southern California, Los Angeles, California 90007, U.S.A.

Shen Xie-chang, Department of Mathematics, Peking University, Beijing, China

Abraham van der Sluis, Mathematisch Instituut, Budapestlaan 6, 3584 CD Utrecht-Uithof, The Netherlands

Jean-Pierre Thiran, Département de Mathématique, Facultés Universitaires de Namur, Rempart de la Vierge, 8, B-5000 Namur, Belgium

Hans-Joachim Töpfer, Institut für Mathematik III der Freien Universität Berlin, Arnim-Allee 2-6, D-1000 Berlin 33, Fed. Rep. Germany

Grace Wahba, Department of Statistics, University of Wisconsin-Madison, 1210 West Dayton Street, Madison, Wisconsin 53706, U.S.A.

G. Alistair Watson, Department of Mathematics, University of Dundee, Dundee DD1 4HN, Scotland, U.K.

Karl Zeller, Mathematisches Institut der Universität Tübingen, Auf der Morgenstelle 10, D-7400 Tübingen 1, Fed. Rep. Germany

PROGRAM OF THE SESSIONS

Monday, February 8

9.00	K. Zeller:	Words of welcome
First morning session.		Chairman: K. Zeller
9.05	G.A. Watson:	A Lagrangian method for multivariate continuous Chebyshev approximation problems
9.40	J.P. Thiran:	Minimal H-sets for multivariate approximation
10.15	M. v. Golitschek:	Approximation of bivariate functions by functions of one variable
Second morning session.		Chairman: M. Reimer
11.10	H.M. Möller:	Eine einfache Methode zur Konstruktion numerischer Differentiations- und Integrationsformeln
11.45	F. Móricz:	The strong approximation by multiple orthogonal series
First afternoon session.		Chairman: G. Meinardus
15.30	K. Zeller:	BOGS procedures in approximation; BOGS remainder in cubature
16.15	W. Dahmen:	Entire functions of affine lineage
Second afternoon session.		Chairman: D.C. Russell
17.10	J.C. Mason:	Near-best approximation in two dimensions by polynomial interpolation and expansion methods
17.45	F. Deutsch:	Which closed convex sets in an incomplete inner product space are Chebyshev?

Tuesday, February 9

First morning session.		Chairman: H.-J. Töpfer
9.00	L.L. Schumaker:	Spaces of piecewise polynomials in two variables
9.35	R.M. Schmidt:	Flächeninterpolation bei unregelmäßig verteilten Daten
Second morning session.		Chairman: W. Schempp
11.10	W. Niethammer	Interpolationsverfahren zur numerischen analytischen Fortsetzung
11.45	S.Xie-Chang:	The recent progress in approximation theory - Part I
First afternoon session.		Chairman: H. Brass
15.30	Ph. Defert:	Approximation by first degree multivariate polynomials
16.05	K. Salkauskas:	The relationship between surface splines of Duchon and Meinguet, the Kriging method of Matheron, and the Backus-Gilbert theory
Second afternoon session.		Chairman: H. Engels
17.00	G. Baszenski:	Bemerkungen zur reduzierten Hermite-Interpolation
17.35	F.J. Delvos:	Remainders in Boolean interpolation

Wednesday, February 10

First morning session.		Chairman: H.S. Shapiro
9.00	J. Meinguet:	Sharp "a priori" error bounds for polynomial approximation in Sobolev spaces

9.35	M.S. Henry:	Multivariate approximation theory: Theoretical error estimates and calculation
10.10	D. Schmidt:	Lipschitz conditions and strong uniqueness for metric projections for almost Chebyshev subspaces of C(X)

Second morning session. Chairman: L.L. Schumaker

11.10	G. Wahba:	Smoothing splines on the sphere with applications in meteorology
11.45	W. Freeden:	Integral formulas of the (unit) sphere and their applications

Thursday, February 11

First morning session. Chairman: M.W. Müller

9.00	W. Schempp:	Drei statt einer reellen Variablen?
9.50	A. le Mehauté:	Constructions of surfaces of class C^k on a domain $\Omega \subset \mathbb{R}^2$, after triangulation

Second morning session. Chairman: M. v. Golitschek

10.45	T. Nishishiraho:	Quantitative theorems on approximation processes of positive linear operators
11.30	M. Gasca:	A generalized Sylvester's identity on determinants and its applications to interpolation problems

First afternoon session. Chairwoman: G. Wahba

15.30	B. Shekhtman:	Some properties of spline projections
16.05	M.F. Barnsley:	Orthogonal polynomials on Julia sets

Second afternoon session.		Chairman: F. Deutsch
17.00	H. Brass:	Ein Beispiel zur Theorie der besten Approximation
17.35	S.Xie-Chang:	The recent progress in approximation theory - Part II

Friday, February 12

First morning session.		Chairman: W. Schempp
9.00	H.S. Shapiro:	When is a vector sum of closed subspaces closed?
9.35	J. Boman:	On the range of the Radon transform and the closure of sums of plane waves
Second morning session.		Chairman: W. Schempp
10.10	A. van der Sluis:	Some remarks on cubature

ARTHUR SARD - IN MEMORIAM

F.J. Delvos and Walter Schempp

Lehrstuhl für Mathematik I der Universität Siegen, Siegen

Professor Arthur Sard was born on July 28, 1909 in New York City, a son of Frederick N. and Maria Belloch Sard. He grew up in New York and spent most of his adult life there.

He graduated in Friends Seminary at New York City and received his B.A. summa cum laude in 1931, his M.A. in 1932, and his Ph.D. in 1936, all from Harvard University.

Dr. Sard was among the first faculty chosen to start Queens College in Flushing, New York, in the fall of 1937 where he taught 1937 until 1970. During the war he was a leading member of the Applied Mathematics Group at Columbia. He retired in 1970 as Professor Emeritus at Queens College and moved to La Jolla to spend five years as a Research Associate in the Mathematics Department of the University of California, San Diego.

In 1975 Professor Sard moved to Binningen (Switzerland) and lectured at a number of European Universities and Mathematical Research Centers. In 1978-1979 he was Visiting Professor at the University of Siegen. In 1978, the Soviet Union's Academy of Sciences invited him to speak as a honored guest. He passed away on August 31, 1980 in Basel (Switzerland).

Professor Arthur Sard was a distinguished mathematician, known internationally for his work in the areas of differential topology and spline approximation. He is noted for

Sard's theorem (Bull. Amer. Math. Soc. 48, 883-890 (1942)) - that the set of critical values of a suitably smooth function has measure zero. Moreover, he was one of the co-inventors of the theory of spline approximation. In this area his starting point was the problem of constructing best quadrature formulae which minimize the norms of certain remainder functionals (Peano-Sard kernels). His name has also become attached to the functional analytic theory of splines and its relations to the probabilistic approach to the calculus of observations. In all, he published 38 research papers and, while at Queens, he also published two monographs - Linear Approximation in 1963 and, in 1971, Book of Splines, jointly with Professor Sol Weintraub.

Priv.-Doz. Dr. F.J. Delvos
Prof. Dr. Walter Schempp
Lehrstuhl für Mathematik I
Universität Siegen
Hölderlinstraße 3
D-5900 Siegen, W. Germany

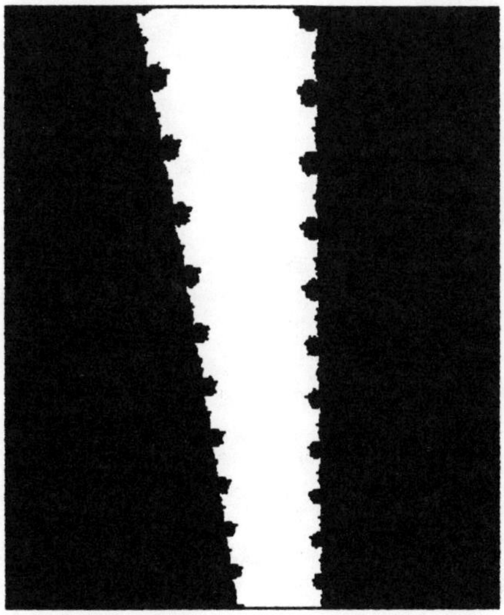

SOME REMARKS ON PRODUCT APPROXIMATIONS

James Angelos and Darrell Schmidt

Montana State University, Department of Mathematics, Bozeman, Montana, USA and Oakland University, Department of Mathematical Sciences, Rochester, Michigan, USA

1. Introduction

Recently there has been considerable interest in various aspects, extensions, and variations of uniform product approximations (see [2,3,6,7] and the references of [3]). This paper concerns convergence and error bounds for polynomial product approximation and a variation similar to appolation considered in [6].

Let $D = I \times J = [a,b] \times [c,d]$ and $F \in C(D)$ where $C(D)$ denotes the space of continuous real-valued functions on D with the uniform norm $\|\cdot\|_D$. For $y \in J$, define $F_y \in C(I)$ by $F_y(x) = F(x,y)$ and let

$$(B_n F_y)(x) = \Sigma_{i=0}^{n} f_i(y)\varphi_i(x) \qquad (1.1)$$

be the best approximation to F_y from the space Π_n of polynomials of degree $\leq n$ with respect to the uniform norm $\|\cdot\|_I$

on $C(D)$. Here $\{\varphi_0,\ldots,\varphi_n\}$ is a basis for Π_n . The coefficient functions f_i are continuous on J (see [7]), and replacing each f_i by its best approximation $S_m f_i$ from the space Π_m of polynomials of degree $\le m$ with respect to the uniform norm $\|\cdot\|_J$ on $C(J)$, we obtain the <u>product approximation</u> of F :

$$(P_{n,m}F)(x,y) = \Sigma_{i=0}^{n}(S_m f_i)(y)\varphi_i(x) . \tag{1.2}$$

Observe that $P_{n,m}F$ is a polynomial in x and y in the tensor product space $\Pi_{n,m} = \{\Sigma_{i=0}^{n}\Sigma_{j=0}^{m}\alpha_{ij}y^j x^i$: each α_{ij} is real$\}$.

Weinstein [7] proved the following density theorem.

<u>Theorem 1.1</u>. Let $F \in C(D)$. For any $\varepsilon > 0$ there exists $N(\varepsilon)$ and for every $n \ge N(\varepsilon)$ there exists $M(\varepsilon,n)$ such that $\|F - P_{n,m}F\|_D < \varepsilon$ for all $n \ge N(\varepsilon)$ and $m \ge M(\varepsilon,n)$.

Henry and Schmidt [3] observed that product approximations depend on the choice of the basis for Π_n and obtained error for various bases. The best global estimates arose for the Lagrange basis functions. That is, we fix $a \le x_0 < x_1 < \ldots < x_n \le b$ and let

$$\varphi_i(x) = \Pi_{\substack{k=0\\k\neq i}}^{n}\{(x-x_k)/(x_i-x_k)\} \tag{1.3}$$

for $i = 0,\ldots,n$. Hereafter, we shall only use these as our basis functions.

<u>Theorem 1.2</u>. Let $x_0,\ldots,x_n$ be the zeros of the $(n+1)$-st degree Chebyshev polynomial (transformed to the interval I) and let $\varphi_i\,(i=0,\ldots,n)$ be given by (1.3). For $F \in C(D)$,

$$\|F - P_{n,m}F\|_D \le \{3 + \tfrac{4}{\pi}\ell n(n+1)\}E_{n,m}(F)$$

where $E_{n,m}(F) = \inf\{\|F-p\|_D : p \in \Pi_{n,m}\}$ is the degree of approximation of F by elements of $\Pi_{n,m}$.

In this paper, we extend Theorem 1.2 to arbitrary Lagrange bases and involve univariate degrees of approximation in the error bound as is done in [6] for blending interpolation and appolation. Furthermore, we show that the error bounds are sharp in an asymptotic sense. By similar techniques, Theorem 1.1 is shown to be sharp in the sense that the dependence of $M(\varepsilon,n)$ on n cannot be removed.

Finally, we propose a hybrid technique (interpolation followed by approximation) which enjoys the same error bounds as product approximation and appolation (approximation followed by interpolation) but is more economical computationally.

2. Product Approximation

We denote by F_x the function in $C(J)$ given by $F_x(y) = F(x,y)$ and by $E_n(g)$, resp. $E_m(h)$, the degree of approximation of $g \in C(I)$, resp. $h \in C(J)$, by polynomials in Π_n, resp. Π_m.

Theorem 2.1. Let $\varphi_i (i=0,\ldots,n)$ be given by (1.3) For $F \in C(D)$,

$$\|F - P_{n,m}F\|_D \le \Lambda_n \max_i E_m(F_{x_i}) + (1+\Lambda_n) \max_y E_n(F_y) \tag{2.1}$$

where $\Lambda_n = \max\{\Sigma_{i=0}^n |\varphi_i(x)| : a \le x \le b\}$ is the Lebesgue constant for the nodes $x_0,\ldots,x_n$.

Proof. For $(x,y) \in D$,

$$|(F-P_{n,m}F)(x,y)|$$

$$\le |(F_y - B_nF_y)(x)| + |\Sigma_{i=0}^n (f_i - S_m f_i)(y)\varphi_i(x)|$$

$$\le E_n(F_y) + \Sigma_{i=0}^n \|f_i - S_m f_i\|_J |\varphi_i(x)|$$

$$\le E_n(F_y) + \Sigma_{i=0}^n \|f_i - S_m F_{x_i}\|_J |\varphi_i(x)| .$$

For fixed $y \in J$, $f_i(y) = (B_nF_y)(x_i)$ and $|(f_i - S_mF_{x_i})(y)|$ $\le |(F_y - B_nF_y)(x_i)| + |(F_{x_i} - S_mF_{x_i})(y)| \le E_n(F_y) + E_m(F_{x_i})$. Thus $\|F_i - S_mF_{x_i}\|_j \le \max_y E_n(F_y) + E_m(F_{x_i})$ and we have

$$|(F - P_{n,m}F)(x,y)|$$

$$\le \max_y E_n(F_y) + \Sigma_{i=0}^n (\max_y E_n(F_y) + E_m(F_{x_i}))|\varphi_i(x)|$$

and (2.1) now follows.

Remark. From the last step of the proof above, we see that a necessary condition for equality to hold in (2.1) is that $E_m(F_{x_0}) = \ldots = E_m(F_{x_i})$. Functions for which equality holds in (2.1) are easily constructed. The next result shows that (2.1) is sharp in a strong asymptotic sense.

Theorem 2.2. For each n select a positive integer m_n where $m_n \to \infty$ as $n \to \infty$ and nodes $a \le x_{n0} < x_{n1} < \ldots < x_{nn} \le b$. Let $\varphi_{ni}(i=0,\ldots,n)$ correspond to these nodes by (1.3). Then there exists $F \in C(D)$ such that

$$\limsup_{n\to\infty} \frac{\|F - P_{n,m}F\|_D}{\Lambda_n \max_i E_{m_n}(F_{x_{ni}}) + (1+\Lambda_n)\max_y E_n(F_y)} = 1 . \quad (2.2)$$

Proof. In the proof, we suppress the subscript on m and the first subscripts on x and φ. We make use of a form of the Uniform Boundedness Principle due to Osgood [4,p.148].

Osgood's Theorem. Let X be a topological space and $\{f_\alpha\}_{\alpha\varepsilon A}$ be a family of real-valued lower semicontinuous functions on X. If $E \subseteq X$ is of second category and $\sup\{f_\alpha(t) : \alpha \in A\} < \infty$ for all $t \in E$, then there is an open set $U \subseteq X$ such that $\sup\{f_\alpha(t) : \alpha \in A, t \in U\} < \infty$.

Let Y_n be the subset of $C(D)$ for which equality holds in (2.1). By the continuity of the operator $P_{n,m}$ [2] and of the operators B_n and S_m [1,p.82], Y_n is closed. By the necessary condition for $F \in Y_n$ (see the remark above), Y_n has an empty interior. Let $E = C(D) \sim \cup_{n=0}^{\infty} Y_n$. By the Baire Category Theorem, E is of second category. Define

$$G_n(F) = \left\{ 1 - \frac{\|F-P_{n,m}F\|_D}{\Lambda_n \max\limits_i E_m(F_{x_i}) + (1+\Lambda_n) \max\limits_y E_n(F_y)} \right\}^{-1}$$

if $F \in C(D) \sim Y_n$ and $G_n(F) = 1$ if $F \in Y_n$. By the continuity of $P_{n,m}$, B_n, and S_m and the definition of G_n, G_n is lower semicontinuous. By the near linerarity properties of B_n and S_m, we have

$$G_n(\alpha F+p) = G_n(F) \tag{2.3}$$

for $F \in C(D)$, $p \in \Pi_{n,m}$, and $\alpha \neq 0$. The proof will be complete if we show that $\{G_n\}_{n=0}^{\infty}$ is not bounded over any open subset of $C(D)$.

We make an initial construction. For fixed n, select $x^* \in J$ such that $\Sigma_{i=0}^{n} |\varphi_i(x^*)| = \Lambda_n$ and let $\sigma_i = \operatorname{sgn}\varphi_i(x^*)$. Select $a \in C(J)$ such that $\|a\|_J = 1$ and $S_m a = 0$. For $0 < \delta < 1$, define $F = F_{n,\delta}$ by

$$F(x,y) = (1-\delta)\sigma_0 a(y)\varphi_0(y) + \Sigma_{i=0}^{n} \sigma_i a(y)\varphi_i(x) . \tag{2.4}$$

Then $E_n(F_y) = 0$ for each $y \in J$, $F_{x_0}(y) = (1-\delta)\sigma_i a(y)$, and $F_{x_i}(y) = \sigma_i a(y)$ $(i=1,\dots,n)$. Thus $\max_i E_m(F_{x_i}) = 1$. Moreover, $P_{n,m}F = 0$ and $\|F-P_{n,m}F\|_D \geq \Lambda_n - \delta\|\varphi_0\|_I$. Thus

$$G_n(F) \geq \left\{1 - \frac{\Lambda_n - \delta\|\varphi_0\|_I}{\Lambda_n}\right\}^{-1} = \frac{\Lambda_n}{\delta\|\varphi_0\|_I} \to \infty \qquad (2.5)$$

as $\delta \to 0^+$.

To see that $\{G_n\}_{n=0}^{\infty}$ is not bounded over any open set, let $F \in C(D)$ and $\varepsilon > 0$. For some n sufficiently large, there exists $p \in \Pi_{n,m}$ such that $\|F-p\|_D < \varepsilon/2$. Letting $\alpha_\delta > 0$ be sufficiently small so that $\|\alpha_\delta F_{n,\delta}\|_D < \varepsilon/2$, we have $\|F - (p+\alpha_\delta F_{n,\delta})\|_D < \varepsilon$. By (2.3) and (2.5)

$$G_n(p+\alpha_\delta F_{n,\delta}) = G_n(F_{n,\delta}) \to \infty$$

as $\delta \to 0^+$. Hence, $\{G_n\}_{n=0}^{\infty}$ is not bounded over any open set and Theorem 2.2 now follows from Osgood's Theorem.

The next theorem shows that product approximations may diverge from the function.

<u>Theorem 2.3</u>. For each n , select a positive integer m_n where $m_n \to \infty$ as $n \to \infty$ and let $x_{n0},\dots,x_{nn}$ be the zeros of the (n+1)-st degree Chebyshev polynomial (transformed to I). Let $\varphi_{n0},\dots,\varphi_{nn}$ be given by (1.3) for these nodes. Then there exists $F \in C(D)$ such that $\limsup_{n\to\infty}\|P_{n,m_n}F\| = \infty$ and thus $P_{n,m_n}F$ does not converge uniformly to F .

<u>Proof</u>. Again we suppress the same subscripts as in Theorem 2.2. We sketch the proof as it is similar to the proof of Theorem 2.1. The key to the proof is that the basis functions φ_i are bounded independent of n , say by M . We take $E = C(D)$ and let $G_n(F) = \|P_{n,m}F\|_D$. Here, E is of second category and G_n is continuous. For fixed n , let x^* and σ_i $(i=0,\dots,n)$ be as in the proof of Theorem 2.2. Select functions $a_i \in C(J)$ such that $\|a_i\|_J = 2$, $S_m a_i = 1$, and

$a_0,\ldots,a_n$ have pairwise disjoint supports. Letting

$$F(x,y) = \Sigma_{i=0}^{n}\sigma_i a_i(y)\varphi_i(x) ,$$

we see that $\|F\|_D \le 2M$ and $(P_{n,m}F)(x,y) = \Sigma_{i=0}^{n}\sigma_i\varphi_i(x)$. So $\|P_{n,m}F\|_D = \Lambda_n \to \infty$ as $n \to \infty$. As in the proof of Theorem 2.2, it can be shown that $\{G_n\}_{n=0}^{\infty}$ is not bounded over any open subset of $C(D)$, and Theorem 2.3 now follows from Osgood's Theorem.

Remark. It is evident that the proof above relies on $\|\varphi_{ni}\|_I$ being bounded while $\|\Sigma_{i=0}^{n}|\varphi_{ni}|\|_I$ is unbounded. Thus the same divergence phenomenon can be demonstrated if we use the mononial bases or the Chebyshev polynomial bases.

3. Product Hybrid Approximation

In this section, we introduce a scheme in which we replace (1.1) with the polynomial

$$(L_nF_y)(x) = \Sigma_{i=0}^{n}F_y(x_i)\varphi_i(x) \tag{3.1}$$

of degree $\le n$ which interpolates F_y at the nodes $x_0,\ldots,x_n$. We then best approximate the coefficient functions F_y by elements of Π_m obtaining the product hybrid approximant

$$(P^*_{n,m}F)(x,y) = \Sigma_{i=0}^{n}(S_mF_{x_i})(y)\varphi_i(x) \tag{3.2}$$

of F . Product hybrid approximations enjoy the same error bounds as product approximation (Theorem 2.1) and appolation [6].

Theorem 3.1. Let $\varphi_i(i=0,\ldots,n)$ be given by (1.3). For $F \in C(D)$,

$$\|F - P^*_{n,m}F\|_D \le \Lambda_n \max_i E_m(F_{x_i}) + (1+\Lambda_n)\max_y E_n(F_y). \tag{3.3}$$

Proof. For $(x,y) \in D$, using [5,p.88] we have

$$
\begin{aligned}
&|(F-P^*_{n,m}F)(x,y)| \\
&\le |(F_y-L_nF_y)(x)| + \Sigma_{i=0}^{n} |(F_{x_i}-S_mF_{x_i})\ (y)|\ |\varphi_i(x)| \\
&\le (1+\Lambda_n)E_n(F_y) + \Lambda_n \max_i E_m(F_{x_i}) .
\end{aligned}
$$

We remark that (3.3) is also asymptotically sharp (see the proof of Theorem 2.2) and that the divergence phenomenon of Theorem 2.3 also holds. The hybrid scheme possesses a considerable computational advantage over both product approximation and appolation. For the latter the expensive aspect is finding the best approximation to F_y for all y in a suitably dense subset of J. In the hybrid scheme, this stage is replaced by interpolations which are essentially free. Moreover, the hybrid scheme possesses error bounds that involve just one Lebesgue constant while tensor product and blending interpolations involve a product of two Lebesgue constants. One drawback to the hybrid scheme is that a density result of the form of Theorem 1.1 does not hold. One needs $L_nF_y \to F_y$ uniformly in x and y to ensure such a result.

References

1. Cheney, E.W., "Introduction to Approximation Theory," McGraw-Hill, New York, 1966.

2. Henry, M.S. and Schmidt, D., Continuity theorems for the product approximation operator, in "Theory of Approximation with Applications" (A.G. Law and B.N. Sahney, Eds.), pp.24-42, Academic Press, New York, 1976.

3. Henry, M.S. and Schmidt, D., Error bounds for polynomial product approximation, J. Approx. Theory 31(1981), pp. 6-21.

4. Larsen, R., "Functional Analysis," Marcel Dekker, Inc., New York, 1973.

5. Rivlin, T.J., "An Introduction to the Approximation of Functions," Dover, New York, 1981.

6. Scherer, R. and Zeller, K., Gestufte Approximation in zwei Variablen, in "Numerical Methods of Approximation Theory" Vol. 5 (L. Collatz, et al, Eds.), pp. 282-288, Birkhäuser Verlag, Basel, 1980.

7. Weinstein, S.E., Approximation of functions of several variables: product Chebyshev approximations I, J. Approx. Theory (1969), pp. 433-447.

James Angelos, Department of Mathematics, Montana State University, Bozeman, Montana 59717, USA.

Darrell Schmidt, Department of Mathematical Sciences, Oakland University, Rochester, Michigan 48063, USA.

APPROXIMATION THEORY ON A SNOWFLAKE

M. F. Barnsley, J. S. Geronimo, A. N. Harrington, and L. D. Dager.

School of Mathematics, Georgia Institute of Technology, Atlanta, Georgia, 30332, U. S. A.

Considered are the monic polynomials orthogonal with respect to the equilibrium measure on the family of Julia sets for the complex mappings $z \mapsto (z-\lambda)^2$ where λ is a parameter. When $\lambda = 2$ they are the Chebychev polynomials on $[0,4]$. For all $\lambda \in C$ an infinite subsequence of the polynomials can be calculated; for $\lambda > 0$ they have the equal oscillation property on the Julia set, and for $\lambda \in (0,2]$ their zeros "interlace" on an underlying tree structure. They are relevant to approximation theory for functions defined on the Julia set.

1. Julia Sets

Let C be the complex plane, $T:C \mapsto C$ be a nonlinear polynomial of the form $T(z) = z^N + K_1 z^{N-1} + \ldots K_N$, and introduce the notation $T^\circ(z) = z$ and $T^{n+1}(z) = T \cdot T^n(z)$ for $n \in \{0,1,2,\ldots\}$. Then the Julia set B for T is the set of points $z \in C$ where $\{T^n(z)\}_1^\infty$ is not a normal family. The functions $\{T^n(z)\}_1^\infty$ form a normal family at $z \in C$ if, when the distance between functions is measured in the metric of the Riemann sphere, there is a neighborhood of z on which an infinite subsequence converges uniformly. B is a bounded, compact, perfect set, with no interior, and it is totally invariant under T. The key early works on Julia sets are [7] and [8], and a good modern reference is [6].

The diversity of appearance of Julia sets is very great; and among those belonging to $T_\lambda(z) = (z-\lambda)^2$ for $0 \le \lambda \le 3$ one finds pictures which look like ferns, clouds, trees, coastlines, and galaxies. Also, in common with the boundaries of natural objects and snowflake curves [9], they share the feature their often intricate shapes are not simplified by magnification. Let B_λ denote the Julia set for $T_\lambda(z)$. The accompanying three pictures show views of $B_{0.75}$, defined by the boundary between the white region and the black region. The first picture shows the whole set, while the second and third pictures show blow-ups of the inset boxes in the respective preceeding pictures. Whereas this Julia set, which we call a cactus, has features which do not change under magnification, other parameter values lead to pictures whose appearance clearly changes under magnification. The table on the next page illustrates some forms of B_λ.

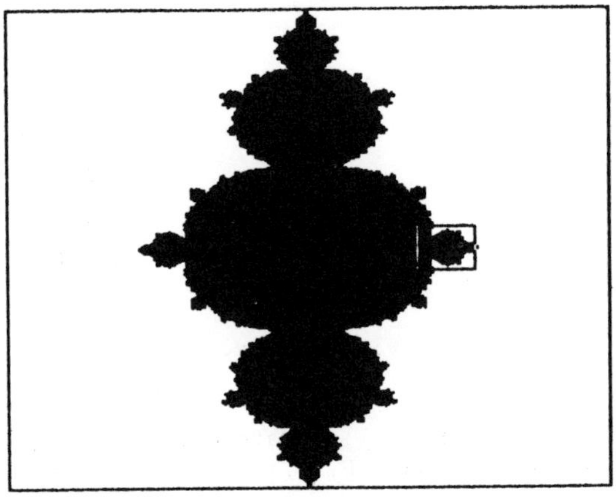

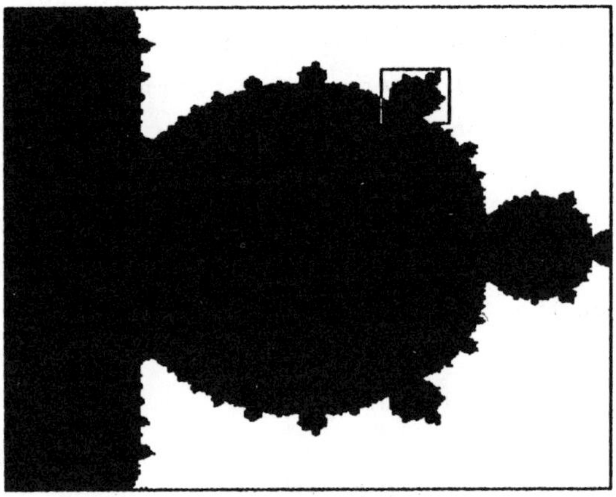

In the next section we show that it is possible to calculate and analyse certain orthogonal polynomials on B. Their properties suggest that they may be useful in the approximation of functions defined on B. Thus, a catalog of Julia sets might be used as the basis for the approximation of functions whose domains are naturally occuring boundaries.

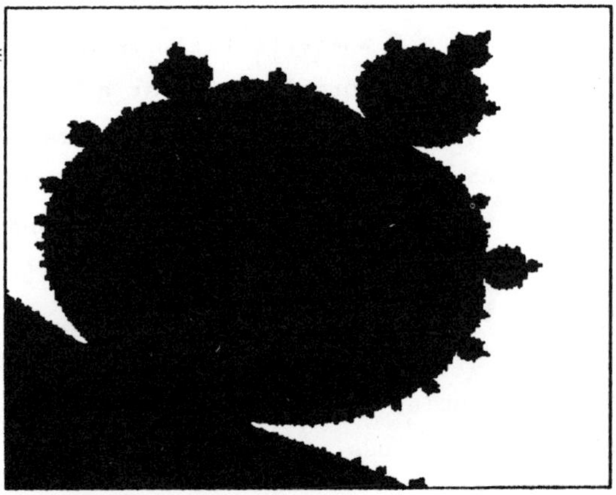

Further details concerning the results mentioned in this paper can be found in [1, 2, 3, 4, 5].

λ	$B\lambda$	Name
0		circle
0.5		leaf
1	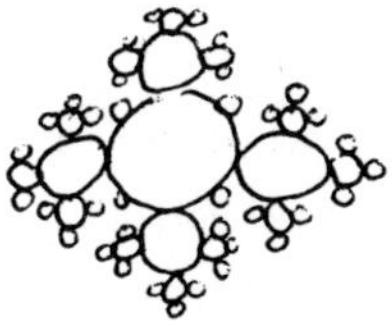	cactus
1.56	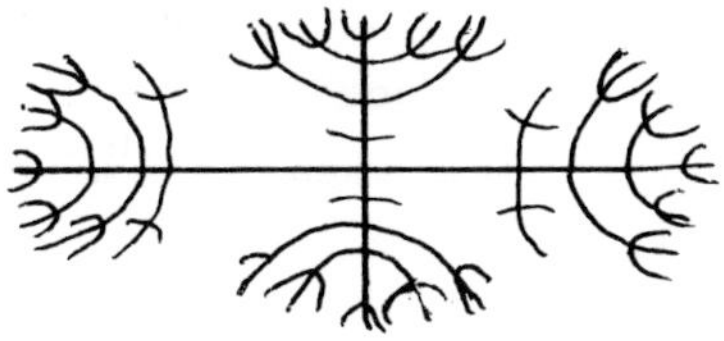	tree
2		interval
3		Cantor set

2. Orthogonal Polynomials

Theorem 1 [2] There exists a unique, balanced, T-invariant probability measure μ on B; that is, if T_j^{-1} for $j \varepsilon\{1,2,\ldots.,N\}$ is a complete assignment of branches of T^{-1}, then for each Borel set S, $\mu(T_j^{-1}S) = \frac{1}{N}\mu(S)$. Moreover, (B, μ, T) is a strongly mixing system.

We now introduce the monic polynomials $\{P_n(z)\}_0^\infty$, where $P_n(z)$ is of degree n, and ${}_B\int P_n(z)\overline{P_m(z)}d\mu(z) = 0$ for $m \neq n$. The following theorem tells how one can calculate an infinite subsequence of the orthogonal polynomials.

Theorem 2 [2] (i) $P_1(z) = z + K_1/N$

(ii) $P_{nN}(z) = P_n(T(z)$ for $n \varepsilon \{0, 1, 2, \ldots\}$

(iii) $P_{N^n}(z) = T^n(z) + K_1/N$ for $n \varepsilon \{0, 1, 2, ..\}$

As an example, consider $T(z) = T_\lambda(z) = (z - \lambda)^2$. Then for $\lambda \geq 2$ we find $P_1(z) = z - \lambda$, $P_2(z) = (z - \lambda)^2 - \lambda$, $P_3(z) = (z-\lambda)(z - \lambda)^2 - \lambda - 1)$, $P_4(z) = ((z - \lambda)^2 - \lambda)^2 - \lambda$. $P_5(z) = (z - \lambda)((z - \lambda)^4 - (2\lambda^2 + 2\lambda - 1)(z - \lambda)^2/(\lambda - 1) + (\lambda^3 - 2\lambda^2 + 2\lambda + 1)/(\lambda - 1))$, and $P_8(z) = (((z-\lambda)^2 - \lambda)^2 - \lambda)^2 - \lambda$. The expressions for $P_1(z)$, $P_2(z)$, $P_4(z)$, $P_8(z), \ldots, P_{2^n}(z), \ldots$ are implied by Theorem 2, and are valid for all $\lambda \varepsilon C$. On the other hand, the formulas for all of the other polynomials, which can calculated explicitly [1] when $\lambda \geq 2$, are not valid for arbitrary $\lambda \varepsilon C$. An illustration of this fact is that the formulas correctly reduce to the Chebychev polynomials on [0, 4] when $\lambda = 2$, but they do not reduce to the polynomials $\{z^\ell\}_{\ell=0}^\infty$ when $\lambda = 0$; such a reduction is valid only when $\ell = 2^n$. A similar situation prevails for arbitrary T, [4].

3. Properties of $\{P_{2^n}(z)\}_0^\infty$ when $T(z) = (z-\lambda)^2$

Theorem 3 [3] Let $0 < \lambda < 2$, $I_0 = [0, \lambda + \frac{1}{2} + \sqrt{\lambda + \frac{1}{4}}]$, and $I_j = T_\lambda^{-1}(I_{j-1})$ for $j \varepsilon \{1, 2, 3, \ldots\}$. Then $\{I_j\}_0^\infty$ is an increasing sequence of trees of analytic arcs with $B_\lambda \subset \overline{\bigcup_{j=0}^\infty I_j} \subset \{z \varepsilon C: \{T_\lambda^N(z)\}_1^\infty$ is bounded$\}$. For infinitely many values of λ with $0 < \lambda < 2$, $B_\lambda = \overline{\bigcup_{j=0}^\infty I_j}$ and $C - B_\lambda$ is connected; in which cases we say that B_λ is tree-like.

Theorem 4 [3] For $\lambda \geq 2$, $P_{2^n}(z)$ has the equal oscillation property on B_λ. For $\lambda > 2$, $P_{2^n}(z)$ with $z \varepsilon B_\lambda$ possesses exactly 2^n maxima and 2^n minima, all of which have the same magnitude. For $0 < \lambda < 2$, $P_{2^n}(z)$ attains its maximum modulus on B_λ at the 2^{n+1} tips of the branches of I_{n+1}

Theorem 5 [3]. The zeros of $P_{2^n}(z)$ are the set of points $\lambda + e_1\sqrt{(\lambda + e_2\sqrt{(\lambda + \ldots \quad \lambda)\ldots))}}$ where each $e_i \varepsilon \{-1, 1\}$ and all distinct sequences are chosen. These zeros are located at the intersection points of the tree I_{n+2} that are not intersection points in I_{n+1}. The zeros of $d\,P_{2^n}(z)/dz$ are the intersection points of branches of the tree I_{n+1}.

When B_λ is tree-like, all these zeros lie on B_λ and the closure of the zeros of $\{P_{2^n}(z)\}_{n=1}^{\infty}$ is B_λ.

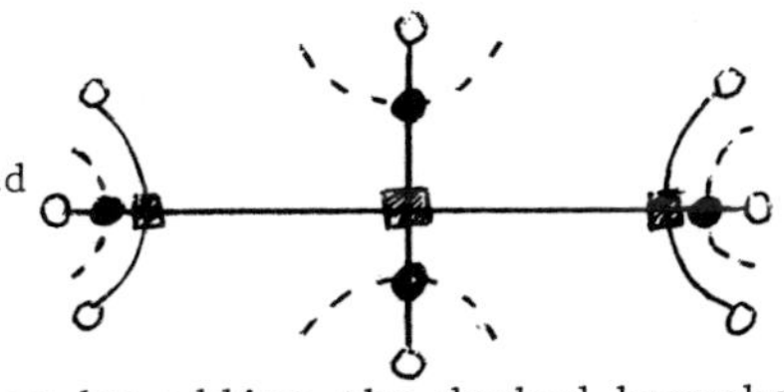

In the drawing we have illustrated part of the content of Theorem 5 for $\lambda \simeq 1$. Shown are the trees I_3 and I_4, where the latter is obtained from the former by adding the dashed branches. The zeros of P_4 are denoted by ● while those of $P_4(z)$ are denoted by ▨ . The symbols O show where $|P_4(z)|$ is maximum for $z \in B_\lambda$

1. Barnsley, M.F., A. N. Harrington, J. S. Geronimo (1981) On the invariant sets of a family of quadratic maps. Submitted to Comm. Math. Phys.

2. Barnsley, M. F., A. N. Harrington, J. S. Geronimo (1982) Orthogonal polynomials associated with invariant measures on Julia sets. Submitted to Bulletin of A.M.S.

3. Barnsley, M. F., A. N. Harrington, J. S. Geronimo (1982) Some treelike Julia sets and Pade´ approximants. Submitted to Comm. Math. Phys.

4. Barnsley, M. F. , A. N. Harrington, J. S. Geronimo (1982) Infinite dimensional Jacobi matrices associated with Julia sets. Submitted to Trans. A.M.S.

5. Barnsley, M. F., A. N. Harrington, J. S. Geronimo (1982) Geometrical and electrical properties of some Julia sets. In preparation

6. Brolin, H. (1965) Invariant sets under iteration of rational functions. Arkiv for Mathematik 6, 103-144.

7. Fatou, P. (1919) Sur les equations fonctionelles. Bull. Soc. Math. France 47, 161-271; Ibidem 48, 33-94, 203-314.

8. Julia, G. (1918) Memoire sur l'iteration des fonctions rationelles. J. de Math. Pur s et Appliquees 8.1, 47-245.

9. Mandlebrot, B., (1977) Fractals: Form, Chance and Dimension. W. H. Freeman, San Francisco.

Prof. Michael F. Barnsley, School of Mathematics, Georgia Institute of Technology, Atlanta, Georgia 30332, U. S. A.

ESTIMATES AND BOUNDS FOR THE DOUBLY PERTURBED EIGENVALUE PROBLEM

M. F. Barnsley and W. Raddatz
School of Mathematics, Georgia Institute of Technology, Atlanta, Georgia 30332, U.S.A.

Considered is the problem of bounding and estimating the lowest eigenvalue of a doubly perturbed self-adjoint linear operator in a Hilbert space, with the aid of Taylor series data for the eigenvalue.

Let A, B, and C be bounded self-adjoint linear operators in a Hilbert space H. We consider $M = A + xB + yC$ where $(x,y)\in\mathbb{R}^2$, and we assume that the lowest portion of the spectrum of M is discrete for $(x,y)\in U$, where U is an open neighborhood of (0,0). Also, the lowest eigenvalues are assumed to be distinct, and we denote them by

$$E(x,y) < F(x,y) < G(x,y) < \ldots \text{ for } (x,y)\in U.$$

These eigenvalues are each regular in some neighborhood of (0,0), and we can write the Taylor series expansion of E as

$$E(x,y) = \sum_{j,k=0}^{\infty} E_{jk}x^j y^k$$

The convexity bound states (see for example [5])

$E(x,y) \le E_{0,0} + E_{1,0}x + E_{0,1}y$ for all $(x,y)\in U$, which uses only the information $\{E_{0,0}, E_{1,0}, E_{0,1}\}$. This suggests the following problems. Given a larger set of Taylor series data, possibly together with information about $F(x,y)$, $G(x,y)$, . . .near $(0,0)$, find best possible bounds and approximations for $E(x,y)$. A motivation for considering such problems comes from atomic physics [2].

One approach is to take $Y = \alpha x$ where $\alpha \in \mathbb{R}$, and set $L = B + \alpha C$. Then we have the one variable problem associated with $H = A + xL$ whose lowest eigenvalue is

$$E(x) = \sum_{j=0}^{\infty} E_j x^j \text{ where } E_j = \sum_{k=0}^{j} \alpha^k E_{j-k,k} .$$

Previously developed techniques for single-parameter situations [3] can now be applied. For example, a best possible upper bound for $E(x)$ can be constructed from the information $\{E_0, E_1, E_2, E_3, F(0,0)\}$.

We have proved [6] that for the set of given information $\{E_0, E_1, E_2, E_3, E_4, E_5, F(0,0), G(0,0)\}$ there are real 3x3 matrices A_3 (diagonal) and L_3 (symmetric tridiagonal) such that $A_3 + xL_3$ matches the given information and hence provides best possible approximations for $E(x,y)$, $F(x,y)$, and $G(x,y)$. Also, if A is diagonal and L is tridiagonal then the approximation for $E(x)$ which is determined from $A_3 + xL_3$ is an upper bound for $E(x)$. More generally, in the latter case, successively larger sets of given information lead to readily calculated best possible upper bounds for $E(x)$. To establish such results, fundamental use is made of the Rayleigh-Ritz variational principle, [4].

We look forward to progress for systems $A + xL$ which are not fundamentally tridiagonal, and to results for $M = A + xB + yC$ which are not obvious extensions of one-variable techniques. For the nonhomogeneous problem $(A + xB + yC)v = w$, v and w in H, such results are available, [1].

[1] Barnsley, M. F. (1978) Approximants for some classes of multivariable functions provided by variational principles. In Multivariate Approximation edited by D. C. Handscomb, Academic Press (New York).

[2] Barnsley, M. F., J. Aguilar (1978) On the approximation of potential energy functions for diatomic molecules. Int. J. Quantum Chemistry 13, 642-677.

[3] Barnsley, M. F. (1981) Bounds for the linearly perturbed eigenvalue problem. In Spectral Theory of Differential Operators, edited by I. W. Knowles and R. T. Lewis, North-Holland.

[4] Hirschfelder, J. O., W. Byers Brown, S. T. Epstein (1964) Recent developments in perturbation theory. In Advances in Quantum Chemistry Vol. 1, Academic Press (New York).

[5] Narnhofer, H., and W. Thirring (1975) Convexity properties for coulomb systems. Acta. Phys. Austr. 41, 281-297.

[6] Raddatz, W. (1982) Doctoral thesis. In preparation.

Professor Michael F. Barnsley, School of Mathematics, Georgia Institute of Technology, Atlanta, Georgia 30332, USA.

REMARKS ON REDUCED HERMITE INTERPOLATION

G. Baszenski, F. J. Delvos, K. Hackenberg

An important class of rectangular finite elements are those of reduced Hermite interpolation type. They have a set of nodes which can be considered as a subset of the interpolation data of a corresponding Hermite tensor product scheme. Those nodes are omitted which do not contribute to the desired properties such as degree of exactness, i.e. the maximal degree of polynomials which are interpolated exactly, or the degree of conformity, i.e. the maximal degree of derivatives which are continuous when interpolating on a rectangular grid using the same scheme several times. This reduction of the number of parameters avoids the computation of unwanted information. Conforming elements of that kind are defined for example in the works of MELKES [5] and WATKINS [7].

The objective of the following paper is to define and to construct a class of two dimensional Hermite interpolation schemes which contains elements of an arbitrary high order of conformity. Moreover we define the elements in a way that no nodes of higher differentiation order than the order of conformity of that element are involved.

For the construction we make use of methods of Boolean interpolation. We begin by introducing two well known elements of that type, both constructed with the help of the following one dimensional two point Hermite interpolation formula.

Definition

Let $H_m : C^m[0,1] \to \Pi_{2m+1}$ be defined by

$$D^i(H_m f)(u) = D^i f(u) \qquad (f \in C^m[0,1];\ i=0,\dots,m;\ u=0,1)$$

where D denotes the differentiation operator with respect to the function's variable x.

Remark

a) Introducing the cardinal polynomials $p_{i,m}^{(u)} \in \Pi_{2m+1}$ which are biorthogonal to the set of interpolation conditions we obtain the following dual representation of $H_m f(x)$:

$$H_m f(x) = \sum_{i=0}^{m} \sum_{u=0}^{1} D^i f(u)\ p_{i,m}^{(u)}(x) \qquad (f \in C^m[0,1]).$$

b) Explicit representations for $p_{i,m}^{(u)}$ are (cf. for instance PHILLIPS [6]):

$$p_{i,m}^{(0)}(x) = \frac{x^i}{i!}\ (1-x)^{m+1} \sum_{s=0}^{m-i} \binom{m+s}{s}\ x^s$$

$$p_{i,m}^{(1)}(x) = (-1)^i\ p_{i,m}^{(0)}(1-x) \qquad (m \in \mathbb{N}_0;\ i=0,\dots,m).$$

We define parametric extensions of H_m so that we are able to use that interpolation scheme for the interpolation of functions defined on $U := [0,1] \times [0,1]$.

Definition

Let $H_m^x := H_m \otimes I : C^{(m,m)}(U) \to \Pi_{2m+1} \otimes C^m[0,1]$ and

$H_m^y := I \otimes H_m : C^{(m,m)}(U) \to C^m[0,1] \otimes \Pi_{2m+1}$.

Then we have for example

$$H_m^x f(x,y) = \sum_{i=0}^{m} \sum_{u=0}^{1} D_x^i f(u,y)\ p_{i,m}^{(u)}(x) \qquad (f \in C^{(m,m)}(U)).$$

H_m and thus H_m^x and H_m^y as well are projections, i.e. idempotent linear operators. $H_m^x f$ interpolates a function f and its normal derivatives up to order m on the edges (O,y) and (1,y) of U, whereas $H_m^y f$ interpolates the appropriate values on (x,O) and (x,1). According to their special tensor product definition the operators H_m^x and H_m^y commute.

Their product $H_m^x H_m^y = H_m \otimes H_m$ is the well known tensor product element of bivariate Hermite interpolation. In terms of Boolean interpolation theory it is the minimal projector which can be constructed using H_m^x and H_m^y. $H_m^x H_m^y$ has only those interpolation properties which are common to H_m^x and H_m^y:

Remark

For each $f \in C^{(m,m)}(U)$ we have

a) $D_x^i D_y^j (H_m^x H_m^y f)(u,v) = D_x^i D_y^j f(u,v) \qquad (i,j=0,\dots,m;\ u,v=0,1).$

b) $H_m^x H_m^y f(x,y) = \sum_{i,j=0}^{m} \sum_{u,v=0}^{1} D_x^i D_y^j f(u,v)\ p_{i,m}^{(u)}(x)\ p_{j,m}^{(v)}(y)$

which is a dual representation for $H_m^x H_m^y f$.

c) Given f of class $C^{(m,m)}$ on a rectangular mesh domain we obtain a piecewise polynomial of global conformity $C^{(m,m)}$ when applying $H_m^x H_m^y$ to f on each rectangular subdomain (this property is called "C^m-conformity" of the interpolation scheme).

d) The remainder projector associated with $H_m^x H_m^y$ is given by

$\overline{H_m^x H_m^y} := I - H_m^x H_m^y = (I-H_m^x) + (I-H_m^y) - (I-H_m^x)(I-H_m^y).$

For $f \in C^{(m,m)}(U_h)$ $(U_h := [O,h]\times[0,h])$ and a linearly transformed projector $H_m^x H_m^y$ which operates on functions defined on U_h we obtain in particular the asymptotic error estimate

$$\left\| \overline{H_m^x H_m^y f} \right\|_\infty = O(h^{2m+2}) \text{ as h tends to zero.}$$

Let us proceed to our second example which is the maximal interpolation scheme in Boolean interpolation that can be constructed using H_m^x and H_m^y. It is called the transfinite Boolean sum

$H_m^x \oplus H_m^y := H_m^x + H_m^y - H_m^x H_m^y$

and was originally introduced by GORDON [3]. Using the fact that H_m^x, H_m^y commute and following Gordon's propositions we are able to state the following properties for $H_m^x \oplus H_m^y$:

Remark

For each $f \in C^{(m,m)}(U)$ we have

a) $D_x^i(H_m^x \oplus H_m^y f)(u,y) = D_x^i f(u,y)$ $\quad (i,j=0,\ldots,m;\ u,v=0,1;\ x,y \in [0,1])$

$D_y^j(H_m^x \oplus H_m^y f)(x,v) = D_y^j f(x,v)$

that is $H_m^x \oplus H_m^y f$ satisfies all of those interpolation conditions which hold for either $H_m^x f$ or $H_m^y f$.

b) The element $H_m^x \oplus H_m^y$ is C^m-conforming (in fact the normal derivatives up to order m of the interpolant along the edges of the domain U are presribed to be equal to those of f which assures global conformity of each derivative of the interpolant).

c) The remainder projector of $H_m^x \oplus H_m^y$ is easily computed to be

$\overline{H_m^x \oplus H_m^y} = \overline{H_m^x}\ \overline{H_m^y}$.

(Gordon showed that any set of commuting projectors generates a Boolean algebra with the binary operations operator product, Boolean sum and the unary remainder operation.)

Therefore we get the asymptotic error behaviour

$$\left\| \overline{H_m^x \oplus H_m^y f} \right\|_\infty = O(h^{4m+4})$$

for f defined on U_h (cf. also WATKINS [8]).

It is our aim to modify the transfinite Boolean sum by appropriate additional projectors to obtain an element of finitely many parameters which still has the same asymptotic error and the same class of conformity. Such an element is defined by replacing the transfinite interpolation data

$D_x^i f(u,y)$, $D_y^j f(x,v)$ $(x,y \in [0,1])$

in the Boolean sum by univariate approximations in a finite-dimensional function vector space. We will choose interpolation polynomials whose interpolation properties are at least those of H_m to construct a so called discrete Boolean sum. The additional univariate scheme is defined as follows:

<u>Definition</u>

Let $0 < x_1 < \ldots < x_n < 1$ be n distinct points.
Let then $H_{m,n} : C^m[0,1] \to \Pi_{2m+n+1}$ be defined by

$$D^i(H_{m,n}f)(u) = D^i f(u) \qquad (i=0,\ldots,m;\ u=0,1) \quad \text{and}$$
$$H_{m,n}f(x_k) = f(x_k) \qquad (k=1,\ldots,n).$$

<u>Remark</u>

Dual functions can be obtained in applying the interpolation method of Newton n times successively to the polynomials $p_{i,m}^{(u)}$. As usual we denote parametric extensions of that projector by $H_{m,n}^x := H_{m,n} \otimes I$, $H_{m,n}^y := I \otimes H_{m,n}$.

<u>Remark</u>

H_m^x, H_m^y, $H_{m,n}^x$, $H_{m,n}^y$ are mutually commuting. (For example we have $H_m^x H_{m,n}^x = H_m^x = H_{m,n}^x H_m^x$.)

We define the discrete Boolean sum projector and denote some of its properties:

<u>Theorem</u>

$$P_{m,n} := H^x_m H^y_{m,n} \oplus H^x_{m,n} H^y_m : C^{(m,m)}(U) \to C^{(m,m)}(U)$$

has the following properties:

a) $P_{m,n} f = H^x_m H^y_{m,n} f + H^x_{m,n} H^y_m f - H^x_m H^y_m f \qquad (f \in C^{(m,m)}(U))$.

(A dual representation formula can easily be established from the above equality.)

b) The interpolation properties are obtained by applying a result of GORDON / CHENEY [4]:

$$D^i_x D^j_y (P_{m,n} f)(u,v) = D^i_x D^j_y f(u,v)$$

$$D^i_x (P_{m,n} f)(u,x_k) = D^i_x f(u,x_k)$$

$$D^j_y (P_{m,n} f)(x_k,v) = D^j_y f(x_k,v)$$

$(f \in C^{(m,m)}(U);\ i,j=0,\dots,m;\ k=1,\dots,n;\ u,v=0,1)$.

c) The function invariance set of $P_{m,n}$ is also calculated by applying a result of GORDON / CHENEY [4]:

$$\mathrm{Im}(P_{m,n}) = \Pi_{2m+1} \otimes \Pi_{2m+n+1} + \Pi_{2m+n+1} \otimes \Pi_{2m+1}.$$

d) Arguing as in the tensor product case we establish that $P_{m,n} f$ is C^m-conforming.

e) Using the calculation rules in a Boolean algebra we find that the remainder projector of $P_{m,n}$ is given by

$$\overline{P_{m,n}} = I - P_{m,n} = \overline{H^x_{m,n}} + \overline{H^y_{m,n}} + \overline{H^x_m}\,\overline{H^y_m} - \overline{H^x_{m,n}}\,\overline{H^y_m} - \overline{H^x_m}\,\overline{H^y_{m,n}}.$$

f) On U_h we therefore obtain the asymptotic error estimate

$$\|\overline{P_{m,n} f}\|_\infty \le F_1 \cdot h^{2m+n+2} + F_2 \cdot h^{4m+4} + F_3 \cdot h^{4m+n+4}$$

$(f \in C^{(2m+n+2,2m+2)} \cap C^{(2m+2,2m+n+2)})$.

We observe that the above estimate is optimal if we choose $n = 2m+2$ so that the powers of h in the first two summands are of equal size. In this special case, however, it is obviously preferable to replace the discretising data $f(x_1),\dots,f(x_{2m+2})$ by an equal number of Hermite conditions in two points

$$f(\tfrac{1}{3}),\ Df(\tfrac{1}{3}),\dots,\ D^m f(\tfrac{1}{3}),\ f(\tfrac{2}{3}),\ Df(\tfrac{2}{3}),\dots,\ D^m f(\tfrac{2}{3})$$

in order to obtain a simpler and more unified structure of interpolation data. We define the discretising projector and derive formulas for its cardinal functions:

Definition & Remark

a) Let $K_m : C^m[0,1] \to \Pi_{4m+3}$ be defined by

$$D^i(K_m f)(\tfrac{u}{3}) = D^i f(\tfrac{u}{3}) \qquad (f \in C^m[0,1];\ i=0,\dots,m;\ u=0,\dots,3).$$

b) The dual polynomials $q_{i,m}^{(u)} \in \Pi_{4m+3}$ in the cardinal representation $K_m f(x) = \sum_{i=0}^{m} \sum_{u=0}^{3} D^i f(\tfrac{u}{3})\ q_{i,m}^{(u)}(x)$

are given by the representations

$$q_{i,m}^{(0)}(x) = \frac{x^i}{i!}\ (1-3x)^{m+1}(1-\tfrac{3}{2}x)^{m+1}(1-x)^{m+1} \sum_{s=0}^{m-i} a_{sm} x^s$$

where $a_{sm} = \sum_{\substack{\kappa+\lambda+\mu=s\\ \kappa,\lambda,\mu\geq 0}} \binom{m+\kappa}{\kappa}\binom{m+\lambda}{\lambda}\binom{m+\mu}{\mu}\ 3^{\lambda+\mu}(\tfrac{1}{2})^{\mu}$,

$$q_{i,m}^{(1)}(x) = \frac{1}{i!}\ (x-\tfrac{1}{3})^i\ x^{m+1}(1-x)^{m+1}(2-3x)^{m+1}(\tfrac{9}{2})^{m+1} \sum_{s=0}^{m-i} b_{sm}(1-3x)^s$$

where $b_{sm} = \sum_{\substack{\kappa+\lambda+\mu=s\\ \kappa,\lambda,\mu\geq 0}} \binom{m+\kappa}{\kappa}\binom{m+\lambda}{\lambda}\binom{m+\mu}{\mu}(-1)^{\lambda+\mu}(\tfrac{1}{2})^{\lambda}$,

$$q_{i,m}^{(2)}(x) = (-1)^i\ q_{i,m}^{(1)}(1-x)\ , \quad q_{i,m}^{(3)}(x) = (-1)^i\ q_{i,m}^{(0)}(1-x)$$

$(m \in \mathbb{N}_0;\ i=0,\dots,m)$.

Proof

We have degree $q_{i,m}^{(0)} = 4m+3$ and

$D^j q_{i,m}^{(0)}(\frac{u}{3}) = 0 \qquad (j=0,\ldots,m;\ u=1,2,3).$

It remains to show $D^j q_{i,m}^{(0)}(0) = \delta_{ij}$.

To prove this we note the series expansions

$$(1-x)^{-m-1} = \sum_{\kappa=0}^{\infty} \binom{m+\kappa}{\kappa} x^{\kappa} \qquad (m \in \mathbb{N}_0;\ -1 < x < 1) ,$$

$$(1-3x)^{-m-1} = \sum_{\lambda=0}^{\infty} \binom{m+\lambda}{\lambda} 3^{\lambda} x^{\lambda} \qquad (m \in \mathbb{N}_0;\ -\tfrac{1}{3} < x < \tfrac{1}{3}) ,$$

$$(1-\tfrac{3}{2}x)^{-m-1} = \sum_{\mu=0}^{\infty} \binom{m+\mu}{\mu} (\tfrac{3}{2})^{\mu} x^{\mu} \qquad (m \in \mathbb{N}_0;\ -\tfrac{2}{3} < x < \tfrac{2}{3}) .$$

Using the Cauchy product formula for converging series we get for each $m \in \mathbb{N}_0$, $i=0,\ldots,m$, $-\frac{1}{3} < x < \frac{1}{3}$ the following identity relation:

$$q_{i,m}^{(0)}(x) = \frac{x^i}{i!} (1-3x)^{m+1} (1-\tfrac{3}{2}x)^{m+1} (1-x)^{m+1} \sum_{s=0}^{m-i} a_{sm} x^s$$

$$= \frac{x^i}{i!} (1-3x)^{m+1} (1-\tfrac{3}{2}x)^{m+1} (1-x)^{m+1} \left\{ \sum_{s=0}^{\infty} a_{sm} x^s - \sum_{s=m-i+1}^{\infty} a_{sm} x^s \right\}$$

$$= \frac{x^i}{i!} (1-3x)^{m+1} (1-\tfrac{3}{2}x)^{m+1} (1-x)^{m+1} \cdot$$

$$\cdot \left\{ \sum_{\kappa=0}^{\infty} \binom{m+\kappa}{\kappa} x^{\kappa} \cdot \sum_{\lambda=0}^{\infty} \binom{m+\lambda}{\lambda} 3^{\lambda} x^{\lambda} \cdot \sum_{\mu=0}^{\infty} \binom{m+\mu}{\mu} (\tfrac{3}{2})^{\mu} x^{\mu} \right.$$

$$\left. - x^{m+1-i} \sum_{s=0}^{\infty} a_{m-i+1+s,m} x^s \right\}$$

$= \frac{x^i}{i!} - x^{m+1} \phi(x)$, where ϕ is an analytical function.

Then we get $D^j q_{i,m}^{(0)}(0) = D^j (\frac{x^i}{i!})|_{x=0} = \delta_{ij} \qquad (j=0,\ldots,m).$

For the representation of $q_{i,m}^{(1)}$ we argue in a similar fashion using the series expansions

$$x^{-m-1} = 3^{m+1} \sum_{\kappa=0}^{\infty} \binom{m+\kappa}{\kappa} (1-3x)^{\kappa} ,$$

$$(1-x)^{-m-1} = (\tfrac{3}{2})^{m+1} \sum_{\lambda=0}^{\infty} \binom{m+\lambda}{\lambda} (-\tfrac{1}{2})^{\lambda} (1-3x)^{\lambda} ,$$

$$(2-3x)^{-m-1} = \sum_{\mu=0}^{\infty} \binom{m+\mu}{\mu} (-1)^{\mu} (1-3x)^{\mu}$$

$(m \in \mathbb{N}_0;\quad |x-\frac{1}{3}| < \frac{1}{3})$. -

The following construction then copies the development of $P_{m,n}$:

Remark

Let $K_m^x := K_m \otimes I$, $K_m^y := I \otimes K_m$.

Then all of the projectors $H_m^x, H_m^y, K_m^x, K_m^y$ commute.

Theorem

For $P_m := H_m^x K_m^y \oplus K_m^x H_m^y$ we have:

a) $P_m f = H_m^x K_m^y f + K_m^x H_m^y f - H_m^x H_m^y f \qquad (f \in C^{(m,m)}(U))$.

b) This yields the following cardinal representation for each $P_m f$ on U:

$$P_m f(x,y) =$$

$$\sum_{i,j=0}^{m} \sum_{u,v=0}^{1}$$

$$\{ D_x^i D_y^j f(u,v) [p_{i,m}^{(u)}(x) q_{j,m}^{(3v)}(y) + q_{i,m}^{(3u)}(x) p_{j,m}^{(v)}(y) - p_{i,m}^{(u)}(x) p_{j,m}^{(v)}(y)]$$

$$+ D_x^i D_y^j f(\tfrac{u+1}{3},v)\ q_{i,m}^{(u+1)}(x) p_{j,m}^{(v)}(y)$$

$$+ D_x^i D_y^j f(u,\tfrac{v+1}{3})\ p_{i,m}^{(u)}(x) q_{j,m}^{(v+1)}(y) \} .$$

c) $D_x^i D_y^j (P_m f)(\frac{u}{3},v) = D_x^i D_y^j f(\frac{u}{3},v)$

$D_x^i D_y^j (P_m f)(v,\frac{u}{3}) = D_x^i D_y^j f(v,\frac{u}{3})$

$(f \in C^{(m,m)}(U);\ i,j=0,\dots,m;\ u=0,\dots,3;\ v=0,1)$.

(c. f. GORDON / CHENEY [4]).

d) $Im(P_m) = \Pi_{2m+1} \otimes \Pi_{4m+3} + \Pi_{4m+3} \otimes \Pi_{2m+1}$.

(c. f. GORDON / CHENEY [4]).

e) P_m yields C^m-conforming elements.

f) $\overline{P_m} = \overline{K_m^x} + \overline{K_m^y} + \overline{H_m^x}\ \overline{H_m^y} - \overline{K_m^x}\ \overline{H_m^y} - \overline{H_m^x}\ \overline{K_m^y}$.

g) On U_h we get the error estimate

$\|\overline{P_m}f\|_\infty = O(h^{4m+4}) \qquad (f \in C^{(4m+4,2m+2)} \cap C^{(2m+2,4m+4)})$.

Example

We give a list of computed errors for the interpolation of $f(x,y) = e^{x\cdot y}$ on $[-1,1]\times[-1,1]$ using $n\times n$ rectangular elements P_m $(n=1,\dots,10;\ m=0,\dots,3)$ (table I).

According to the limited machine accuracy some of the list entries are not evaluated.

Table I. Interpolation error for $f(x,y) = e^{x\cdot y}$.

n	$\|\overline{P_0}f\|_\infty$	$\|\overline{P_1}f\|_\infty$	$\|\overline{P_2}f\|_\infty$	$\|\overline{P_3}f\|_\infty$
1	5.49e-1	4.67e-2	1.58e-3	2.84e-5
2	6.67e-2	5.04e-4	1.46e-6	2.18e-9
3	2.02e-2	3.58e-5	2.34e-8	8.11e-12
4	8.11e-3	4.96e-6	1.10e-9	(7.11e-13)
5	3.87e-3	1.02e-6	9.70e-11	-
6	2.07e-3	2.73e-7	1.32e-11	
7	1.21e-3	8.81e-8	2.64e-12	
8	7.50e-4	3.27e-8	(9.81e-13)	
9	4.90e-4	1.36e-8	-	
10	3.34e-4	6.13e-9		

We conclude by presenting plots of the function $e^{x \cdot y}$ as well as the C^0- and C^1-conforming errors on four adjacent rectangles. The interpolation nodes are marked by points in figure 1.

Fig. 1.
$f(x,y)=e^{x \cdot y}$

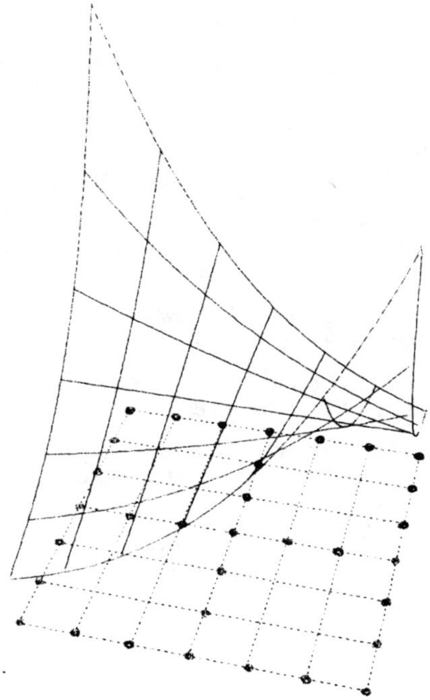

Fig. 2. $25 \cdot \overline{P_0} f$

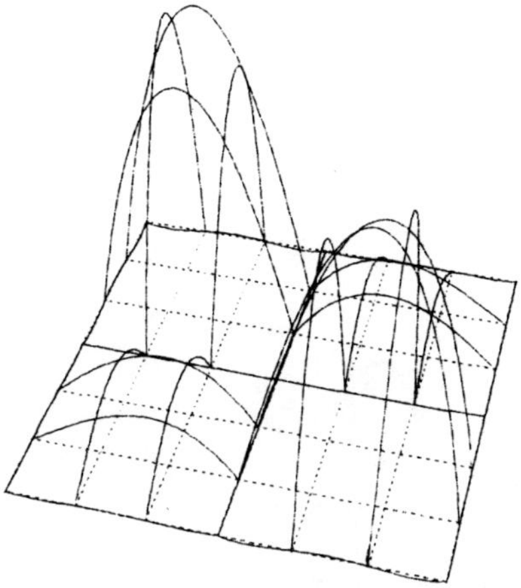

Fig. 3. $2500 \cdot \overline{P_1} f$

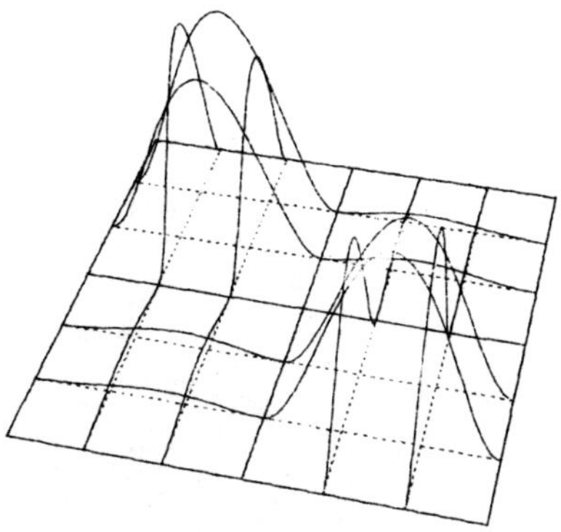

References

1. Baszenski, G., Posdorf, H., Delvos, F.J. (1980) Representation formulas for conforming bivariate interpolation In: Approximation theory III, ed. E. W. Cheney, 193-198.

2. Coons, S.A. (1964) Surfaces for computer aided design of space forms. Project MAC, design div., dept of mech. engineering, MIT.

3. Gordon, W.J. (1969) Distributive lattices and approximation of multivariate functions. Proc. Symp Madison (Wisc.), 223-277.

4. Gordon, W.J., Cheney, E.W. (1977) Bivariate and multivariate interpolation with noncommutative projectors. ISNM 40, 381-387.

5. Melkes, F. (1972) Reduced piecewise bivariate Hermite interpolations. Num Math 19, 326-340.

6. Phillips, G.M. (1973) Explicit forms for certain Hermite approximations. BIT 13, 177-180.

7. Watkins, D.S. (1976) On the construction of conforming rectangular plate elements. Int J Num methods engineering 10, 925-933.

8. Watkins, D.S. (1977) Error bounds for polynomial blending function methods. SIAM J Numer Anal 14 No 4, 721-734.

Günter Baszenski, Klaus Hackenberg, Rechenzentrum der Ruhr-Universität, Universitätsstr. 150, 4630 Bochum 1 (Fed. Rep. of Germany).

Dr. Franz-Jürgen Delvos, Lehrstuhl für Mathematik I, Gesamthochschule Siegen, Hölderlinstr. 3, 5900 Siegen 21 (Fed. Rep. of Germany).

EIN BEISPIEL ZUR THEORIE DER BESTEN APPROXIMATION

Helmut Braß

1. Das Resultat

Das Beispiel des Titels ist die beste Approximation im Tschebyscheffschen Sinne der Funktion

$$f_p(x_1,\dots,x_d) := x_1^{p_1} x_2^{p_2} \dots x_d^{p_d} \qquad p_\nu \in \mathbb{N}$$

aus dem Raum P_m^d der Polynome in d Variabeln vom Gesamtgrad $m := p_1+p_2+\dots+p_d-1$. Grundgebiet ist dabei das Simplex

$$Q_s := \{(x_1,\dots,x_d);\ -1 \leq x_1 \leq x_2 \leq \dots \leq x_d \leq 1\}.$$

Für andere Grundgebiete ist das Problem behandelt von EHLICH/ ZELLER [1] und SHAPIRO [7](achsenparalleler Würfel) sowie von GEARHART [2] und REIMER [4](Kreis), für die Kugel scheint das Problem schwieriger zu sein. In einer chinesisch geschriebenen Arbeit, von deren Existenz ich erst nach Fertigstellung dieser Arbeit erfuhr, hat LIANG [3] auch schon das Gebiet Q_s betrachtet, allerdings nur für d = 2, und in diesem Spezialfall Satz 1 bewiesen. Es ist dabei zu beachten, daß das Problem für d = 2 im Prinzip auf den Kreisfall zurückgeführt werden kann, durch

$$x_1 = 2y_1^2 - 1 \quad , \qquad x_2 = -\,2y_2^2 + 1$$

wird der Einheitskreis $y_1^2 + y_2^2 \leq 1$ auf Q_s abgebildet, und damit wird aus einem Proximum für $f(y_1,y_2) = y_1^{2p}\, y_2^{2q}$ ein Proximum für $f(x_1,x_2) = x_1^p\, x_2^q$ erzeugt. Die hier konstruierten Lösungen sind aber verschieden von denen, die man durch Transformation aus den GEARHARTschen oder den REIMERschen erhalten kann.

Satz 1. Es gilt

$$\inf_{v\in P_m^d} \| f_p - v \| = 2^{-m} \qquad (\| f \| := \sup_{x\in Q_s} |f(x)|)$$

und ein Proximum v_p wird gegeben durch

$$(1) \qquad f_p(x_1,\ldots,x_d) - v_p(x_1,\ldots,x_d) = \prod_{\lambda=1}^{d} \prod_{\nu=q_{\lambda-1}}^{q_\lambda - 1} (x_\lambda - \xi_\nu),$$

$$q_\lambda = p_1 + \ldots + p_\lambda + 1$$

dabei sind ξ_ν die der Größe nach geordneten Nullstellen des Tschebyscheffpolynoms T_{m+1}.

Das Beispiel verdient auch deswegen Interesse, weil es in gewisser Weise extrem singulär ist. Um diese Aussage zu präzisieren, muß an einige Begriffe und Resultate aus der Theorie der besten Approximation von Funktionen aus $C[Q]$ durch Elemente eines n-dimensionalen Unterraumes V erinnert werden:

Definition. Das Paar $(M;\sigma)$ heißt Extremalsignatur (ES) für V, wenn $M \subset Q$ endlich ist und σ eine auf M definierte Funktion mit $|\sigma(x)|=1$ für alle $x\in M$ ist, die noch der folgenden Bedingung genügt: Es gibt Zahlen $b_\nu \geq 0$ mit

$$\sum_{x_\nu \in M} b_\nu \sigma(x_\nu) v(x_\nu) = 0, \qquad \sum |b_\nu| \neq 0$$

für alle $v \in V$.

Eine primitive ES (PES) ist eine ES$(M;\sigma)$, für die $(M';\sigma|_{M'})$ mit $M' \subsetneq M$ keine ES ist.

Satz. v_0 ist genau dann Proximum für f, wenn eine PES $(M;\sigma)$ existiert mit

$$f(x)-v_0(x) = \|f-v_0\| \, \sigma(x) \qquad \text{für alle } x \in M.$$

Es ist dies die klassische Kennzeichnung der Proxima im Raum C[Q] in der von RIVLIN/SHAPIRO [6] gegebenen Form. Es kann viele Proxima geben, jedoch gehört nach einem Satz von RICE [5] zu allen die gleiche PES. Im Regelfall hat sie n+1 Punkte, die Singularität in unserem Beispiel ist darin zu sehen, daß die PES besonders wenig Punkte hat. Es ist nämlich leicht zu sehen, daß es in $V = P_m^d$ keine PES mit $|M| < m+2$ gibt, die einzigen PES mit $|M| = m+2$ bestehen aus m+2 Punkten auf einer Geraden, wobei benachbarten Punkten entgegengesetzte σ - Werte zugeordnet sind. Wie sich zeigen wird, liegt gerade dieser Typ von PES unserem Beispiel zugrunde.

Der folgende Satz (im wesentlichen bei TÖPFER [8]) zeigt, daß die Menge $V_0[f]$ der Proxima um so größer sein kann, je weniger Punkte die PES enthält.

Satz. Es gilt

$$\dim V_0[f] \leq n+1 - |M|$$

wenn $(M;\sigma)$ eine zu dem Approximationsproblem gehörende PES ist.

(Die Dimension der konvexen Menge $V_0[f]$ wird definiert als Dimension des von V_0 [f] erzeugten affinen Raumes). Der folgende Satz zeigt nun, daß im hier betrachteten Beispiel die maximale Menge von Proxima tatsächlich vorliegt. Es gilt (man beachte

$n = \dim P_m^d = \binom{m+d}{d}$) :

<u>Satz 2.</u> Sei $V_o[f_p]$ die Menge der Proxima des Problems aus Satz 1. Dann gilt

$$\dim V_o = \binom{m+d}{d} - m - 1 ,$$

sofern alle p_i positiv sind.

<u>2. Beweis von Satz 1</u>

$\xi_\nu = -\cos \frac{2\nu-1}{2(m+1)} \pi$ bedeuten stets die Nullstellen von T_{m+1}.

<u>Lemma 1.</u> Für $s = 1,...,m$ gilt

$$\left| \prod_{\nu=1}^{s} \frac{u-\xi_\nu}{v-\xi_\nu} \right| < 1$$

sofern $v \geq \xi_{s+1}$ und $-1 \leq u < v \leq 1$.

<u>Beweis.</u> Fall 1: $u \geq \xi_s$. Dann ist

$$0 \leq u - \xi_\nu < v - \xi_\nu \qquad \nu = 1,...,s$$

und also

$$\left| \frac{u-\xi_\nu}{v-\xi_\nu} \right| < 1 ,$$

woraus die Behauptung folgt.

Fall 2: $u < \xi_s$. Dann ist

$$\left|\prod_{\nu=1}^{s}\frac{u-\xi_\nu}{v-\xi_\nu}\right| \le \left|\prod_{\nu=1}^{s}\frac{u-\xi_\nu}{\xi_{s+1}-\xi_\nu}\right| \le \left|\prod_{\substack{\nu=1\\ \nu\neq s+1}}^{m+1}\frac{u-\xi_\nu}{\xi_{s+1}-\xi_\nu}\right|$$

$$= \left|\frac{T_{m+1}(u)}{(u-\xi_{s+1})\,T'_{m+1}(\xi_{s+1})}\right| <$$

$$\frac{1}{|\xi_{s+1}-\xi_s|\;|T'_{m+1}(\xi_{s+1})|} = \frac{\sin\frac{2s+1}{2(m+1)}\pi}{(m+1)[\cos\frac{2s-1}{2(m+1)}\pi - \cos\frac{2s+1}{2(m+1)}\pi]} =$$

$$\frac{1}{2(m+1)}\left[\operatorname{ctg}\frac{\pi}{2(m+1)} + \operatorname{ctg}\frac{s\pi}{2(m+1)}\right] \le \frac{1}{m+1}\operatorname{ctg}\frac{\pi}{2(m+1)} \le \frac{2}{\pi} .$$

<u>Beweis von Satz 1.</u> Zunächst ist klar, daß durch (1) wirklich ein $v_p \in P_m^d$ bestimmt ist.

Die Funktion auf der rechten Seite von (1) werde mit F bezeichnet. Es ist $F(x,x,\dots,x)=2^{-m}T_{m+1}(x)$, somit bilden die m+2 Extremstellen von T_{m+1} eine PES auf der Geraden $x_1=x_2=\dots=x_d$.

Es ist also nur noch $\|F\|=2^{-m}$ zu zeigen. Dieser Beweis wird durch Induktion nach d geführt. Der Induktionsanfang d=1 führt auf $F = 2^{-m}\,T_{m+1}$ und ist somit klar. Beim Induktionsschritt sind zwei Fälle zu unterscheiden:

Fall 1: Für ein $\lambda\in[2,d]$ sei $x_\lambda < \xi_{q_{\lambda-1}}$. Dann ist

$$|F(x_1,\dots,x_{\lambda-1},x_\lambda,x_{\lambda+1},\dots,x_d)| < |F(x_1,\dots,x_{\lambda-1},x_{\lambda-1},x_{\lambda+1},\dots,x_d)|,$$

und rechts kann die Induktionsvoraussetzung angewandt werden.

Fall 2: Für alle $\lambda\in[2,d]$ sei $x_\lambda \ge \xi_{q_{\lambda-1}}$. Man schreibt

$$F(x_1,\ldots,x_d) = 2^{-m}\, T_m(x_d) \prod_{\lambda=1}^{d-1} G_\lambda(x_\lambda, x_{\lambda+1}) ,$$

$$G_\lambda(x_\lambda, x_{\lambda+1}) = \prod_{\nu=1}^{q_\lambda -1} \frac{x_\lambda - \xi_\nu}{x_{\lambda+1} - \xi_\nu} .$$

Auf jeden Faktor G_λ kann Lemma 1 angewandt werden, sofern $x_\lambda < x_{\lambda+1}$. Ist $x_\lambda = x_{\lambda+1}$, so ist $G_\lambda = 1$ und die zu erweisende Ungleichung ebenfalls richtig. Man erkennt noch, daß F seine Extremwerte auf Q_s nur an den Stellen der PES annimmt.

3. Beweis von Satz 2

Es genügt offenbar, zu beweisen: Hat $P \in P_m^d$ die Eigenschaft $P(x,x,\ldots,x)=0$, so existiert ein $\varepsilon > 0$ mit

(2) $$\|F + \varepsilon P\| = 2^{-m} .$$

Denn die Gesamtheit dieser Polynome hat gerade die in Satz 2 angegebene Dimension, wie man so einsieht (diesen einfachen Beweis verdanke ich Herrn Möller): Es kommt offenbar nur darauf an, diejenigen Polynome zu erfassen, die längs einer Geraden verschwinden. Es wird übersichtlich, wenn man die Gerade $x_1=x_2=\ldots=x_{d-1}=0$ wählt. Schreibt man alle Monome hin, dann erkennt man, daß nur die m+1 Potenzen von x_d nicht zur betrachteten Menge gehören, damit hat man die gewünschte Dimension.

Der Beweis von (2) wird übersichtlicher, wenn man sich auf den Fall d=3 beschränkt und $x_1=x$, $x_2=y$, $x_3=z$ setzt.

Lemma 2. Es sei $T_{m+1}(\eta)=1$, aber $\eta \neq -1$. Dann ist $F_x(\eta,\eta,\eta) > 0$; $F_x(\eta,\eta,\eta) + F_y(\eta,\eta,\eta) > 0$.

<u>Beweis.</u> Man erkennt unschwer, daß es genügt, für

$$U(x) := \prod_{\nu=1}^{s} (x-\xi_\nu) \qquad s=1,\ldots,m$$

$\operatorname{sgn} U'(\eta) = \operatorname{sgn} U(\eta)$ zu beweisen. Ist $\eta < 1$, so geht man aus von

$$0 = \frac{T'_{m+1}(\eta)}{T_{m+1}(\eta)} = \sum_{\nu=1}^{m+1} \frac{1}{\eta-\xi_\nu} .$$

Hier sind die ersten Summanden positiv, die anderen negativ, also muß

$$\frac{U'(\eta)}{U(\eta)} = \sum_{\nu=1}^{s} \frac{1}{\eta-\xi_\nu} > 0$$

sein. Wegen $1-\xi_\nu > 0$ ist diese letzte Beziehung auch für $\eta=1$ richtig.

<u>Lemma 3.</u> η habe die Bedeutung aus Lemma 2. Es gibt ein $\varepsilon_1 > 0$ derart, daß für alle Punkte einer Umgebung U von (η,η,η) in Q_s gilt:

$$0 \leq F(x,y,z) \leq 2^{-m} - \varepsilon_1(y-x) - \varepsilon_1(z-y).$$

<u>Beweis.</u> Durch zweimalige Anwendung des Mittelwertsatzes folgt

$$F(x,y,z) = F(y,y,z) + (x-y)\, F_x(\theta,y,z) =$$

$$F(z,z,z) + (y-z)\, [F_x(\zeta,\zeta,z)+F_y(\zeta,\zeta,z)] +$$

$$(x-y)\, F_x(\theta,y,z)$$

mit $x \leq \theta \leq y$ und $y \leq \zeta \leq z$. Aus Stetigkeitsgründen gilt in einer passenden Umgebung U wegen Lemma 2

$$F_x(x,y,z) \geq \varepsilon_1 , \quad F_x(x,y,z) + F_y(x,y,z) \geq \varepsilon_1 ,$$

woraus wegen $F(z,z,z) \leq 2^{-m}$ die rechte Seite der behaupteten Ungleichung abzulesen ist. Die Gültigkeit der linken Hälfte läßt sich durch Verkleinerung von U immer erreichen.

Beweis von Satz 2. Durch Division mit Rest und Berücksichtigung von $P(z,z,z)=0$ erhält man die Existenz von Polynomen P_1 und P_2 mit

$$P(x,y,z) = (y-x)P_1(x,y,z) + (z-y)\, P_2\, (x,y,z).$$

Wegen Lemma 3 folgt hiermit, daß $F+\varepsilon P \leq 2^{-m}$ und sogar $|F+\varepsilon P| \leq 2^{-m}$ in U gilt, wenn ε genügend klein gewählt ist. So verfährt man an allen Extremstellen, nach der Bemerkung am Schluß des Beweises von Satz 1 kommen dafür nur die Stellen (η,η,η) mit $|T_{m+1}(\eta)| = 1$ in Frage. Ist $T_{m+1}(\eta) = -1$ und $\eta \neq -1$, so sind die erforderlichen Modifikationen fast trivial, ist $\eta = -1$, so geht man aus von

$$F(x,y,z) = F(-1,-1,-1) + (x+1)\, F_x(\ldots) + (y+1)\, F_y(\ldots) + (z+1) F_z(\ldots).$$

Da $\sup F < 2^{-m}$ außerhalb der Vereinigung der Umgebungen U gilt, kann (2) gegebenenfalls durch weitere Verkleinerung von ε erzwungen werden.

4. Ein weiteres Proximum

Das Proximum v_p aus Satz 1 hat irrationale Koeffizienten. Es möge daher ohne Beweis noch ein Proximum v_q mit rationalen Koeffizienten angegeben werden. Wegen der größeren Kompliziertheit beschränken wir uns auf $d = 3$ ($d = 2$ folgt bei der Spezialisierung $x = -1$) und schreiben wieder $x_1=x$, $x_2=y$, $x_3=z$. U_ν bedeute das Tschebyscheff-Polynom zweiter Art. Damit gilt

$$x^r y^s z^t - v_q(x,y,z) = 2^{-r-s-t+1}\{T_r(x)T_s(y)T_t(z)$$

$$+ (z-1)(y+1)\ U_{t-1}(z)\ U_{s-1}(y)T_r(x)$$

$$+ (z-1)(x+1)\ U_{t-1}(z)\ U_{r-1}(x)T_s(y)$$

$$+ (y-1)(x+1)\ U_{s-1}(y)\ U_{r-1}(x)T_t(z)\}.$$

5. Literatur

[1] Ehlich, H./ Zeller, K. (1966) Čebyšev-Polynome in mehreren Veränderlichen. Math. Z. 93, 142-143.

[2] Gearhart, W.B. (1973)Some Chebyshev approximations by polynomials in two variables.

[3] Liang, Xie-zhang (1979) Polynome mit kleinster Abweichung von Null in gewissen mehrdimensionalen Gebieten (Chinesisch). Numer. Math. Nanking 2, 189-193.

[4] Reimer, M. (1977) On multivariate polynomials of least deviation from zero on the unit ball. Math. Z. 153, 51-58.

[5] Rice, J. R. (1963) Tchebycheff approximation in several variables. Transactions AMS 109, 444-465.

[6] Rivlin, T. J./ Shapiro, H. S. (1961) A unified approach to certain problems of approximation and minimization. J. Soc. Indust. Appl. Math. 9, 670-699.

[7] Shapiro, H. S. (1967) Some theorems on Čebyšev approximation II. J. Math. Analysis Appl. 17, 262-268.

[8] Töpfer, H.-J. (1965) Über die Tschebyscheffsche Approximationsaufgabe bei nicht erfüllter Haarscher Bedingung. Berichte des Hahn-Meitner-Instituts für Kernforschung Berlin HMI- B40.

Prof. Dr. Helmut Braß, Lehrstuhl E für Mathematik, Technische Universität Braunschweig, Pockelsstr. 14, 3300 Braunschweig, West-Germany.

ON SPACES OF PIECEWISE POLYNOMIALS WITH BOUNDARY CONDITIONS. I. RECTANGLES

C. K. Chui and L. L. Schumaker
Center for Approximation Theory
Texas A&M University
College Station, Texas, USA

1. Introduction

In the past few years there has been considerable interest in spaces of piecewise polynomials defined on partitions of a plane region Ω. While the study of such spaces for general partitions is extremely difficult, considerable progress has been made for special kinds of partitions--see e.g. [1-8, 10-11] and references therein.

Despite the fact that they are of interest in data fitting, function approximation, and in finite element analysis, spaces of piecewise polynomials satisfying boundary conditions do not seem to have been studied. The purpose of this paper is to examine such spaces in the case where Ω is a rectangle which has been partitioned into subrectangles. Results for triangulations will appear elsewhere.

We devote the remainder of this introduction to notation. Let

(1.1) $$\Delta_k = \{a = x_0 < x_1 < \ldots . < x_{k+1} = b\}$$

and

(1.2) $$\tilde{\Delta}_{\tilde{k}} = \{\tilde{a} = \tilde{x}_0 < \tilde{x}_1 < \ldots . < \tilde{x}_{\tilde{k}+1} = \tilde{b}\} \quad .$$

Then Δ_k and $\tilde{\Delta}_{\tilde{k}}$ together define a partition of the rectangle $\Omega = [a,b] \otimes [\tilde{a},\tilde{b}]$ into subrectangles

$$(1.3) \qquad \Omega_{ij} = [x_i, x_{i+1}) \otimes [\tilde{x}_j, \tilde{x}_{j+1}) \quad , \quad \begin{array}{l} i = 0,1,\ldots,k. \\ j = 0,1,\ldots,\tilde{k}. \end{array}$$

We denote this partition of Ω by $\Delta_{k\tilde{k}}$.

Suppose now that d and μ are non-negative integers. Then we define

$$(1.4) \qquad \mathcal{S}_d^\mu (\Delta_{k\tilde{k}}) = \{s \in C^\mu(\Omega) : s|_{\Omega_{ij}} \in \pi_d^2 \ , \ \begin{array}{l} i=0,\ldots,k \\ j=0,\ldots,\tilde{k} \end{array}\}$$

where π_d^2 denotes the space of polynomials in x and $\tilde{x}$ of <u>total degree</u> d. $\mathcal{S}_d^\mu(\Delta_{k\tilde{k}})$ can be considered as a kind of spline space--for results on its dimension, see e.g. [1,2].

In this paper we are interested in subspaces of $\mathcal{S}_d^\mu(\Delta_{k\tilde{k}})$ which satisfy boundary conditions. Let α, β, $\tilde{\alpha}$, $\tilde{\beta}$ be non-negative integers, and define

$$(1.5) \qquad \mathcal{S} := \mathcal{S}_d^{\mu,\alpha,\beta,\tilde{\alpha},\tilde{\beta}}(\Delta_{k\tilde{k}}) = \{s \in \mathcal{S}_d^\mu(\Delta_{k\tilde{k}}) \ : \ D_x^i s(a,\tilde{x})=0,$$
$$i = 0,1,\ldots,\alpha \text{ and } D_x^j s(b,\tilde{x})=0, \ j=0,1,\ldots,\beta \text{ for all } \tilde{a} \le \tilde{x} \le \tilde{b}$$
$$\text{while } D_{\tilde{x}}^p s(x,\tilde{a}) = 0, \ p = 0, \ldots,\tilde{\alpha} \text{ and } D_{\tilde{x}}^q s(x,\tilde{b}) = 0, \ q = 0,1,\ldots,\tilde{\beta} \text{ for all } a \le x \le b\} .$$

The aim of this paper is to compute the dimension of $\mathcal{S}$ and to construct a local basis for it. We begin in section 2 with some preliminary results for one-dimensional splines. Our main results are in sections 3 and 4. Several special cases are discussed in section 5, and we conclude the paper with remarks and references.

2. One dimensional splines

Suppose π_q denotes the space of polynomials of degree q in one variable, and suppose Δ_k is a partition as in (1.1) for the interval [a,b]. Then given any negative integer μ, we define

$$(2.1)\qquad \mathscr{S}_q^{\mu}(\Delta_k) = \{s \in C^{\mu}[a,b] : \; s|_{[x_i,x_{i+1})} \in \pi_q, \; i=0,1,\ldots k\}.$$

This is the usual space of polynomial splines of degree q with $(q - \mu)$ tuple knots at the points $x_1,\ldots,x_k$.

In this section we are interested in subspaces of $\mathscr{S}_q^{\mu}(\Delta_k)$ which satisfy boundary conditions at a and b. Let $\alpha, \beta < q$ be non-negative integers, and let

$$(2.2)\qquad \mathscr{S}_q^{\mu,\alpha,\beta}(\Delta_k) = \{s \in \mathscr{S}_q^{\mu}(\Delta_k): D^i s(a) = 0, i = 0,1,\ldots,\alpha \text{ and } D^j s(b) = 0, j = 0,1,\ldots,\beta\}.$$

We begin this section with two elementary lemmas concerning this space of splines.

<u>Lemma 2.1</u> Suppose $q \leq \mu$. Then

$$(2.3)\qquad n_q := \dim \mathscr{S}_q^{\mu,\alpha,\beta}(\Delta_k) = (q - \alpha - \beta - 1)_+ \;,$$

and a basis is given by the polynomials

$$(2.4)\qquad \phi_i(x) = (x-a)^{\alpha+1}(b-x)^{\beta+1} x^{i-1} \;, \; i = 1,2,\ldots,q-\alpha-\beta-1.$$

<u>Proof</u>: Since $q \leq \mu$, clearly the spline space reduces to π_q. Now a polynomial which satisfies the boundary conditions must have the form $p(x) = (x-a)^{\alpha+1}(b-x)^{\beta+1} g(x)$, where $g \in \pi_{q-\alpha-\beta-2}$. The result follows. ∎

<u>Lemma 2.2</u> Suppose $q > \mu$. Then

(2.5) $$n_q := \dim \mathcal{S}_q^{\mu,\alpha,\beta}(\Delta_k) = [(k+1)(q-\mu) + \mu-\alpha-\beta-1]_+ \quad .$$

Moreover, if $\{N_i^{q+1}\}_{i=1}^{q+k(q-\mu)+1}$ is the basis of normalized B-spline spanning $\mathcal{S}_q^{\mu}(\Delta)$ (cf. [9], p. 116), then

(2.6) $$\{N_{i+\alpha+1}^{q+1}\}_{i=1}^{n_q} \text{ form a basis for } \mathcal{S}_q^{\mu,\alpha,\beta}(\Delta_k).$$

Finally, there exist points $a < t_1 < \ldots. < t_{n_q} < b$ such that

(2.7) $$\det\,(N_{i+\alpha+1}^{q+1}(t_j))_{i,j=1}^{n_q} \neq 0 \quad .$$

<u>Proof</u>: The space $\mathcal{S}_q^{\mu}(\Delta_k)$ has dimension $n := (k+1)(q-\mu) + \mu + 1$. Now by Theorem 4.67 of [9], there exist points $a = \tau_1 = \ldots = \tau_{\alpha+1} < \tau_{\alpha+2} < \ldots. < \tau_{n-\beta} = \ldots = \tau_n = b$ such that

$$\det(N_i^{q+1}(\tau_j))_{i,j=1}^{n} \neq 0 \quad .$$

The corresponding matrix has the form

$$\begin{bmatrix} A & \vdots & 0 & \vdots & 0 \\ \cdots & & \cdots & & \cdots \\ B & \vdots & C & \vdots & D \\ \cdots & & \cdots & & \cdots \\ 0 & \vdots & 0 & \vdots & G \end{bmatrix}$$

where A and G are $\alpha+1$ by $\alpha+1$ and $\beta+1$ by $\beta+1$ non-singular triangular matrices, respectively. We conclude that the space $\mathcal{S}_q^{\mu,\alpha,\beta}(\Delta_k)$ has dimension $n_q = n - \alpha - \beta - 2 = (k+1)(q-\mu) + \mu + 1 - \alpha - \beta - 2$, and that the B-splines $N_{\alpha+2}^{q+1},\ldots,N_{n-\beta-1}^{q+1}$ form a basis with

$$\det(N_i^{q+1}(\tau_j))_{i,j=\alpha+2}^{n-\beta-1} \neq 0 \quad .$$

Setting $t_i = \tau_{i+\alpha+1}$, $i = 1,\ldots,n_q$, we obtain (2.7). ∎

The basis given in Lemma 2.2 is made up of B-splines of degree q. For the construction of a basis for multi-dimensional spline spaces satisfying boundary conditions, it is important to have a basis for the one-dimensional spline space $\mathcal{S}_q^{\mu,\alpha,\beta}(\Delta_k)$ consisting of elements of lowest possible degree. The following lemma gives such a basis.

Lemma 2.3 Suppose $\mu - \alpha - \beta - 1 \geq 0$. Then there exist functions $\{\phi_i\}_1^\infty$ and points $\{t_i\}_1^\infty$ in $[a,b]$ such that for all $q \geq \alpha + \beta + 2$

(2.8) $\{\phi_i\}_1^{n_q}$ is a basis for $\mathcal{S}_q^{\mu,\alpha,\beta}(\Delta_k)$

with

(2.9) $\det\,(\phi_i(t_j))_{i,j=1}^{n_q} \neq 0.$

Moreover,

(2.10) $\phi_i \in \pi_{\alpha+\beta+1+i}\ ,\ i = 1,\ldots,\mu - \alpha - \beta - 1$

while

(2.11) $\phi_{n_{q-1}+1}, \ldots, \phi_{n_q}$ are B-splines of degree q

for $q \geq \mu+1$. The number of ϕ_i's of exact degree q is given by

(2.12) $$\nabla n_q = n_q - n_{q-1} = \begin{cases} 0\ ,\ q = 1, \ldots, \alpha+\beta+1 \\ 1\ ,\ q = \alpha+\beta+2, \ldots, \mu \\ k+1,\ q = \mu+1, \ldots \quad . \end{cases}$$

Proof: By Lemma 2.1, $n_q = 0$ for $q = 1,2,\ldots,\alpha+\beta+1$, while if $\alpha+\beta+2 \leq q \leq \mu$, then $n_q = q-\alpha-\beta-1$ and $\mathcal{S}_q^{\mu,\alpha,\beta}(\Delta_k)$ is spanned by the polynomials $\phi_1, \ldots, \phi_{q-\alpha-\beta-1}$ defined in (2.4). This establishes the lemma for $q = 1,2,\ldots,\mu$. We now proceed by induction on q. Suppose the lemma has been established for q-1. To advance the induction, we note that $\mathcal{S}_q^{\mu,\alpha,\beta}(\Delta_k)$ is spanned by $\{N_{i+\alpha+1}^{q+1}\}_1^{n_q}$

while $\mathcal{S}_{q-1}^{\mu,\alpha,\beta}(\Delta_k) \subseteq \mathcal{S}_q^{\mu,\alpha,\beta}(\Delta_k)$. Thus, we can substitute $\phi_1,\ldots,\phi_{n_{q-1}}$ for all but k+1 of these B-splines to obtain a new basis $\phi_1,\ldots,\phi_{n_q}$ for $\mathcal{S}_q^{\mu,\alpha,\beta}(\Delta_k)$. Assertions (2.11) and (2.12) follow for q, and the induction argument is complete. ■

<u>Lemma 2.4</u> Suppose $\mu-\alpha-\beta-1 < 0$, and let r be the unique integer such that

$$(r-1)(k+1) \le \alpha+\beta+1-\mu < r(k+1) \quad . \tag{2.13}$$

Then there exist functions $\{\phi_i\}_1^\infty$ and points $\{t_i\}_1^\infty$ in $[a,b]$ such that the statements (2.8), (2.9), and (2.11) hold for each $q \ge \mu + r$. The number of ϕ_i's of <u>exact degree</u> q is given by

$$\nabla n_q = n_q - n_{q-1} = \begin{cases} 0 \quad , \; q = 1,2,\ldots,\mu+r-1 \\ (k+1)r+\mu-\alpha-\beta-1, \; q = \mu+r \\ k+1, \; q = \mu+r+1,\ldots \quad . \end{cases} \tag{2.14}$$

<u>Proof</u>: By Lemmas 2.1 and 2.2, $n_q = 0$ for $q = 1,2,\ldots,\mu+r-1$. For $q = \mu+r$ we have $n_{\mu+r} = (k+1)r+\mu-\alpha-\beta-1$, and a basis is given by B-splines $\phi_i = N_{i+\alpha+1}^{\mu+r+1}$, $i=1,\ldots,n_{\mu+r}$ of degree $\mu+r$. Now we proceed by induction on q. Assuming the lemma has been established for q-1, we note that by Lemma 2.2, $\mathcal{S}_q^{\mu,\alpha,\beta}(\Delta_k)$ has a basis of B-splines $\{N_{i+\alpha+1}^{q+1}\}_1^{n_q}$ of degree q, while $\phi_1,\ldots,\phi_{n_{q-1}}$ form a basis for $\mathcal{S}_{q-1}^{\mu,\alpha,\beta}(\Delta_k) \subseteq \mathcal{S}_q^{\mu,\alpha,\beta}(\Delta_k)$. Substituting the ϕ_i's for all but k+1 of these B-splines, we obtain the result for q. ■

<u>3. Two dimensional splines with $d \le \mu$</u>

In this section we discuss the spline spaces defined in (1.5) in the simple case where the degree d is no larger than the smoothness μ.

Theorem 3.1 Suppose $d \le \mu$. Then

$$(3.1) \qquad \dim \mathcal{S}_d^{\mu,\alpha,\beta,\tilde\alpha,\tilde\beta}(\Delta_{k\tilde k}) = \frac{(d-\sigma+2)(d-\sigma+1)_+}{2} ,$$

where $\sigma = \alpha+\beta+\tilde\alpha+\tilde\beta+4$.

Proof: The condition $d \le \mu$ implies that $\mathcal{S}_d^{\mu} = \pi_d^2$. Now if $p \in \pi_d^2$ is a polynomial which satisfies the boundary conditions, it must have the form $p(x,\tilde x) = (x-a)^{\alpha+1}(b-x)^{\beta+1}(\tilde x-\tilde a)^{\tilde\alpha+1}(\tilde b-\tilde x)^{\tilde\beta+1} \cdot g(x,\tilde x)$, where $g \in \pi_{d-\sigma}^2$. The dimensionality of $\pi_{d-\sigma}^2$ is the number given in (3.1). A basis for this space is given by

$$(3.2) \qquad \{(x-a)^{\alpha+1}(b-x)^{\beta+1}(\tilde x-c)^{\tilde\alpha+1}(d-\tilde x)^{\tilde\beta+1}\, x^i \tilde x^{j-i}\}_{i=0,j=0}^{j,d-\sigma} .$$

4. Two dimensional splines with $d > \mu$

Throughout this section we assume that $d > \mu$. Our construction of a basis for the spline space $\mathcal{S}$ defined in (1.5) will involve taking tensor products of one-dimensional splines. Given $k,\tilde k,\Delta_k,\Delta_{\tilde k},\mu,\alpha,\beta,\tilde\alpha,\tilde\beta$, as in (1.5), suppose that r, n_q, $\{\phi_i\}_1^\infty$ and $\{t_i\}_1^\infty$ are the objects defined in Lemmas 2.1 - 2.4 for splines in the one variable x. Similarly, let $\tilde r,\tilde n_q,\{\tilde\phi_i\}_1^\infty$ and $\{\tilde t_i\}_1^\infty$ be the corresponding objects associated with splines in the variable $\tilde x$. Then $\{\phi_i\}_1^{n_q}$ and $\{\tilde\phi_i\}_1^{\tilde n_q}$ are bases for $\mathcal{S}_q^{\mu,\alpha,\beta}(\Delta_k)$ and $\mathcal{S}_q^{\mu,\tilde\alpha,\tilde\beta}(\Delta_{\tilde k})$, respectively.

In order to construct a basis for the space $\mathcal{S}$ of two-dimensional splines, it is natural to consider the tensor-product splines $\phi_i(x)\ \tilde\phi_j(\tilde x)$ with total degree at most d. The following theorem shows that this set of tensor-product splines does form a basis for $\mathcal{S}$.

Theorem 4.1 Suppose $d > \mu$. Then

$$(4.1) \qquad \dim \mathcal{S}_d^{\mu,\alpha,\beta,\tilde\alpha,\tilde\beta}(\Delta_{k\tilde k}) = \sum_{q=1}^{d} \tilde n_{d-q}\ \nabla n_q .$$

Moreover, a basis for this space is given by the tensor-product splines

(4.2) $$\bigcup_{q=1}^{d} \{\phi_i(x)\tilde{\phi}_j(\tilde{x})\}_{i=n_{q-1}+1}^{n_q}\ ,\ {}_{j=1}^{\tilde{n}_{d-q}}.$$

Proof: We continue to use the notation $\mathcal{S}$ for the space of splines in (1.5). Each of the splines listed in (4.2) is clearly of total degree at most d and satisfies all boundary conditions. The total number of splines given in (4.2) is $N:=\sum_{q=1}^{d} \tilde{n}_{d-q}\ \nabla n_q$. Since the splines (4.2) are obviously linearly independent, it follows that the dimension of $\mathcal{S}$ is at least N.

To show that the dimension can be no larger than N, consider the N linear functionals

(4.3) $$\lambda_{ij}f = f(t_i,\tilde{t}_j), \qquad \begin{aligned} i &= n_{q-1}+1,\dots,n_q \\ j &= 1,\dots,\tilde{n}_{d-q} \\ q &= 1,\dots,d. \end{aligned}$$

By a well known lemma (cf. Lemma 3.3 in [8]), N will be an upper bound for the dimension of $\mathcal{S}$ if we can show that

(4.4) $$\lambda_{ij}s = s(t_i,\tilde{t}_j) = 0 \text{ for all } i,j,q \text{ as in (4.3) implies s is identically zero.}$$

To show this, it suffices to show that (4.4) implies

(4.5) $$s(t_i,\cdot) = 0 \quad , \ i = 1,2,\dots,n_d \quad .$$

Indeed, since for each $\tilde{a} \le \tilde{x} \le \tilde{b}$, $s(\cdot,\tilde{x})$ is in the n_d-dimensional spline space $\mathcal{S}_d^{\mu,\alpha,\beta}(\Delta_k)$, coupling (4.5) with (2.9) implies that

(4.6) $$s(\cdot,\tilde{x}) \equiv 0 \quad \text{for all } \tilde{a} \le \tilde{x} \le \tilde{b}$$

and it follows that s is identically zero.

It remains to show that (4.4) implies (4.5). We give the proof only for the case where $\mu-\alpha-\beta-1 \geq 0$ since the other case is similar. In this case $n_1 = \ldots = n_{\alpha+\beta+1} = 0$, and so we need to show that

$$(4.7) \qquad s(t_i,\cdot) = 0 \quad , \quad i = n_{q-1}+1,\ldots,n_q$$

for $q = \alpha+\beta+2,\ldots,d$. This we do by induction on q. To get the induction started, we consider the case $q = \alpha+\beta+2$.

If s satisfies (4.6), then as observed above, we are done. If not, then there must be some $\tilde{a} \leq \tilde{x} \leq \tilde{b}$ such that $s(\cdot,\tilde{x})$ is a nontrivial spline of degree q. This implies that the one-dimensional spline $s(t_i,\cdot)$ must be of degree at most d-q. But for any i as in (4.7),

$$(4.8) \qquad s(t_i,\tilde{t}_j) = 0 \quad , \quad j = 1,2,\ldots,\tilde{n}_{d-q} \quad .$$

By our choice of the $\tilde{t}_j$'s, this implies that (4.7) must hold for $q = \alpha+\beta+2$.

To advance the induction, suppose that (4.7) has been established for q. Then if (4.6) does not hold, we conclude that $s(\cdot,\tilde{x})$ is a nontrivial spline of degree q+1 in x for some $\tilde{a} \leq \tilde{x} \leq \tilde{b}$. This implies that $s(t_i,\cdot)$ is of degree at most d-q-1, and condition (4.8) with q replaced by q+1 implies that (4.7) must hold for q+1. This completes the proof of the theorem. ∎

5. Uniform boundary conditions

In this section we give two corollaries of Theorem 4.1 for the special case where $\alpha = \tilde{\alpha} = \beta = \tilde{\beta}$. In this case we use the simpler notation $\mathcal{S}_d^{\mu,\alpha}(\Delta_{k\tilde{k}}) = \mathcal{S}_d^{\mu,\alpha,\alpha,\alpha,\alpha}(\Delta_{k\tilde{k}})$.

Corollary 5.1 For all nonnegative $d \geq 0$ and $\mu > 0$,

(5.1) $$\dim \mathcal{S}_d^{\mu,0}(\Delta_{k\tilde{k}}) = \tfrac{1}{2}[k\tilde{k}(d-2\mu)_+(d-2\mu-1) + (k+\tilde{k})(d-\mu-2)_+(d-\mu-1) + (d-3)_+(d-2)] .$$

Proof: By Lemmas 2.1 and 2.2, we know that

$$n_q = \begin{cases} 0 & , q = 0,\ 1 \\ (q-1) & , q = 2,\dots,\mu \\ (\mu-1) + (k+1)(q-\mu) & , q = \mu+1,\dots, \end{cases}$$

and a similar formula holds for $\tilde{n}_q$ with k replaced by $\tilde{k}$. There are two cases.

Case 1: Suppose $d > 2\mu+1$. Then by Theorem 4.1,

$$\dim \mathcal{S} = \sum_{q=2}^{\mu} [(\mu-1) + (\tilde{k}+1)(d-\mu-q)_+ + (k+1) \sum_{q=\mu+1}^{d-\mu-1} [\mu-1 + (\tilde{k}+1)(d-q-\mu)] + (k+1) \sum_{q=d-\mu}^{d-2} (d-q-1) .$$

Combining terms, we obtain (5.1).

Case 2: Suppose $d \le 2\mu+1$. Then Theorem 4.1 implies

$$\dim \mathcal{S} = \sum_{q=2}^{d-\mu-1} [(\mu-1) + (\tilde{k}+1)(d-q-\mu)] + \sum_{q=d-\mu}^{\mu} (d-q-1) + (k+1) \sum_{q=\mu+1}^{d-2} (d-q-1) .$$

This leads to (5.1). ■

Corollary 5.2 Suppose $k,\tilde{k} \ge \mu \ge 0$. Then for all $d > 2\mu+1$,

(5.2) $$\dim \mathcal{S}_d^{\mu,\mu}(\Delta_{k\tilde{k}}) = \tfrac{1}{2}[k\tilde{k}(d-2\mu-1)(d-2\mu) + (k+\tilde{k})(d-2\mu-1)\cdot \cdot(d-4\mu-2) + (d-2\mu-2)(d-4\mu-3) - 2\mu(d-3\mu-2)] \quad ,$$

and the dimension is 0 if $d \leq 2\mu+1$.

<u>Proof</u>: Lemma 2.4 with $r = \tilde{r} = 1$ gives the value of ∇n_q and $\tilde{n}_{d-q}$ needed in Theorem 4.1. The result follows after some algebra. ■

6. Remarks

1. In dealing with one-dimensional splines, most authors (cf. [9]) prefer to work with <u>order</u> rather than <u>degree</u>. Here we have chosen to work with degree, however, since we believe it is more convenient for functions of several variables.

2. Because of the symmetry of the problem, the dimension of the spline space $\mathcal{S}$ in Theorem 4.1 can also be written as $\sum_{q=1}^{d} n_{d-q} \; \nabla\tilde{n}_q$.

3. It is quite easy to construct a dual basis for the basis constructed in Theorem 4.1. In particular, if $\{\phi_i^*\}_1^\infty$ and $\{\tilde{\phi}_i^*\}_1^\infty$ are dual linear functionals corresponding to the functions $\{\phi_i\}_1^\infty$ and $\{\tilde{\phi}_i\}_1^\infty$ of Lemmas 2.3 and 2.4, then a dual basis for the basis in (4.2) is given by $\{\phi_i^* \; \tilde{\phi}_j^*\}_{i,j \in E}$, where E is the set of indices in (4.2).

4. One might expect that the dimensionality statement of Theorem 4.1 could be obtained by looking at the dimensionality of $\mathcal{S}_d^\mu(\Delta_{k\tilde{k}})$ and subtracting the number of conditions required to enforce the boundary conditions. While this works in some isolated cases, it does not work in general. This can be seen already for the simple case where $k = \tilde{k} = 0$. In this case $\mathcal{S}_d^\mu(\Delta_{00})$ has dimension $(d+1)(d+2)/2$ while the number of boundary conditions necessary to force an element in this space to be zero on the boundary is $4d$. The difference is $(d^2-5d+2)/2$ while the correct

dimension of $\mathcal{S}_d^{\mu,0}(\Delta_{00})$ is actually $(d^2-5d+6)/2$.

5. It is also easily shown that one cannot find the dimension of $\mathcal{S}_d^{\mu,0}(\Delta_{k\tilde{k}})$ from $\mathcal{S}_d^{\mu-1,0}(\Delta_{k\tilde{k}})$ simply by subtracting the number of linear functionals required to make an element of the second spline space be a member of the first.

6. Lemmas 2.3 and 2.4 show that the one-dimensional spline space $\mathcal{S}_q^{\mu,\alpha,\beta}(\Delta_k)$ has a basis of polynomials and splines of lowest possible degree, and that this basis can be obtained by substituting a basis for $\mathcal{S}_{q-1}^{\mu,\alpha,\beta}(\Delta_k)$ in place of the B-spline basis of degree d for $\mathcal{S}_q^{\mu,\alpha,\beta}(\Delta_k)$. This substitution must be done with care--one cannot retain an arbitrary set of k+1 B-splines in the basis.

References

1. Chui, C.K. and R.H. Wang, Bases of bivariate spline spaces with cross-cut grid partitions, J. Math. Res. and Exp. 2 (1982), 1-4.
2. Chui, C.K. and R.H. Wang, On smooth multivariate spline functions, CAT Report #3, Texas A&M Univ., 1981.
3. Chui, C.K. and R.H. Wang, On a bivariate B-spline basis, CAT Report #7, Texas A&M Univ., 1981.
4. Chui, C.K. and R.H. Wang, Multivariate spline spaces, J. Math. Anal. and Appl., to appear.
5. Fredrickson, P., Triangular spline interpolation, Report #670, Whitehead Univ., Canada, 1970.
6. Heindl, G., Interpolation and approximation by piecewise quadratic C1-functions of two variables, in Multivariate Approximation Theory, ed. by W. Schempp and K. Zeller, Birkhauser, Basel, 1979, 146-161.
7. Morgan, J. and R. Scott, A nodal basis for C1 piecewise polynomials of degree $n \geq 5$, Math. Comp. 29 (1975), 736-740.
8. Schumaker, L.L., On the dimension of spaces of piecewise polynomials in two variables, in Multivariate Approximation Theory, ed. by W. Schempp and K. Zeller, Birkhauser, Basel, 1979, 396-412.
9. Schumaker, L.L., Spline Functions: Basic Theory, Wiley, N.Y., 1981.
10. Strang, G., The dimension of piecewise polynomials and one-sided approximation, Springer-Verlag Lecture Notes 365, 1974, 144-152.
11. Zwart, P., Multi-variate splines with non-degenerate partitions, SIAM J. Numer. Anal. 10 (1973), 665-673.

SOME REMARKS ON MULTIVARIATE B-SPLINES

Wolfgang Dahmen and Charles A. Micchelli

In a recent paper [1], C. de Boor and K. Höllig introduced the following notion of a multivariate B-spline. Let P denote the orthogonal projection of R^n onto R^s, $n \geq s$. For any convex polyhedral body $B \subset R^n$ satisfying

$$(1) \qquad vol_{n-s}(\{u \in B: \ Pu = x\}) < \infty \qquad \text{for} \quad x \in R^s,$$

we define the B-spline $M_B(x)$, $x \in R^s$, by setting

$$(2) \qquad M_B(x) = vol_{n-s}(\{u \in B : \ Pu = x\})$$

or, equivalently, by requiring that

$$(3) \qquad \int_B \phi(Pu)du = \int_{R^s} M_B(x)\phi(x)dx$$

for all $\phi \in C_o(R^s)$, the set of all locally supported continuous functions on R^s. Note that the left hand side of (3) makes sense for any Lebesgue measurable set B and any linear map P from R^n into R^s, even for $n < s$. In general we interprete the right hand side of (3) in distributional sense and express the dependency of the corresponding distributions on P by writing $M_B(x|P)$.

Instances of such B-splines have been studied earlier. In fact, choosing B to be an n-dimensional convex cone generated by $u^1,\dots,u^n \in R^n$ (2), (3) defines the multivariate truncated power function [2, 4]. In this case condition (1) is equivalent to requiring that $0 \notin [Pu^1,\dots,Pu^n]$, the convex hull of the points Pu^i, $i=1,\dots,n$. Furthermore, for $B = \sigma = [v^o,\dots,v^n] \subset R^n$ one obtains the well-known simplicial B-spline whose structure is by now well understood [2, 4]. In particular, it has the nice property that it is completely determined by its knots in R^s, namely

$$(4) \qquad M(x|Pv^0,\dots,Pv^n) = M_\sigma(x)/\mathrm{vol}_n(\sigma)$$

does not depend on the special choice of the simplex $\sigma = [v^0,\dots,v^n]$ but only on its projected vertices $Pv^i = x^i \in R^s$. This characteristic feature is also expressed by the following recurrence relation due to the second named author, [4]. For $\mathrm{vol}_s([x^0,\dots,x^n]) > 0$, $n > s+1$ one has

$$(5) \qquad M(x|x^0,\dots,x^n) = \frac{n}{n-s} \sum_{j=0}^{n} \lambda_j M(x|x^0,\dots,x^{j-1},x^{j+1},\dots,x^n)$$

whenever

$$(6) \qquad x = \sum_{j=0}^{n} \lambda_j x^j, \qquad \sum_{j=0}^{n} \lambda_j = 1.$$

Moreover, for $D_z f = \sum_{i=1}^{s} z_i \frac{\partial f}{\partial x_i}$ one has [2, 4]

$$(7) \qquad D_z M(x|x^0,\dots,x^n) = n \sum_{j=0}^{n} \mu_j M(x|x^0,\dots,x^{j-1},x^{j+1},\dots,x^n)$$

whenever

$$(8) \qquad z = \sum_{j=0}^{n} \mu_j x^j, \qquad \sum_{j=0}^{n} \mu_j = 0.$$

The relations (5), (7) as well as the analogous formulas for the truncated powers, [2, 4], may be viewed as special cases of the following theorem given in [1]. To this end, let us denote by B_i a typical (n-1)-dimensional face with outer normal n^i of a given convex body $B \subset R^n$ while b^i will always stand for an arbitrary point in the flat spanned by B_i. Finally, $u \cdot v = \sum u_i v_i$ is the usual inner product of two vectors where it is understood that if $u \in R^n$, $v \in R^m$, $m < n$, v is to be extended to an n-vector by appending to it n-m zero components.

THEOREM 1. (de Boor, Höllig)

i) $M_B(x) = \frac{1}{n-s} \sum_i (b^i-u)\cdot n^i \, M_{B_i}(x)$, for all $u \in R^n$ such that $Pu = x$.

ii) $D_z M_B(x) = -\sum_i v\cdot n^i \, M_{B_i}(x)$, for all $v \in R^n$ such that $Pv = z$.

The derivation of Theorem 1 given in [1] was based on Stoke's theorem and did not explicitly use any of the known results for the simplicial B-spline.

The purpose of this note is to briefly point out two facts: Theorem 1 may be in turn viewed as a simple consequence of the relations (5), (7) for the simplicial B-spline. Second, inner products of the B-splines (2) also admit a recursive representation as in the case of the simplicial B-splines [3]. This latter fact may be of particular importance for eventual applications of the B-splines (2) in a finite element setting, say.

To derive Theorem 1 from (5), (7), we observe first that it is sufficient to confirm Theorem 1 for $B = \sigma = [v^0,\ldots,v^n]$. In fact, choosing a collection T of n-simplices such that $B = \cup\{\sigma \in T\}$ where $\sigma \cap \sigma'$ is either empty or a common face of σ, $\sigma' \in T$ we may write, in view of (3),

$$M_B(x) = \sum_{\sigma \in T} M_\sigma(x).$$

Once Theorem 1 has been proven for simplices we may apply the recursions to each of the summands $M_\sigma(x)$, $D_z M_\sigma(x)$ choosing the same b^i for any common face of two simplices. Observe now that every lower order simplicial B-spline in the resulting expansion which does not correspond to an (n-1)-face of B has to occur exactly twice since it has to be a common face of two adjacent elements of T. Consequently, by definition of b^i and n^i, the corresponding coefficients differ only in sign. Hence all such terms cancel and only the terms corresponding to faces of B are left which proves the assertion for arbitrary B.

So it remains to translate (5),(7) into the terminology of Theorem 1. Let $B = \sigma = [v^0,\dots,v^n]$, $\sigma_i = [v^0,\dots,v^{i-1},v^{i+1},\dots,v^n]$, $x^i = Pv^i$ and suppose b^i is in the flat spanned by σ_i having outer normal n^i, $i=0,\dots,n$.

Recall that for $u \in R^n$ the barycentric coordinates $\lambda_i(u)$ of u with respect to the simplex σ are given by

$$u = \sum_{j=0}^{n} \lambda_j(u) v^j, \qquad 1 = \sum_{j=0}^{n} \lambda_j(u).$$

In particular, for $u \in \sigma$ the coefficients $\lambda_i(u)$ may be represented as $\lambda_i(u) = \mathrm{vol}_n([u,\sigma_i])/\mathrm{vol}_n(\sigma)$. But since $\mathrm{vol}_n([u,\sigma_i]) = \mathrm{vol}_{n-1}(\sigma_i)(b^i-u)\cdot n^i/n$, for $u \in \sigma$, we may write for all $u \in R^s$

$$(9) \qquad \lambda_i(u) = \mathrm{vol}_{n-1}(\sigma_i)(b^i-u)\cdot n^i/(n\ \mathrm{vol}_n(\sigma)).$$

Hence, recalling that $x^i = Pv^i$, we obtain

$$(10) \qquad x = Pu = \sum_{j=0}^{n} \lambda_j(u) x^j, \quad 1 = \sum_{j=0}^{n} \lambda_j(u).$$

Thus, combining (4), (5) with (9) proves Theorem 1 i) for $B = \sigma$.

Moreover, for any $v \in R^n$ such that $Pv = z$, the coefficients $\mu_i = D_z\lambda_i(u)$ satisfy by (10) the relations (8) whence by (9) Theorem 1 ii) follows too.

Following the lead of [3] we are now ready to state

<u>THEOREM 2.</u> For any two polyhedral bodies $B \subset R^n$, $C \subset R^m$, $n \geq m > s$, let B_i, C_i denote the $(n-1)$, $(m-1)$-faces of B, C with outer normals n^i, m^i, respectively. For b^i, c^i in the flats

spanned by B_i, C_i, respectively, and an arbitrary $z \in R^m$ one has

$$\text{(11)} \qquad \int_{R^s} M_B(x)M_C(x)dx = \frac{1}{n+m-s}\Big(\sum_i (b^i-z)\cdot n^i \int_{R^s} M_{B_i}(x)M_C(x)dx + \sum_i (c^i-z)\cdot m^i \int_{R^s} M_B(x)M_{C_i}(x)dx\Big).$$

Perhaps the simplest way to prove the above assertion is to follow exactly the line of arguments in [3], by appropriately using barycentric coordinates as above. We prefer, however, to adopt a different point of view which takes advantage of the greater generality of our present setting. In fact, this allows us to point out that (11) is actually a special case of Theorem 1 i) because the inner product of two B-splines may be viewed as a certain higher order B-spline evaluated at zero. The reason for this, as we shall show below, is that the class of B-splines defined by (3) is closed under convolution.

LEMMA 1. For any two measurable sets $B \subset R^n$, $C \subset R^m$, both satisfying (1), and two linear maps P, Q from R^n, R^m into R^s, respectively, one has

$$\int_{R^s} M_B(x-y|P)M_C(y|Q)dy = M_{B\times C}(x|P \oplus Q), \qquad x \in R^s,$$

where $P \oplus Q$ is the direct sum of P and Q.

Proof. let $\phi \in C_o(R^s)$, then by (3)

$$\int_{R^s} \phi(x)\Big(\int_{R^s} M_B(x-y|P)M_C(y|Q)dy\Big)dx$$

$$= \int_{R^s} M_C(y|Q) \int_B \phi(y+Pu)du\, dy = \int_C \int_B \phi(Pu + Qv)dudv$$

$$= \int_{R^s} \phi(x)\, M_{B\times C}(x|P \oplus Q)dx.$$

Proof of Theorem 2. By Lemma 1 we have

$$\int_{R^s} M_B(x|P)\, M_C(x|Q)\,dx = M_{B\times C}(0|(-P)\oplus Q).$$

To make use of this identity we first observe that Theorem 1 remains valid for any linear map $P: R^n \to R^s$ with the interpretation that $M_{B_i}(x) = M_{B_i}(x|P_i)$ where P_i is the restriction of P to the flat spanned by B_i. Note that the (n+m-1)- faces of $B\times C$ have the form $B_i\times C$ or $B\times C_i$. Moreover, with b^i, c^i, n^i, m^i as before, (n^i,O), $(0,m^i) \in R^{n+m}$ are outer normals of $B_i\times C$, $B\times C_i$, respectively, while for $a \in C$, $w \in B$ we have that (b^i,a) and (w,c^i) belong to the flat spanned by $B_i\times C$, $B\times C_i$, respectively. For $n \geq m$, say, and some $z \in R^m$, let

$$u^O = (z,0,z), \qquad 0 \in R^{n-m}.$$

If P, Q are orthogonal projections from R^n, R^m onto R^s, then

$$((-P)\oplus Q)(u^O) = Qz - Pz = 0.$$

Thus, applying Theorem 1 i) to $M_{B\times C}(0|(-P)\oplus Q)$ yields

$$\int_{R^s} M_B(x)M_C(x)\,dx = \frac{1}{n+m-s}\Big(\sum_i (b^i-z)\cdot n^i\, M_{B_i\times C}(0|(-P)_i \oplus Q)$$

$$+ \sum_i (c^i-z)\cdot m^i\, M_{B\times C_i}(0|(-P)\oplus Q_i)\Big).$$

The assertion follows now from applying Lemma 1 to each of the summands above.

The above proof shows that the inner product representation in |3| for simplicial B-splines may be considered as a special case of the recurrence relation for a B-spline corresponding to a simploid, that is, a cross product of simplices.

Let us also observe that it is possible to keep the number of lower order B-splines occurring in the right hand side of (11) small by judiciously choosing the vector z. In fact, when z can be chosen as a common vertex of B and C the coefficients $(b^i-z)\cdot n^i$, $(c^i-z)\cdot m^i$ vanish if $z \in B_i$, $z \in C_i$, respectively.

Finally let us mention that the class of B-splines is also closed under pointwise multiplication. In fact,

$$\int_{R^s}\int_{R^t} M_B(x|P)M_C(y|Q)\phi(x,y)\,dxdy = \int_B\int_C \phi(Pu,Qv)\,dudv$$

$$= \int_{R^t}\int_{R^s} M_{B\times C}((x,y)\,|\,(P,Q))\phi(x,y)\,dxdy,$$

where we define $(P,Q): R^{n+m} \to R^{s+t}$ by $(P,Q)(x,y) = (Px,Qy)$. Thus we have the identity

$$M_B(x|P)M_C(y|Q) = M_{B\times C}((x,y)\,|\,(P,Q)).$$

REFERENCES

[1] C. de Boor, K. Höllig, Recurrence relations for multivariate B-splines, to appear in Proc. Amer. Math. Soc.

[2] W. Dahmen, On multivariate B-splines, SIAM J. Numer. Anal. 17 (1980), 179-191.

[3] W. Dahmen, C.A. Micchelli, Computation of inner products of multivariate B-splines. Numer. Funct. Anal. And Optimiz. 3 (1981), 367-375.

[4] C.A. Micchelli, On a numerically efficient method for computing multivariate B-splines, in Multivariate Approximation Theory, W. Schempp and K. Zeller, eds., Birkhäuser Basel, 1979, 211-248.

Wolfgang Dahmen
Fakultät für Mathematik
Universität Bielefeld
Universitätsstraße
4800 Bielefeld
West-Germany

Charles A. Micchelli
IBM Thomas J. Watson Research Center
P.O. Box 218
Yorktown Heights, New York 10598
U.S.A.

ON DISCRETE TRIVARIATE BLENDING INTERPOLATION

F.J. Delvos
Lehrstuhl für Mathematik I
University of Siegen, Siegen (West Germany)

In this paper we will discuss some trivariate polynomial interpolation schemes which are related to the method of transfinite trivariate blending function interpolation introduced by GORDON [6] . In particular we will derive explicit remainders for these interpolation schemes.

1. Trivariate blending

First we briefly recall the method of trivariate polynomial blending function interpolation [6] . Let $C(Q_3)$ be the algebra of continuous real valued functions f defined on the cube $Q_3 = [0,H]^3$ with side H . We consider parametrically extended polynomial Lagrange interpolation projectors associated with the sets of distinct interpolation points in $[0,H]$:

$$\left\{x_{i_1,1} : i_1 \in \mathbb{N}\right\}, \quad \left\{x_{i_2,2} : i_2 \in \mathbb{N}\right\}, \quad \left\{x_{i_3,3} : i_3 \in \mathbb{N}\right\}.$$

These interpolation projectors are defined by

$$P_1^{m_1}(f)(x_1,x_2,x_3) = \sum_{i_1=1}^{a_1(m_1)} f(x_{i_1,1},x_2,x_3) f_{i_1,1}^{m_1}(x_1,x_2,x_3) \quad ,$$

$$P_2^{m_2}(f)(x_1,x_2,x_3) = \sum_{i_2=1}^{a_2(m_2)} f(x_1,x_{i_2,2},x_3) f_{i_2,2}^{m_2}(x_1,x_2,x_3) \quad ,$$

$$P_3^{m_3}(f)(x_1,x_2,x_3) = \sum_{i_3=1}^{a_3(m_3)} f(x_1,x_2,x_{i_3,3}) f_3^{m_3}(x_1,x_2,x_3) \quad .$$

The blending functions are univariate Lagrange polynomials:

$$f_{i_u,u}^{m_u}(x_1,x_2,x_3) = \prod_{\substack{k_u=1 \\ k_u \neq i_u}}^{a_u(m_u)} (x_u - x_{k_u,u})(x_{i_u,u} - x_{k_u,u})^{-1} \quad ,$$

$$a_u(m_u) \in \mathbb{N} \ , \quad a_u(m_u) < a_u(m_u+1) \quad (m_u \in \mathbb{N},\ u=1,2,3) \quad .$$

The projectors $P_u^{m_u}$ ($m_u \in \mathbb{N}$, $u=1,2,3$) generate a maximal Boolean algebra $\mathbb{P}''$ of commuting projectors on $C(Q_3)$ which which contains also the product projectors

$$P_k^{m_k} P_l^{m_l} \ (m_k, m_l \in \mathbb{N} \ , \ 1=k<l\leq 3), \quad P_1^{m_1} P_2^{m_2} P_3^{m_3} \ (m_1,m_2,m_3 \in \mathbb{N})$$

as well as Boolean sums of these projectors.

The projector of trivariate blending is defined as

$$B_3 = P_1^1 \oplus P_2^1 \oplus P_3^1 = I - (I-P_1^1)(I-P_2^1)(I-P_3^1)$$

$$= P_1^1 + P_2^1 + P_3^1 - P_1^1P_2^1 - P_1^1P_3^1 - P_2^1P_3^1 + P_1^1P_2^1P_3^1$$

where I is the identity on $C(Q_3)$. B_3 possesses the following interpolation properties.

<u>Theorem 1</u> ([6])

For any $f \in C(Q_3)$ the trivariate blending interpolant $B_3(f)$ satisfies the interpolation conditions

$$B_3(f)(x_{i_1,1},x_2,x_3) = f(x_{i_1,1},x_2,x_3) \quad (i_1=1,..,a_1(1)) \ ,$$

$$B_3(f)(x_1,x_{i_2,2},x_3) = f(x_1,x_{i_2,2},x_3) \quad (i_2=1,..,a_2(1)) \ ,$$

$$B_3(f)(x_1,x_2,x_{i_3,3}) = f(x_1,x_2,x_{i_3,3}) \quad (i_3=1,\ldots,a_3(1)) .$$

The parametrically extended remainder projectors are defined as

$$R_u^{m_u} = I - P_u^{m_u} \quad (m_u \in \mathbb{N} , u=1,2,3) .$$

Then the remainder projector of trivariate blending has the simple form

$$R_3 = I - B_3 = R_1^1 R_2^1 R_3^1 .$$

More general we have

$$I - P_1^{n_1} \oplus P_2^{n_2} \oplus P_3^{n_3} = R_1^{n_1} R_2^{n_2} R_3^{n_3} \quad (n_u \in \mathbb{N} , u=1,2,3) .$$

To formulate error bounds we use the notation

$$M_{1,2,3}^{n_1,n_2,n_3}(f)$$
$$= ||D_{x_1}^{a_1(n_1)} D_{x_2}^{a_2(n_2)} D_{x_3}^{a_3(n_3)}(f)||_\infty (a_1(n_1)!a_2(n_2)!a_3(n_3)!)^{-1} .$$

Using standard results concerning the error of univariate Lagrange interpolation the following error bounds can be derived.

<u>Theorem 2</u>([6])
Assume that $f \in C^{a_1(n_1),a_2(n_2),a_3(n_3)}(Q_3)$. Then

$$||f - P_1^{n_1} \oplus P_2^{n_2} \oplus P_3^{n_3}(f)||_\infty$$
$$\leq M_{1,2,3}^{n_1,n_2,n_3}(f)\, H^{a_1(n_1)+a_2(n_2)+a_3(n_3)} .$$

This result has some simple corollaries .

Corollary 1

The order of convergence for trivariate blending is given by

$$r = a_1(1) + a_2(1) + a_3(1) \quad , \text{ i. e.}$$

$$\|f - B_3(f)\|_\infty = \mathcal{O}(H^r) \quad \text{as} \quad H \longrightarrow 0 \quad .$$

Let $\mathbb{P}_{s,3}$ denote the linear space of trivariate polynomials of total degree $\leq s$.

Corollary 2

The invariance set $\mathrm{im}(B_3)$ of B_3 [6] contains the space of trivariate polynomials of degree $\leq r-1$:

$$\mathbb{P}_{r-1,3} \subset \mathrm{im}(B_3) \quad .$$

2. Discrete trivariate blending

First we recall the methods of discretization proposed by COMAN [2] and GORDON [5] . Coman's method is based on replacing the projectors P_1^1, P_2^1, P_3^1 by the projectors

$$P_1^1P_2^3P_3^2 \quad , \quad P_1^2P_2^1P_3^3 \quad , \quad P_1^3P_2^2P_3^1 \quad .$$

This yields the Boolean sum projector

$$B_3' = P_1^1P_2^3P_3^2 \oplus P_1^2P_2^1P_3^3 \oplus P_1^3P_2^2P_3^1 \in \mathbb{P}'' \quad .$$

It is easily verified that Coman's projector B_3' has the explicit form

$$\begin{aligned} B_3' = {} & P_1^1P_2^3P_3^2 + P_1^2P_2^1P_3^3 + P_1^3P_2^2P_3^1 \\ & - P_1^1P_2^1P_3^2 - P_1^1P_2^2P_3^1 - P_1^2P_2^1P_3^1 \\ & + P_1^1P_2^1P_3^1 \quad . \end{aligned}$$

It follows from the general theory of Boolean sum interpolation that the set of interpolation points of B_3' is the union of the sets of interpolation points of the product projectors $P_1^1P_2^3P_3^2$, $P_1^2P_2^1P_3^3$, $P_1^3P_2^2P_3^1$.

Theorem 3

For any $f \in C(Q_3)$ Coman's interpolant $B_3'(f)$ satisfies the interpolation conditions

$$B_3'(f)(x_{i_1,1},x_{i_2,2},x_{i_3,3}) = f(x_{i_1,1},x_{i_2,2},x_{i_3,3})$$

$$\begin{aligned}(i_1 &= 1,..,a_1(1) , & i_2 &= 1,..,a_2(2) , & i_3 &= 1,..,a_3(3) & \text{or}\\ i_1 &= 1,..,a_1(3) , & i_2 &= 1,..,a_2(1) , & i_3 &= 1,..,a_3(2) & \text{or}\\ i_1 &= 1,..,a_1(2) , & i_2 &= 1,..,a_2(3) , & i_3 &= 1,..,a_3(1) &) .\end{aligned}$$

Next we consider Gordon's method of discrete trivariate blending. Gordon's method is based on replacing the projectors P_1^1, P_2^1, P_3^1 by the projectors

$$\begin{aligned} P_1^1(P_2^2P_3^3 \oplus P_2^3P_3^2) &= P_1^1P_2^2P_3^3 \oplus P_1^1P_2^3P_3^2 ,\\ P_2^1(P_3^2P_1^3 \oplus P_3^3P_1^2) &= P_1^3P_2^1P_3^2 \oplus P_1^2P_2^1P_3^3 ,\\ P_3^1(P_1^2P_2^3 \oplus P_1^3P_2^2) &= P_1^2P_2^3P_3^1 \oplus P_1^3P_2^2P_3^1 . \end{aligned}$$

This yields the Boolean sum projector

$$B_3'' = P_1^1P_2^2P_3^3 \oplus P_1^1P_2^3P_3^2 \oplus P_1^3P_2^1P_3^2 \oplus P_1^2P_2^1P_3^3 \oplus P_1^2P_2^3P_3^1 \oplus P_1^3P_2^2P_3^1 \in \mathbb{P}'' .$$

Using standard techniques of Boolean sum interpolation the following explicit representation for Gordon's projector B_3'' can be derived (see also [5]) :

$$\begin{aligned} B''_3 = \; & P_1^1P_2^2P_3^3 + P_1^1P_2^3P_3^2 + P_1^3P_2^1P_3^2 + P_1^2P_2^1P_3^3 + P_1^2P_2^3P_3^1 + P_1^3P_2^2P_3^1 \\ & - P_1^1P_2^2P_3^2 \qquad - P_1^2P_2^1P_3^2 \qquad - P_1^2P_2^2P_3^1 \\ & - P_1^1P_2^1P_3^3 \qquad - P_1^1P_2^3P_3^1 \qquad - P_1^3P_2^1P_3^1 \\ & + P_1^1P_2^1P_3^1 \quad . \end{aligned}$$

Concerning the set of interpolation points Theorem 3 has the following counterpart.

<u>Theorem 4</u>
For any $f \in C(Q_3)$ Gordon's interpolant $B''_3(f)$ satisfies the interpolation conditions

$$B''_3(f)(x_{i_1,1},x_{i_2,2},x_{i_3,3}) = f(x_{i_1,1},x_{i_2,2},x_{i_3,3})$$

$$(i_1 = 1,..,a_1(s(1)),\ i_2 = 1,..,a_2(s(2)),\ i_3 = 1,..,a_3(s(3))$$

$$\text{for any permutation } s = \begin{pmatrix} 1 & 2 & 3 \\ s(1) & s(2) & s(3) \end{pmatrix} \text{ of } \{1,2,3\}) .$$

Gordon's projector is greater than Coman's projector :

$$B''_3 \geq B'_3 \quad , \text{ i. e. } \quad B''_3\, B'_3 = B'_3 = B'_3\, B''_3$$

(see [5,6]) .

We will consider now a method of discrete trivariate blending which is based on replacing P_1^1, P_2^1, P_3^1 by the projectors

$$\begin{aligned} P_1^1(P_2^3P_3^1 \oplus P_2^2P_3^2 \oplus P_2^1P_3^3) &= P_1^1P_2^3P_3^1 \oplus P_1^1P_2^2P_3^2 \oplus P_1^1P_2^1P_3^3 \quad , \\ P_2^1(P_1^3P_3^1 \oplus P_1^2P_3^2 \oplus P_1^1P_3^3) &= P_1^3P_2^1P_3^1 \oplus P_1^2P_2^1P_3^2 \oplus P_1^1P_2^1P_3^3 \quad , \\ P_3^1(P_1^3P_2^1 \oplus P_1^2P_2^2 \oplus P_1^1P_2^3) &= P_1^3P_2^1P_3^1 \oplus P_1^2P_2^2P_3^1 \oplus P_1^1P_2^3P_3^1 \quad . \end{aligned}$$

This yields the Boolean sum projector

$$B_{5,3} = P_1^1P_2^1P_3^3 \oplus P_1^1P_2^2P_3^2 \oplus P_1^1P_2^3P_3^1 \oplus P_1^2P_2^1P_3^2 \oplus P_1^2P_2^2P_3^1 \oplus P_1^3P_2^1P_3^1 \in \mathbb{P}''$$

which has the explicit form

$$\begin{aligned} B_{5,3} &= P_1^1P_2^1P_3^3 + P_1^1P_2^2P_3^2 + P_1^1P_2^3P_3^1 + P_1^2P_2^1P_3^2 + P_1^2P_2^2P_3^1 + P_1^3P_2^1P_3^1 \\ &- 2(P_1^1P_2^1P_3^2 + P_1^1P_2^2P_3^1 + P_1^2P_2^1P_3^1) \\ &+ P_1^1P_2^1P_3^1 . \end{aligned}$$

In [3,4] we have introduced projectors $B_{q,3}$ of trivariate Boolean interpolation schemes for any q :

$$B_{q,3} = \bigoplus_{m_1+m_2+m_3=q} P_1^{m_1}P_2^{m_2}P_3^{m_3} \quad (q \in \mathbb{N} , \quad q \geq 3).$$

This projector possesses the explicit representation

$$B_{q,3} = \sum_{\substack{m_1+m_2+m_3 \\ =q}} P_1^{m_1}P_2^{m_2}P_3^{m_3} - 2\sum_{\substack{m_1+m_2+m_3 \\ =q-1}} P_1^{m_1}P_2^{m_2}P_3^{m_3} + \sum_{\substack{m_1+m_2+m_3 \\ =q-2}} P_1^{m_1}P_2^{m_2}P_3^{m_3} .$$

In particular we have

$$B_{3,3} = P_1^1P_2^1P_3^1 ,$$

$$B_{4,3} = P_1^1P_2^1P_3^2 + P_1^1P_2^2P_3^1 + P_1^2P_2^1P_3^1 - 2\, P_1^1P_2^1P_3^1 .$$

Concerning the set of interpolation points of $B_{q,3}$ Theorems 3 and 4 have the following counterpart.

Theorem 5 ([3,4])
For any $f \in C(Q_3)$ the trivariate Boolean interpolant $B_{q,3}(f)$ satisfies the interpolation conditions

$$B_{q,3}(f)(x_{i_1,1},x_{i_2,2},x_{i_3,3}) = f(x_{i_1,1},x_{i_2,2},x_{i_3,3})$$

$$(i_1 = 1,..,a_1(m_1),\ i_2 = 1,..,a_2(m_2),\ i_3 = 1,..,a_3(m_3)$$
$$\text{for any } m_1,m_2,m_3 \in \mathbb{N} \text{ such that } m_1+m_2+m_3=q) .$$

We note that the projectors $B_{5,3}$, B_3' , B_3'' satisfy $B_{5,3} \leq B_3' \leq B_3''$ in the sense of lattice-theoretical ordering.

3. Remainder formulas

In this section we will derive remainder formulas for the trivariate Boolean interpolation schemes presented in section 2 . First we consider Gordon's interpolation scheme. Using the representation formula for B_3'' and the parametrically extended remainder projectors we have

$$\begin{aligned} &B_3'' \\ &= (I-R_1^1)(I-R_2^2)(I-R_3^3) + (I-R_1^1)(I-R_2^3)(I-R_3^2) + (I-R_1^2)(I-R_2^1)(I-R_3^3) \\ &\quad + (I-R_1^3)(I-R_2^1)(I-R_3^2) + (I-R_1^2)(I-R_2^3)(I-R_3^1) + (I-R_1^3)(I-R_2^2)(I-R_3^1) \\ &\quad - (I-R_1^1)(I-R_2^2)(I-R_3^2) - (I-R_1^2)(I-R_2^1)(I-R_3^2) - (I-R_1^2)(I-R_2^2)(I-R_3^1) \\ &\quad - (I-R_1^1)(I-R_2^1)(I-R_3^3) - (I-R_1^1)(I-R_2^3)(I-R_3^1) - (I-R_1^3)(I-R_2^1)(I-R_3^1) \\ &\quad + (I-R_1^1)(I-R_2^1)(I-R_3^1) . \end{aligned}$$

After some calculations we obtain the following result.

<u>Theorem 6</u>

The remainder projector of Gordon's scheme for discrete trivariate blending is given by

$$
\begin{aligned}
R_3'' &= I - B_3'' \\
&= R_1^1R_2^2R_3^3 + R_1^1R_2^3R_3^2 + R_1^3R_2^1R_3^2 + R_1^2R_2^1R_3^3 + R_1^2R_2^3R_3^1 + R_1^3R_2^2R_3^1 \\
&\quad - R_1^1R_2^1R_3^3 - R_1^1R_2^2R_3^2 - R_1^1R_2^3R_3^1 - R_1^2R_2^1R_3^2 - R_1^2R_2^2R_3^1 - R_1^3R_2^1R_3^1 \\
&\quad + R_1^1R_2^1R_3^1 \\
&\quad - R_1^2R_2^3 - R_1^3R_2^2 - R_1^2R_3^3 - R_1^3R_3^2 - R_2^2R_3^3 - R_2^3R_3^2 \\
&\quad + R_1^2R_2^2 + R_1^2R_3^2 + R_2^2R_3^2 \\
&\quad + R_1^3 + R_2^3 + R_3^3 \quad .
\end{aligned}
$$

Using Theorem 2 with the obvious conventions $a_u(0) = 0$ and $P_u^0 = 0$, $R_u^0 = I$, $D_{x_u}^0 = I$ we have for sufficiently smooth functions

$$
\| R_1^{n_1} R_2^{n_2} R_3^{n_3}(f) \|_\infty \le M_{1,2,3}^{n_1,n_2,n_3}(f)\; H^{a_1(n_1)+a_2(n_2)+a_3(n_3)}
$$

with

$$
\begin{aligned}
&M_{1,2,3}^{n_1,n_2,n_3}(f) \\
&= \| D_{x_1}^{a_1(n_1)} D_{x_2}^{a_2(n_2)} D_{x_3}^{a_3(n_3)}(f) \|_\infty (a_1(n_1)!\,a_2(n_2)!\,a_3(n_3)!)^{-1}
\end{aligned}
$$

$$(n_1, n_2, n_3 \in \mathbb{Z}_+) \quad .$$

These inequalities together with the explicit remainder formula for Gordon's scheme of discrete trivariate blending function interpolation yield the following corollary .

Corollary 3

The order of convergence for Gordon's scheme is given by

$$r'' = \min\{a_1(3),\ a_2(3),\ a_3(3),\ a_1(2)+a_2(2),\ a_1(2)+a_3(2),\ a_2(2)+a_3(2),\ a_1(1)+a_2(1)+a_3(1)\} \quad ,$$

i. e. $\|f - B_3''(f)\|_\infty = \mathcal{O}(H^{r''})$ as $H \to 0$.

Moreover the invariance set of B_3'' contains the space of trivariate polynomials of degree $\leq r''-1$:

$$\mathbb{P}_{r''-1,3} \subset \operatorname{im}(B_3'') \quad .$$

Next we consider Coman's scheme for discrete trivariate blending. B_3' possesses the alternative representation

$$\begin{aligned} B_3' &= (I-R_1^1)(I-R_2^3)(I-R_3^2) + (I-R_1^2)(I-R_2^1)(I-R_3^3) + (I-R_1^3)(I-R_2^2)(I-R_3^1) \\ &- (I-R_1^1)(I-R_2^1)(I-R_3^2) - (I-R_1^1)(I-R_2^2)(I-R_3^1) - (I-R_1^2)(I-R_2^1)(I-R_3^1) \\ &+ (I-R_1^1)(I-R_2^1)(I-R_3^1) \end{aligned}$$

which yields the following

Theorem 7

The remainder of Coman's scheme has the form

$$\begin{aligned} R_3' &= I - B_3' \\ &= R_1^1R_2^3R_3^2 + R_1^2R_2^1R_3^3 + R_1^3R_2^2R_3^1 \\ &- R_1^1R_2^1R_3^2 - R_1^1R_2^2R_3^1 - R_1^2R_2^1R_3^1 \\ &+ R_1^1R_2^1R_3^1 \\ &- R_1^1R_2^3 - R_1^3R_2^2 - R_1^2R_3^3 - R_1^3R_3^1 - R_2^1R_3^3 - R_2^3R_3^2 \\ &+ R_1^2R_2^1 + R_1^1R_3^2 + R_2^2R_3^1 \\ &+ R_1^3 + R_2^3 + R_3^3 \quad . \end{aligned}$$

Corollary 4

The order of convergence for Coman's scheme is given by

$$r' = \min\{a_1(3),\ a_2(3),\ a_3(3),\ a_1(2)+a_2(1),\ a_1(1)+a_3(2),\ a_2(2)+a_3(1)\ a_1(1)+a_2(1)+a_3(1)\},$$

i. e. $\|f - B_3'(f)\|_\infty = \mathcal{O}(H^{r'})$ as $H \to 0$.

The invariance set of B_3' contains the space of trivariate polynomials of degree $\le r'-1$:

$$\mathbb{P}_{r'-1,3} \subset \mathrm{im}(B_3') \quad .$$

Our next objective is to determine the remainder projector for our scheme of discrete trivariate blending defined by $B_{5,3}$. We will prove a more general result.

Theorem 8

The remainder projector for trivariate Boolean interpolation defined by $B_{q,3}$ is given by

$$\begin{aligned} R_{q,3} &= I - B_{q,3} \\ &= \sum_{1\le k\le 3} R_k^{q-2} + \sum_{1\le k<l\le 3}\ \sum_{\substack{m_k+m_l\\=q-2}} R_k^{m_k} R_l^{m_l} + \sum_{\substack{m_1+m_2+m_3\\=q-2}} R_1^{m_1} R_2^{m_2} R_3^{m_3} \\ &\quad - \sum_{1\le k<l\le 3}\ \sum_{\substack{m_k+m_l\\=q-1}} R_k^{m_k} R_l^{m_l} - 2\sum_{\substack{m_1+m_2+m_3\\=q-1}} R_1^{m_1} R_2^{m_2} R_3^{m_3} \\ &\quad + \sum_{\substack{m_1+m_2+m_3\\=q}} R_1^{m_1} R_2^{m_2} R_3^{m_3} \quad . \end{aligned}$$

Proof: Using the representation formula for $B_{q,3}$ and the parametrically extended remainder projectors we can conclude

$$
\begin{aligned}
B_{q,3}
&= \overline{\sum_{m_1+m_2+m_3=q}}(I-R_1^{m_1})(I-R_2^{m_2})(I-R_3^{m_3}) \\
&\quad - 2\overline{\sum_{m_1+m_2+m_3=q-1}}(I-R_1^{m_1})(I-R_2^{m_2})(I-R_3^{m_3}) \\
&\quad + \overline{\sum_{m_1+m_2+m_3=q-2}}(I-R_1^{m_1})(I-R_2^{m_2})(I-R_3^{m_3}) \\
&= \binom{q-1}{2}I - 2\binom{q-2}{2}I + \binom{q-3}{2}I + \dots \\
&= I - \overline{\sum_{m_1+m_2+m_3=q}}(R_1^{m_1}+R_2^{m_2}+R_3^{m_3}) \\
&\quad + 2\overline{\sum_{\substack{m_1+m_2+m_3\\=q-1}}}(R_1^{m_1}+R_2^{m_2}+R_3^{m_3}) - \overline{\sum_{\substack{m_1+m_2+m_3\\=q-2}}}(R_1^{m_1}+R_2^{m_2}+R_3^{m_3}) + \dots \\
&= I - \Big(\sum_{m_1=1}^{q-2}(q-m_1-1)R_1^{m_1} + \dots\Big) \\
&\quad + 2\Big(\sum_{m_1=1}^{q-3}(q-m_1-2)R_1^{m_1} + \dots\Big) - \Big(\sum_{m_1=1}^{q-4}(q-m_1-3)R_1^{m_1} + \dots\Big) + \dots \\
&= I - (R_1^{q-2} + 2R_1^{q-3} + \dots) + 2(R_1^{q-3} + \dots) + \dots \\
&= I - (R_1^{q-2} + R_2^{q-2} + R_3^{q-2}) \\
&\quad + \overline{\sum_{m_1+m_2+m_3=q}}(R_1^{m_1}R_2^{m_2}+R_1^{m_1}R_3^{m_3}+R_2^{m_2}R_3^{m_3}) \\
&\quad - 2\overline{\sum_{m_1+m_2+m_3=q-1}}(R_1^{m_1}R_2^{m_2}+R_1^{m_1}R_3^{m_3}+R_2^{m_2}R_3^{m_3}) \\
&\quad + \overline{\sum_{m_1+m_2+m_3=q-2}}(R_1^{m_1}R_2^{m_2}+R_1^{m_1}R_3^{m_3}+R_2^{m_2}R_3^{m_3}) + \dots
\end{aligned}
$$

$$= I - (R_1^{q-2} + R_2^{q-2} + R_3^{q-2})$$

$$+ \left(\sum_{m_3=1}^{q-2} \sum_{m_1+m_2=q-m_3} R_1^{m_1} R_2^{m_2} + \dots\right) - 2\left(\sum_{m_3=1}^{q-3} \sum_{m_1+m_2=q-1-m_3} R_1^{m_1} R_2^{m_2} + \dots\right)$$

$$+ \left(\sum_{m_3=1}^{q-4} \sum_{m_1+m_2=q-2-m_3} R_1^{m_1} R_2^{m_2} + \dots\right) + \dots$$

$$= I - (R_1^{q-2} + R_2^{q-2} + R_3^{q-2})$$

$$+ \left(\sum_{m_1+m_2=q-1} R_1^{m_1} R_2^{m_2} + \sum_{m_1+m_2=q-2} R_1^{m_1} R_2^{m_2} + \dots\right)$$

$$- 2\left(\sum_{m_1+m_2=q-2} R_1^{m_1} R_2^{m_2} + \dots\right) + \dots$$

$$= I$$

$$- (R_1^{q-2} + R_2^{q-2} + R_3^{q-2})$$

$$+ \left(\sum_{\substack{m_1+m_2\\=q-1}} R_1^{m_1} R_2^{m_2} + \sum_{\substack{m_1+m_3\\=q-1}} R_1^{m_1} R_3^{m_3} + \sum_{\substack{m_2+m_3\\=q-1}} R_2^{m_2} R_3^{m_3}\right)$$

$$- \left(\sum_{\substack{m_1+m_2\\=q-2}} R_1^{m_1} R_2^{m_2} + \sum_{\substack{m_1+m_2\\=q-2}} R_1^{m_1} R_3^{m_3} + \sum_{\substack{m_2+m_3\\=q-2}} R_2^{m_2} R_3^{m_3}\right)$$

$$- \sum_{\substack{m_1+m_2+m_3\\=q}} R_1^{m_1} R_2^{m_2} R_3^{m_3} + 2 \sum_{\substack{m_1+m_2+m_3\\=q-1}} R_1^{m_1} R_2^{m_2} R_3^{m_3} - \sum_{\substack{m_1+m_2+m_3\\=q-2}} R_1^{m_1} R_2^{m_2} R_3^{m_3}$$

$$= B_{q,3}$$

whence Theorem 8 is proved.

For $q = 5$ we obtain the special result :

Theorem 9

The remainder of discrete trivariate blending defined by $B_{5,3}$ possesses the representation

$$
\begin{aligned}
R_{5,3} &= I - B_{5,3} \\
&= R_1^1R_2^1R_3^3 + R_1^1R_2^2R_3^2 + R_1^1R_2^3R_3^1 + R_1^2R_2^1R_3^2 + R_1^2R_2^2R_3^1 + R_1^3R_2^1R_3^1 \\
&- 2(R_1^1R_2^1R_3^2 + R_1^1R_2^2R_3^1 + R_1^2R_2^1R_3^1) \\
&+ R_1^1R_2^1R_3^1 \\
&- (R_1^1R_2^3 + R_1^2R_2^2 + R_1^3R_2^1 + R_1^1R_3^3 + R_1^2R_3^2 + R_1^3R_3^1 + R_2^1R_3^3 + R_2^2R_3^2 + R_2^3R_3^1) \\
&+ (R_1^1R_2^2 + R_1^2R_2^1 + R_1^1R_3^2 + R_1^2R_3^1 + R_2^1R_3^2 + R_2^2R_3^1) \\
&+ R_1^3 + R_2^3 + R_3^3 \quad .
\end{aligned}
$$

Corollaries 3 and 4 have the following counterpart.

Corollary 6

The order of convergence for $B_{5,3}$ is given by

$$
\begin{aligned}
r_5 = \min \{ & a_1(3),\ a_2(3),\ a_3(3), \\
& a_1(1)+a_2(2),\ a_2(2)+a_1(1), \\
& a_1(1)+a_3(2),\ a_3(2)+a_1(1), \\
& a_2(1)+a_3(2),\ a_3(2)+a_2(1), \\
& a_1(1)+a_2(1)+a_3(1) \} \quad ,
\end{aligned}
$$

i. e $\quad ||f - B_{5,3}(f)||_\infty = \mathcal{O}(H^{r_5})$ as $H \longrightarrow 0$.

The invariance set of $B_{5,3}$ contains the space of trivariate polynomials of degree $\leq r_5-1$:

$$\mathbb{P}_{r_5-1,3} \subset \operatorname{im}(B_{5,3}) \quad .$$

4. The number of interpolation points

The numbers of interpolation points for the different schemes of discrete trivariate blending are just the dimensions of the ranges of the projectors B_3'' , B_3' , $B_{5,3}$ respectively .

The "orthogonal" decomposition

$$\begin{aligned} B_3'' = {} & P_1^1P_2^2(P_3^3-P_3^2) + P_1^1P_2^3(P_3^2-P_3^1) \\ & + (P_1^3-P_1^2)P_2^1P_3^2 + (P_1^2-P_1^1)P_2^1P_3^3 \\ & + P_1^2(P_2^3-P_2^2)P_3^1 + P_1^3(P_2^2-P_2^1)P_3^1 + P_1^1P_2^1P_3^1 \end{aligned}$$

yields the following result .

Theorem 10

The number of interpolation points for Gordon's scheme of discrete trivariate blending is given by

$$\begin{aligned} d'' = {} & \dim(\mathrm{im}(B_3'')) \\ = {} & a_1(1)(a_2(2)(a_3(3) - a_3(2)) + a_2(3)(a_3(2) - a_3(1))) \\ & + a_2(1)(a_3(2)(a_1(3) - a_1(2)) + a_3(3)(a_1(2) - a_1(1))) \\ & + a_3(1)(a_1(2)(a_2(3) - a_2(2)) + a_1(3)(a_2(2) - a_2(1))) \\ & + a_1(1)a_2(1)a_3(3) \quad . \end{aligned}$$

The projector B_3' possesses the orthogonal decomposition

$$B_3' = P_1^1(P_2^3-P_2^1)P_3^2 + P_1^2P_2^1(P_3^3-P_3^1) + (P_1^3-P_1^1)P_2^2P_3^1 + P_1^1P_2^1P_3^1$$

which implies the following theorem .

Theorem 11

The number of interpolation points for Coman's scheme of discrete trivariate blending is given by

$$\begin{aligned} d' = {} & a_1(1)a_3(2)(a_2(3)-a_2(1)) + a_1(2)a_2(1)(a_3(3)-a_3(1)) \\ & + (a_1(3)-a_1(1))a_2(2)a_3(1) + a_1(1)a_2(1)a_3(1) \quad . \end{aligned}$$

For the projector $B_{5,3}$ the orthogonal decomposition

$$\begin{aligned} B_{5,3} &= P_1^1P_2^1(P_3^3-P_3^2) + P_1^1P_3^1(P_2^3-P_2^2) + P_2^1P_3^1(P_1^3-P_1^2) \\ &+ P_1^1P_3^2(P_2^2-P_2^1) + P_1^2P_2^1(P_3^2-P_3^1) + P_2^2P_3^1(P_1^2-P_1^1) + P_1^1P_2^1P_3^1 \end{aligned}$$

yields the following

<u>Theorem 12</u>

The number of interpolation points for the scheme of discrete trivariate blending defined by $B_{5,3}$ is given by

$$\begin{aligned} d_5 = \; & a_1(1)(a_2(1)(a_3(3)-a_3(2)) + a_3(2)(a_2(2)-a_2(1))) \\ &+ a_2(1)(a_1(2)(a_3(2)-a_3(1)) + a_3(1)(a_1(3)-a_1(2))) \\ &+ a_3(1)(a_1(1)(a_2(3)-a_2(2)) + a_2(2)(a_1(2)-a_1(1))) \\ &+ a_1(1)a_2(1)a_3(1) \quad . \end{aligned}$$

Finally we will assume

$$a_1(m) = a_2(m) = a_3(m) = a(m) \quad (m \in \mathbb{N}) \quad ,$$
$$a(m) = m\,a(1) = m\,n \quad (n \in \mathbb{N}) \quad .$$

In this case the orders of convergence for the different schemes of discrete trivariate blending are equal to the order of convergence for transfinite trivariate blending :

$$r = r'' = r' = r_5 = 3n \quad (n \in \mathbb{N}) \quad .$$

The numbers of interpolation points for the different schemes of discrete trivariate blending are given by

$$d'' = 16\,n^3 \quad , \quad d' = 13\,n^3 \quad , \quad d_5 = 10\,n^3 \quad .$$

In this connection it is interesting to note that the order of convergence for $P_1^3P_2^3P_3^3$ is also $3\,n$ while the number of interpolation points is given by $27\,n^3$.

Possible distributions of the interpolation points for the simplest case $n = 1$ are indicated in the following figures:

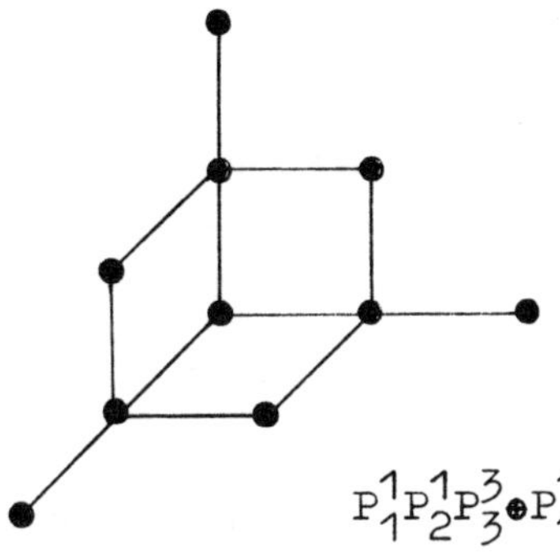

$$P_1^1P_2^1P_3^3 \oplus P_1^1P_2^2P_3^2 \oplus P_1^1P_2^3P_3^1 \oplus P_1^2P_2^1P_3^2 \oplus P_1^2P_2^2P_3^1 \oplus P_1^3P_2^1P_3^1$$

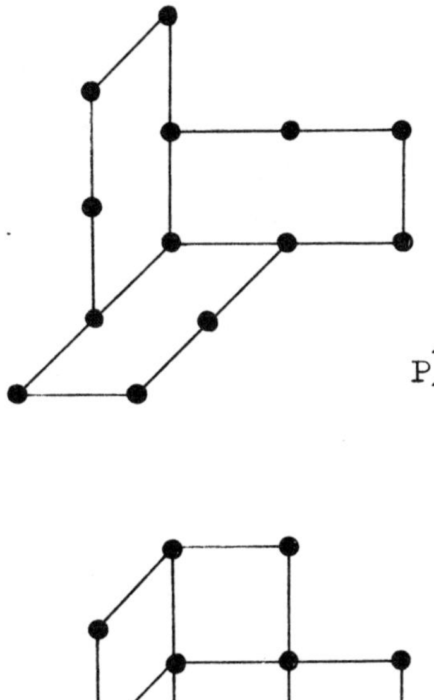

$$P_1^1P_2^3P_3^2 \oplus P_1^2P_2^1P_3^3 \oplus P_1^3P_2^2P_3^1$$

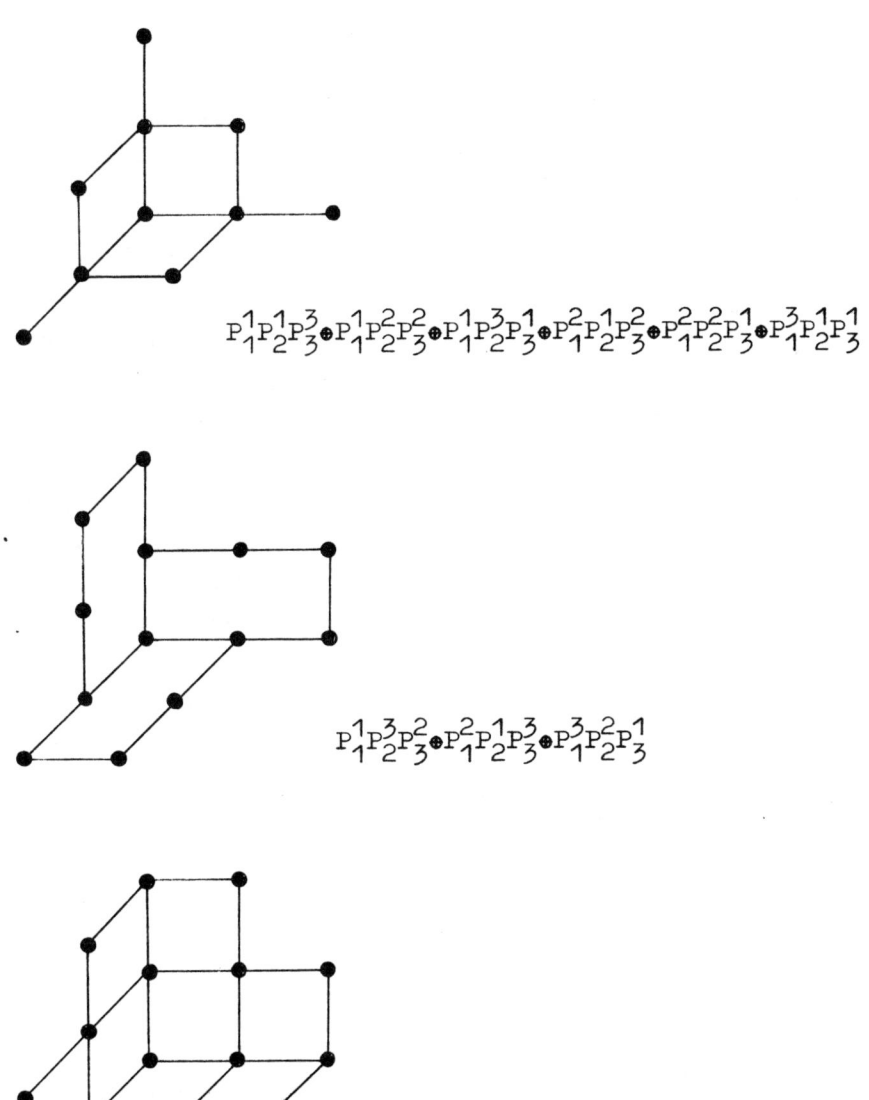

$$P_1^1P_2^2P_3^3 \oplus P_1^1P_2^3P_3^2 \oplus P_1^3P_2^1P_3^2 \oplus P_1^2P_2^1P_3^3 \oplus P_1^2P_2^3P_3^1 \oplus P_1^3P_2^2P_3^1$$

References

1 E. W. CHENEY , W. J. GORDON : Bivariate and multivariate interpolation with noncommutative projectors . In "Linear spaces and Approximation" (ed. by P. L. Butzer and B.Sz. Nagy), ISNM 40 , 1977 , pp. 381 - 387 .

2. Gh. COMAN : Multivariate approximation schemes and the approximation of linear functionals. Mathematica 16 , 1974, pp. 229 - 249 .

3. F. J. DELVOS : d-variate Boolean interpolation . J. of Approximation Theory, 34 , 1982, pp. 99-114 .

4. F. J. DELVOS , H. POSDORF : Boolean trivariate interpolation. To appear in "Proceedings of the international conference on functions, series, operators" , Budapest 1980

5. W. J. GORDON : Distributive lattices and the approximation of multivariate functions . In "Approximation with special emphasis on spline functions" (ed. by I. J. Schoenberg) . Academic Press, New York , 1969 , pp. 223 - 277 .

6. W. J. GORDON : Blending function methods of bivariate and multivariate interpolation and approximation . SIAM J. Numer. Anal. 8, 1971, pp. 158 .

Priv.-Doz. Dr. F. J. Delvos

Lehrstuhl für Mathematik I
University of Siegen
Hölderlinstraße 3

D-5900 Siegen , West Germany

ON PRECISION SETS OF INTERPOLATION PROJECTORS

F. J. Delvos , W. Schempp

Lehrstuhl für Mathematik I
University of Siegen , Siegen
West Germany

0. Introduction

Let (X,d) be a compact metric space and let $C(X)$ denote the real Banach algebra of all continuous functions defined on X . For any continuous projector P on $C(X)$ its precision set $\mathrm{prec}(P)$ is defined by

$$\mathrm{prec}(P) = \{y \in X : \hat{y}P = \hat{y}\}$$

where $\hat{y} \in C(X)'$ denotes the point evaluation at y , i. e. , the Dirac measure with carrier $\{y\}$:

$$\hat{y}(f) = f(y) \quad (f \in C(X)) .$$

P is called an interpolation projector iff

(i) $\mathrm{prec}(P) \neq \emptyset$

(see Lancaster [9]) . We will call P an interpolator iff P is an interpolation projector and if the following two additional conditions are satisfied :

(ii) $P(1) = 1$;

(iii) $\hat{x}P \in \mathrm{lin}\{\hat{y} : y \in \mathrm{prec}(P)\}$ for all $x \in X$;

where lin stands for the linear hull.

This paper presents some general results concerning products and Boolean sums of commuting interpolators . In this connection we also correct a statement in [7] on precision sets of commuting projectors.

1. Interpolators with finite and transfinite precision sets

In many applications interpolation projectors are such that prec(P) is finite, i.e., we have

$$\mathrm{prec}(P) = \{y_1,\dots,y_m\} \quad \text{with} \quad m > 1 \quad .$$

For $i = 1,\dots,m$ put

$$A_i = \mathrm{prec}(P)\setminus\{y_i\} \quad .$$

Then the functions $h_1,\dots,h_m$ defined by

$$h_i(x) = d(x,A_i)(d(x,y_i)+d(x,A_i))^{-1} \quad (i=1,\dots,m)$$

are continuous on X and satisfy the cardinality relations :

$$h_i(y_j) = \delta_{i,j} \quad (i,j = 1,\dots,m) \quad .$$

The functions $g_1,\dots,g_m$ are defined by

$$g_i = P(h_i) \quad (i=1,\dots,m) \quad .$$

Since $\mathrm{prec}(P) = \{y_1,\dots,y_m\}$ they also satisfy the cardinality relations :

$$g_i(y_j) = \delta_{i,j} \quad (i,j = 1,\dots,m) \quad .$$

Moreover, condition (ii) yields the equality

$$\sum_{i=1}^{m} g_i = 1 \quad .$$

Next we will show that condition (iii) yields the representation

$$P = \sum_{i=1}^{m} \hat{y}_i \otimes g_i$$

where $\hat{y}_i \otimes g_i$ denotes the (dyadic) tensorproduct :

$$\hat{y}_i \otimes g_i(f) = f(y_i)\,g_i \qquad (f \in C(X)) .$$

For this purpose suppose that $g \in \mathrm{Im}(P)$ and define G by

$$G = \sum_{i=1}^{m} g(y_i) g_i .$$

Then we have

$$G(y_i) = g(y_i) \qquad (i=1,\ldots,m)$$

and in view of condition (iii) for any $x \in X$ there are real numbers $a_1(x) , \ldots , a_m(x)$ such that

$$g(x) = P(g)(x) = \sum_{i=1}^{m} a_i(x) g(y_i)$$

$$= \sum_{i=1}^{m} a_i(x) G(y_i) = P(G)(x) = G(x)$$

whence

$$g = \sum_{i=1}^{m} g(y_i) g_i$$

follows. For arbitrary $f \in C(X)$ put $g = P(f)$. Since

$$g(y_i) = P(f)(y_i) = f(y_i) \qquad (i=1,\ldots,m)$$

we obtain

$$P(f) = \sum_{i=1}^{m} f(y_i) g_i .$$

Thus we have proved the following

Theorem 1

Assume that P is an interpolator with finite precision set

$$\operatorname{prec}(P) = \{y_1,\ldots,y_m\} .$$

Then there exists a unique cardinal basis $\{g_1,\ldots,g_m\}$ of $\operatorname{Im}(P)$ such that

$$P = \sum_{i=1}^{m} \hat{y}_i \otimes g_i \quad , \quad \sum_{i=1}^{m} g_i = 1 .$$

The existence of an interpolator P for a given precision set $\{y_1,\ldots,y_m\}$ is obtained from Shepard's method of "metric interpolation" [8] :

Theorem 2

Define for $x \in X$

$$w_k(x) = \prod_{\substack{j=1 \\ j\neq k}}^{m} d(x,y_j) \qquad (k=1,\ldots,m) ,$$

$$w(x) = \sum_{k=1}^{m} w_k(x) ,$$

$$v_k(x) = w_k(x)/w(x) \qquad (k=1,\ldots,m) .$$

Then

$$S = \sum_{j=1}^{m} \hat{y}_j \otimes v_j$$

is an interpolator with precision set $\operatorname{prec}(S) = \{y_1,\ldots,y_m\}$.

It is interesting to note that in the case of Theorem 1 the precision set of the interpolator P is also its carrier . By definition the carrier $F = \operatorname{carr}(P)$ of P is the smallest closed subset F of X such that $P(f) = 0$ whenever $f(x) = 0$ for all $x \in F$ ([2]) .
In general merely the relation

$$\mathrm{prec}(P) \subset \mathrm{carr}(P)$$

holds.

Theorem 3
Assume that P is an interpolator on $C(X)$. Then we have

$$\mathrm{prec}(P) = \mathrm{carr}(P) .$$

Proof: For any $f \in C(X)$ put

$$X_f = \{ y \in X : P(f)(y) = f(y) \} .$$

X_f is closed. Since

$$\mathrm{prec}(P) = \bigcap_{f \in C(X)} X_f$$

the precision set of P is also closed. Now condition (iii) yields

$$\mathrm{carr}(P) \subset \mathrm{prec}(P)$$

which completes the proof of Theorem 2 .

As instances of interpolators with transfinite precision sets we consider parametrically extended polynomial Lagrange interpolators [6] :

$$P^1_m(f)(x_1,x_2) = \sum_{i=1}^{m} f(y^1_i,x_2)\, L^1_{i,m}(x_1) ,$$

$$P^2_n(f)(x_1,x_2) = \sum_{j=1}^{n} f(x_1,y^2_j)\, L^2_{j,n}(x_2) ;$$

here we assume that $(y^1_i)_{i \in \mathbb{N}}$ and $(y^2_j)_{j \in \mathbb{N}}$ are sequences of distinct numbers in $[0,1]$ and

$$L^1_{i,m}(x_1) = \prod_{\substack{r=1\\ r\neq i}}^{m} (x_1-y^1_r)(y^1_i-y^1_r)^{-1} ,$$

$$L^2_{j,n}(x_2) = \prod_{\substack{s=1\\ s\neq j}}^{n} (x_2-y^2_s)(y^2_j-y^2_s)^{-1} .$$

It is easily seen that P^1_m and P^2_n are interpolators on $C([0,1]^2)$ whose precision sets are given by

$$\text{prec}(P^1_m) = \{y^1_1,\dots,y^1_m\}\times[0,1] ,$$

$$\text{prec}(P^2_n) = [0,1]\times\{y^2_1,\dots,y^2_n\} .$$

It is an important fact that P^1_m and P^2_n are commuting projectors which implies that the product $P^1_mP^2_n$ and the Boolean sum $P^1_m \oplus P^2_n = P^1_m + P^2_n - P^1_mP^2_n$ are also projectors on $C(X)$. Thus, it is a natural question wether $P^1_mP^2_n$ and $P^1_m \oplus P^2_n$ are also interpolators. This will be discussed in a more general setting in the following sections.

2. Products of interpolators

First we will determine the kernel of interpolators .

Theorem 4

Suppose that P is an interpolator on $C(X)$. Then we have

$$\text{Ker}(P) = \{f \in C(X) : f|_{\text{prec}(P)} = 0\} ;$$

i. e., $\text{Ker}(P)$ is an ideal in $C(X)$ (see also [1]).

Proof: This is an immediate consequence of condition (iii) and the definition of prec(P) .

Theorem 5

Let P and Q be commuting interpolators on C(X) . Then

$$\mathrm{prec}(P) \cap \mathrm{prec}(Q) \neq \emptyset .$$

Proof: Assume that

$$\mathrm{prec}(P) \cap \mathrm{prec}(Q) = \emptyset .$$

Define $h \in C(X)$ by

$$h(x) = d(x,\mathrm{prec}(P))\,(\,d(x,\mathrm{prec}(P)) + d(x,\mathrm{prec}(Q))\,)^{-1} .$$

Then we have

$$h|_{\mathrm{prec}(P)} = 0 , \qquad h|_{\mathrm{prec}(Q)} = 1 .$$

In account of condition (ii) and Theorem 4 we obtain

$$P(h) = 0 , \qquad Q(h) = 1 .$$

In view of condition (ii) we can conclude

$$1 = P(1) = PQ(h) = Q(P(h)) = 0$$

which is impossible.

Theorem 6

The product PQ of two commuting interpolators on C(X) is also an interpolator on C(X) .

Proof: Condition (ii) is trivially satisfied for $PQ = QP$.

Since

$$\mathrm{prec}(P) \cap \mathrm{prec}(Q) \subset \mathrm{prec}(PQ) \tag{2.1}$$

an application of Theorem 5 yields condition (i) , i. e. , PQ is an interpolation projector . Since P and Q are interpolators we can conclude for any $x \in X$:

$$\hat{x}P \in \mathrm{lin}\{\hat{y} : y \in \mathrm{prec}(P)\} ,$$

$$\hat{y}Q \in \mathrm{lin}\{\hat{z} : z \in \mathrm{prec}(Q)\}$$

whence

$$\hat{x}PQ \in \mathrm{lin}\{\hat{z} : z \in \mathrm{prec}(Q)\} \qquad (x \in X)$$

follows. Since $PQ = QP$ we also have

$$\hat{x}PQ \in \mathrm{lin}\{\hat{y} : y \in \mathrm{prec}(P)\} \qquad (x \in X) .$$

In account of (2.1) we obtain

$$\hat{x}PQ \in \mathrm{lin}\{\hat{y}: y \in \mathrm{prec}(P)\} \cap \mathrm{lin}\{\hat{z}: z \in \mathrm{prec}(Q)\}$$

$$= \mathrm{lin}\{\hat{u} : u \in \mathrm{prec}(P) \cap \mathrm{prec}(Q)\} ,$$

i. e. ,

$$\hat{x}PQ \in \mathrm{lin}\{\hat{v} : v \in \mathrm{prec}(PQ)\} .$$

Thus, condition (iii) is also true for PQ and the proof of Theorem 6 is complete.

<u>Theorem 7</u>
The precision set of the product of two commuting interpolators P and Q is given by

$$\mathrm{prec}(PQ) = \mathrm{prec}(P) \cap \mathrm{prec}(Q) .$$

Proof: In the proof of Theorem 6 we have shown that

$$\hat{x}\,PQ \in \operatorname{lin}\{\hat{v} : v \in \operatorname{prec}(P) \cap \operatorname{prec}(Q)\}$$

which implies

$$\operatorname{prec}(PQ) \subset \operatorname{prec}(P) \cap \operatorname{prec}(Q) \quad .$$

Now an application of (2.1) completes the proof.

Theorem 7 was first established for (weakly) commuting projectors by W.J. Gordon and J.A. Wixom [7] . In view of Theorem 3 and Theorem 7 we have the

Corollary 1

The carrier of the product of two commuting interpolators P and Q is given by

$$\operatorname{carr}(PQ) = \operatorname{carr}(P) \cap \operatorname{carr}(Q) \quad .$$

We conclude this section by considering the product $P_m^1 P_n^2$ of the parametrically extended Lagrange projectors P_m^1 and P_n^2 :

$$P_m^1 P_n^2(f)(x_1,x_2) = \sum_{i=1}^{m} \sum_{j=1}^{n} f(y_i^1,y_j^2)\, L_{i,m}^1(x_1)\, L_{j,n}^2(x_2) \quad .$$

The precision set $\operatorname{prec}(P_m^1 P_n^2)$ and thus the carrier of $P_m^1 P_n^2$ is given by

$$\begin{aligned} &\operatorname{prec}(P_m^1 P_n^2) \\ &= \operatorname{prec}(P_m^1) \cap \operatorname{prec}(P_n^2) \\ &= \{(y_i^1,y_j^2) : i=1,..,m ; j=1,...,n\} \end{aligned}$$

as is well known.

3. Boolean sums of interpolators

Assume that P and Q are commuting interpolators on $C(X)$. The formula

$$P \oplus Q = P + Q - PQ$$

implies that the precision set and the carrier of PQ satisfy the relations

$$\operatorname{prec}(P \oplus Q) \supset \operatorname{prec}(P) \cup \operatorname{prec}(Q) \quad , \tag{3.1}$$

$$\operatorname{carr}(P \oplus Q) \subset \operatorname{carr}(P) \cup \operatorname{carr}(Q) \quad . \tag{3.2}$$

Theorem 8

The Boolean sum $P \oplus Q$ of two commuting interpolators P and Q on $C(X)$ is also an interpolator on $C(X)$.

Proof: The validity of condition (i) for $P \oplus Q$ follows from (3.1). Also condition (ii) is immediately verified.
Assume now that $x \in X$ is arbitrary. In account of Theorem 7 we have

$$\begin{aligned} \hat{x} P \oplus Q &= \hat{x} P + \hat{x} Q - \hat{x} PQ \\ &\in \operatorname{lin}\{\hat{y} : y \in \operatorname{prec}(P)\} + \operatorname{lin}\{\hat{z} : z \in \operatorname{prec}(Q)\} \\ &= \operatorname{lin}\{\hat{u} : u \in \operatorname{prec}(P) \cup \operatorname{prec}(Q)\} \end{aligned}$$

whence

$$\hat{x} P \oplus Q \in \operatorname{lin}\{\hat{u} : u \in \operatorname{prec}(P \oplus Q)\}$$

follows. This completes the proof of Theorem 8.

Our next purpose is to improve the relation (3.1) .

Theorem 9

The precision set of the Boolean sum $P \oplus Q$ of the two commuting interpolators P and Q is given by

$$\mathrm{prec}(P \oplus Q) = \mathrm{prec}(P) \cup \mathrm{prec}(Q) \quad .$$

Proof: Since $P \oplus Q$ is an interpolator it follows from Theorem 2 and (3.2) that

$$\begin{aligned} &\mathrm{prec}(P \oplus Q) \\ &= \mathrm{carr}(P \oplus Q) \\ &\subset \mathrm{carr}(P) \cup \mathrm{carr}(Q) \\ &= \mathrm{prec}(P) \cup \mathrm{prec}(Q) \quad , \end{aligned}$$

i. e. ,

$$\mathrm{prec}(P \oplus Q) \subset \mathrm{prec}(P) \cup \mathrm{prec}(Q) \quad .$$

In account of (3.1) the proof of Theorem 9 is complete.

Corollary 1 has the following counterpart.

Corollary 2

The carrier of the Boolean sum $P \oplus Q$ of the two commuting interpolators P and Q is given by

$$\mathrm{carr}(P \oplus Q) = \mathrm{carr}(P) \cup \mathrm{carr}(Q) \quad .$$

As a simple instance we consider $P_m^1 \oplus P_n^2$ which is the projector of bivariate polynomial Lagrange blending :

$$\begin{aligned} &P_m^1 P_n^2(f)(x_1,x_2) \\ &= \sum_{i=1}^{m} f(y_i^1,x_2)\, L_{i,m}^1(x_1) + \sum_{j=1}^{n} f(x_1,y_j^2)\, L_{j,m}^2(x_2) \\ &\quad - \sum_{i=1}^{m} \sum_{j=1}^{n} f(y_i^1,y_j^2)\, L_{i,m}^1(x_1)\, L_{j,n}^2(x_2) \quad . \end{aligned}$$

Its precision set is given by

$$\mathrm{prec}(P_m^1 \oplus P_n^2)$$

$$= \mathrm{prec}(P_m^1) \cup \mathrm{prec}(P_n^2)$$

$$= \{(y_i^1, x_2) : 1 \le i \le m \; ; \; 0 \le x_2 \le 1\} \cup \{(x_1, y_j^2) : 0 \le x_1 \le 1 \; ; \; 1 \le j \le n\}$$

as is well known.

Theorem 9 does not hold for arbitrary commuting projectors P and Q on $C(X)$. This can be seen as follows. Observe that for any interpolator P on $C(X)$ the associated remainder projector

$$\overline{P} = I - P$$

is not an interpolator since

$$\overline{P}(1) = 0$$

and therefore

$$\mathrm{prec}(\overline{P}) = \emptyset \quad .$$

On the other hand we have for $P \neq I$

$$\mathrm{prec}(P \oplus \overline{P}) = \mathrm{prec}(I) = X \quad ,$$

$$\mathrm{prec}(P) \cup \mathrm{prec}(\overline{P}) = \mathrm{prec}(P) \neq X \quad .$$

Thus , the property of commuting projectors and their precision sets as stated in [7] is not true in general. It is an open question wether Theorem 9 is true for interpolation projectors.

4. Lattices of interpolators

Let $\mathbb{P}$ be a collection of commuting interpolators on $C(X)$. We will construct a maximal distributive lattice of commuting interpolators on $C(X)$ which contains the generator $\mathbb{P}$. We apply the same method as in [5] . Define

$$\mathbb{P}' = \{ Q : Q \text{ is an interpolator on } C(X) , \; QP = PQ \text{ for all } P \in \mathbb{P} \} .$$

Note that

$$I \in \mathbb{P}' , \quad \mathbb{P} \subset \mathbb{P}' .$$

In general $\mathbb{P}'$ is not closed with respect to the operator product. Thus we consider the smaller set

$$\mathbb{P}'' = \{ R \in \mathbb{P}' : QR = RQ \text{ for all } Q \in \mathbb{P}' \} .$$

It is obvious that

$$\mathbb{P} \subset \mathbb{P}'' \subset \mathbb{P}' .$$

<u>Lemma 1</u>
For any two interpolators $R_1 , R_2 \in \mathbb{P}''$ we have

$$R_1R_2 = R_2R_1 \in \mathbb{P}'' ,$$

$$R_1 \bullet R_2 = R_1 + R_2 - R_1R_2 \in \mathbb{P}'' .$$

Proof: Since $R_1 , R_2 \in \mathbb{P}'$ we have for any $P \in \mathbb{P}$:

$$PR_1R_2 = R_1PR_2 = R_1R_2P .$$

Morever, since R_1 and R_2 are both members of $\mathbb{P}'' \subset \mathbb{P}'$ we also have

$$R_1R_2 = R_2R_1 .$$

In view of Theorem 6 we obtain that R_1R_2 is an interpolator such that

$$R_1 R_2 \in \mathbb{P}' \quad .$$

Assume now that $Q \in \mathbb{P}'$ is arbitary. The definition of $\mathbb{P}''$ yields

$$QR_1R_2 = R_1QR_2 = R_1R_2Q \quad ,$$

i. e. , we have

$$R_1R_2 \in \mathbb{P}'' \quad (R_1, R_2 \in \mathbb{P}'') \quad .$$

The relation

$$R_1 \oplus R_2 \in \mathbb{P}'' \quad (R_1, R_2 \in \mathbb{P}'')$$

is proved similarly with the aid of Theorem 8 .

We proceed with the following isomorphism theorem .

<u>Theorem 10</u>
The set $\mathbb{P}''$ is a distributive lattice with respect to the the partial ordering

$$P \leq Q \quad \Leftrightarrow \quad PQ = QP = P \quad .$$

Moreover, $(\mathbb{P}'' , \leq)$ is isomorphic to the lattice $(\{ \mathrm{prec}(P) : P \in \mathbb{P}'' \}, \subseteq)$ of subsets of X by the mapping

$$P \longrightarrow \mathrm{prec}(P) \quad .$$

<u>Proof:</u> In view of Lemma 1 we have

$$\inf\{P,Q\} = PQ \in \mathbb{P}'' \quad ,$$
$$\sup\{P,Q\} = P \oplus Q \in \mathbb{P}'' \quad ,$$

i. e. $(\mathbb{P}'', \leq)$ is a lattice. Moreover, the equality

$$P \oplus (QR) = (P \oplus Q)(P \oplus R) \quad (P,Q,R \in \mathbb{P}'')$$

is easily proved which implies that $(\mathbb{P}'', \leq)$ is distributive. In account of Theorems 7 and 9 we have only to show that

$$\mathrm{prec}(P_1) = \mathrm{prec}(P_2) \Rightarrow P_1 = P_2 \quad (P_1, P_2 \in \mathbb{P}'') .$$

Assume that $\mathrm{prec}(P_1) = \mathrm{prec}(P_2)$. Then we have

$$\mathrm{prec}(P_1P_2) = \mathrm{prec}(P_1) = \mathrm{prec}(P_2)$$

which implies

$$P_1(f)(x) = P_1P_2(f)(x) = P_2P_1(f)(x) = P_2(f)(x) = f(x)$$

$$(x \in \mathrm{prec}(P_1) = \mathrm{prec}(P_2) = \mathrm{prec}(P_1P_2)) .$$

Now an application of Theorem 3 yields

$$f - P_2(f) \in \mathrm{Ker}(P_1) , \quad f - P_1(f) \in \mathrm{Ker}(P_2) ,$$

i. e.,

$$P_1(f) = P_1P_2(f) = P_2P_1(f) = P_2(f) \quad (f \in C(X))$$

which completes the proof of Theorem 10 .

In $\mathbb{P}''$ the intersection of all sublattices of $\mathbb{P}''$ which contain $\mathbb{P}$ is the smallest lattice $\mathbb{P}^+$ generated by $\mathbb{P}$. $\mathbb{P}^+$ is also given by a construction due to Cheney and Gordon [3] . Define

$$\mathbb{P}_0 = \mathbb{P} ,$$

$$\mathbb{P}_{n+1} = \{ P_1P_2 : P_1, P_2 \in \mathbb{P}_n \} \cup \{ P_1 \oplus P_2 : P_1, P_2 \in \mathbb{P}_n \}$$

$$(n = 0,1,2,\dots) .$$

Then the following result is easily established.

<u>Theorem 11</u>

The lattice $\mathbb{P}^+$ is given by

$$\mathbb{P}^+ = \bigcup_{n \geq 0} \mathbb{P}_n .$$

It is an immediate consequence of this characterization that

$$\operatorname{prec}(P) \subset \bigcup_{Q \in \mathbb{P}} \operatorname{prec}(Q) \quad \text{for all} \quad P \in \mathbb{P}^+ \quad .$$

Another application of Theorems 11, 7 and 9 yields the following

<u>Theorem 12</u>

Assume that the

$$\mathbb{P}_0^+ = \{P \in \mathbb{P}^+ : \operatorname{prec}(P) \text{ is finite}\}$$

is not empty. Then $\mathbb{P}_0^+$ is a sublattice of $\mathbb{P}^+$.

We construct some simple examples using the parametrically extended Lagrange projectors of section 1 .

<u>Example 1</u>

$$\{P_1^1 , P_1^2\}^+ = \{P_1^1, P_1^2, P_1^1P_1^2, P_1^1 \oplus P_1^2\} \quad ,$$

$$\{P_1^1 , P_1^2\}_0^+ = \{P_1^1P_1^2\} \quad .$$

<u>Example 2</u>

$$\{P_1^1, P_2^1, P_1^2, P_2^2\}^+$$

$$= \{P_1^1, P_2^1, P_1^2, P_2^2,$$

$$P_1^1P_1^2, P_1^1P_2^2, P_2^1P_1^2, P_2^1P_2^2, P_1^1 \oplus P_1^2, P_1^1 \oplus P_2^2, P_2^1 \oplus P_1^2, P_2^1 \oplus P_2^2,$$

$$P_1^1 \oplus P_2^1P_1^2, P_1^1 \oplus P_2^1P_2^2, P_1^2 \oplus P_1^1P_2^2, P_1^2 \oplus P_2^1P_2^2,$$

$$P_1^1P_2^2 \oplus P_2^1P_1^2, P_2^1P_2^2 \oplus (P_1^1 \oplus P_1^2)\}$$

$$\left\{P_1^1,\ P_2^1,\ P_1^2,\ P_2^2\right\}_0^+$$

$$= \left\{P_1^1P_1^2,\ P_1^1P_2^2,\ P_2^1P_1^2,\ P_2^1P_2^2,\ P_1^1P_2^2 \oplus P_2^1P_1^2\right\} \quad .$$

References

1 G. BIRKHOFF : The algebra of multivariate interpolation. In: "Constructive approaches to mathematical models" (Eds.: C.W. Coffman and G. Fix). Academic Press (1979), New York, pp. 347-363 .

2. E. W. CHENEY : Projection operators in approximation theory. In: "Studies in functional analysis" (Ed.: R.G. Bartle). The mathematical association of America (1980), Washington, pp. 50-80 .

3. E. W. CHENEY and W.J. GORDON : Bivariate and multivariate interpolation with noncommutative projectors. In : "Linear spaces and approximation" (Eds.: P.L. Butzer and B.Sz. Nagy). ISNM 40 (1977), pp. 381-387 .

4. F.J. DELVOS : d-variate Boolean interpolation. J. of Approximation Theory 34 (1982), pp. 99-114 .

5. F.J. DELVOS and H. POSDORF : Generalized Biermann interpolation. Resultate der Mathematik (1982), pp.

6. W.J. GORDON : Distributive lattices and the approximation of multivariate functions. In : "Approximation with special emphasis on spline functions" (Ed.: I.J. Schoenberg). Academic Press (1969), New York, pp. 223-277 .

7. W.J. GORDON and J.A. WIXOM : Pseudo-harmonic interpolation on convex domains. SIAM J. Numer. Anal. 11 (1974), pp. 909-933 .

8. W.J. GORDON and J.A. WIXOM : Shephard's method of "metric interpolation" to bivariate and multivariate interpolation. Mathematics of computation 32 (1978), pp. 253-264 .

9. P. LANCASTER : Composite methods for generating surfaces. In : "Polynomial and spline approximation" (Ed.: B.N. Sahney). D. Reidel Publishing Company (1979), pp. 91-102 .

Priv.-Doz. Dr. F.J. Delvos

o.Prof. Dr. W. Schempp

Lehrstuhl für Mathematik I
University of Siegen

Hölderlinstraße 3
D-5900 Siegen 21

(West-Germany)

WHICH CLOSED CONVEX SUBSETS OF AN INNER PRODUCT SPACE ARE CHEBYSHEV?

Frank Deutsch

ABSTRACT. In a certain class of convex sets (including the subspaces of finite-codimension), a precise characterization is given of those convex sets which are Chebyshev.

1. INTRODUCTION

It is well-known that one of the most fundamental and important facts about Hilbert space is that every closed convex subset is a "Chebyshev set." That is, each point of the space has a unique nearest point in the convex set. This fact can be used, for example, to give simple proofs of the Fréchet-Riesz representation theorem and the projection theorem.

However, in an inner product space which is <u>not</u> complete, this fact is no longer true. Indeed, there are closed subspaces of codimension one which are not Chebyshev. This raises the following question: "Which closed convex subsets of an incomplete inner product space are Chebyshev?"

One reason this seems to be an important question is the following. In many applications of the approximation of functions which arise in the engineering sciences, the "natural" space to work in is the space of continuous functions with the L_2-norm. For example, consider the space $C_2[a,b]$ consisting of all real-valued continuous functions x on $[a,b]$ with the norm $\|x\|^2 = \int_a^b |x(t)|^2 dt$. Then $C_2[a,b]$ is an incomplete inner product space (its completion is $L_2[a,b]$), and

$$C = \{x \in C_2[a,b] \mid \int_a^b x(t)dt \leq 0, \int_a^b tx(t)dt \leq 0\}$$

is a closed convex cone in $C_2[a,b]$. Is C a Chebyshev set?

If we regard C as a subset of the larger Hilbert space $L_2[a,b]$, then C is not closed and hence certainly not Chebyshev in $L_2[a,b]$. On the other hand, if we replace C by its closure $\tilde{C}$ in $L_2[a,b]$, then $\tilde{C}$ is Chebyshev in $L_2[a,b]$. However, $\tilde{C}$ contains many discontinuous functions and it is not a priori clear that the nearest point in $\tilde{C}$ to a continuous function is continuous, i.e. in C.

We can summarize this briefly. When approximating by a closed convex subset of an incomplete inner product space, the Hilbert space theory is usually not helpful. What we need is a useful criterion which tells us when a closed convex subset of an inner product space is Chebyshev.

It is well-known that each finite-dimensional subspace of an inner product space is Chebyshev. In fact, any closed convex subset of a finite-dimensional subspace of an inner product space is Chebyshev. What we would like is a useful condition which characterizes when a given closed convex subset is Chebyshev. Today I would like to give a partial answer to this problem. Specifically, if a convex set belongs to a certain class of convex cones (which includes the subspaces of finite-codimension), then we can give a simple characterization of when this set is Chebyshev. Detailed proofs, along with somewhat more general results and related work, will appear in [2].

2. THE MAIN RESULTS

Throughout this section, X will denote some given (real) inner product space and X^* its dual space of all bounded linear functionals on X.

A functional $x^* \in X^*$ is said to have a representer $x \in X$ provided that

$$x^*(y) = \langle y,x \rangle \quad \text{for all} \quad y \in X.$$

(If X is complete, i.e. a Hilbert space, then every $x^* \in X^*$ has a representer in X by virtue of the Fréchet-Riesz representation theorem. However, if X is not complete, there always exist $x^* \in X^*$ which do not have representers in X.)

A functional $x^* \in X^*$ is said to <u>attain its norm</u> if there is an $x \in X$ with $\|x\| = 1$ such that $x^*(x) = \|x^*\|$.

THEOREM 1. x^* attains its norm if and only if x^* has a representer in X.

THEOREM 2. Let $\{x_1^*, x_2^*, \ldots, x_n^*\}$ be a linearly independent set in X^*, let $c_1, c_2, \ldots, c_n$ be n real numbers, and let

$$C = \bigcap_{i=1}^{n} \{x \in X \mid x_i^*(x) \le c_i\}$$

or

$$C = \bigcap_{1}^{n} \{x \in X \mid x_i^*(x) = c_i\}.$$

Then C is a Chebyshev set if and only if each x_i^* has a representer in X.

As an application of Theorem 2, consider the example of the Introduction: $X = C_2[a,b]$ and

$$C = \bigcap_{i=1}^{2} \{x \in X \mid x_i^*(x) \le 0\},$$

where

$$x_1^*(x) := \int_a^b x(t)dt = \langle x, x_1 \rangle \quad , \quad x_1(t) \equiv 1,$$

and

$$x_2^*(x) := \int_a^b tx(t)dt = \langle x, x_2 \rangle \, , \quad x_2(t) \equiv t.$$

Since x_i^* has the representer $x_i \in X$ $(i = 1,2)$, it follows by Theorem 2 that C is Chebyshev.

I suspect that there is a result analogous to Theorem 2 valid for <u>any</u> closed convex set C in terms of the functionals which define supporting hyperplanes to C. Thus far the proof has eluded me.

3. A PARTIAL GENERALIZATION

Let X be any normed linear space, and let X^* denote the Banach space of all bounded linear functionals on X. A subset C of X is called <u>proximinal</u> (resp. <u>Chebyshev</u>) if each $x \in X$ has at least (resp. exactly) one nearest point $y_0 \in C$: $\|x - y_0\| = \inf_{y \in C} \|x - y\|$. Owing to the strict convexity of an inner product space, a convex subset of an inner product space is proximinal if and only if it is Chebyshev.

In one direction, Theorem 2 can be generalized to any normed linear space.

THEOREM 3. Let X be a normed linear space, $\{x_1^*, x_2^*, \ldots, x_n^*\}$ a linearly independent set in X^*, c_i $(i = 1,2,\ldots,n)$ real numbers, and

$$C = \bigcap_{i=1}^{n} \{x \in X \mid x_i^*(x) \le c_i\}$$

or

$$C = \bigcap_{i=1}^{n} \{x \in X \mid x_i^*(x) = c_i\}.$$

If C is proximinal, then each x_i^* attains its norm.

Thus a <u>necessary</u> condition that C be proximinal is that each x_i^* attain its norm. In Theorem 2, we have shown that in an inner product space, this condition is also sufficient. Also, the condition is sufficient if X is any reflexive Banach space since in this case every closed convex subset is proximinal. Blatter and Cheney [1] and Pollul [3; Lemma 2.6] have essentially shown that in the space c_0 the condition is sufficient (for subspaces of finite-codimension).

However, the condition is <u>not</u> sufficient in general. In fact, in the space $C[0,1]$ of real-valued continuous functions with the supremum norm, the subspace of codimension 2:

$$C = \bigcap_{i=1}^{2} \{x \in C_2[0,1] \mid x_i^*(x) = 0\},$$

where

$$x_1^*(x) := x(0)$$

$$x_2^*(x) := \int_0^1 x(t)\,dt,$$

has the property that both x_i^* attain their norms (at $e(t) \equiv 1$). However, the function $x(t) = t$ has no nearest point in C.

REFERENCES

1. J. Blatter and E.W. Cheney, On the existence of extremal projections, J. Approximation Theory, 6(1972), 72-79.
2. F. Deutsch, Representers of linear functionals, norm-attaining functionals, and best approximation by cones and linear varieties in inner product spaces, J. Approximation Theory, to appear.
3. W. Pollul, Reflexivität und Existenz-Tielraüme in der linearen Approximationstheorie, Dissertation, Bonn, 1972.

Department of Mathematics
The Pennsylvania State University
University Park, PA 16802

Interpolation methods for numerical analytic continuation

Michael Eiermann, Wilhelm Niethammer
Institut für Praktische Mathematik,
Universität Karlsruhe, W.-Germany

1. Introduction

Given a function f by its power series

(1.1) $$f(z) = \sum_{j=0}^{\infty} u_j z^j \quad ,$$

convergent for $z \in D_R := \{z \in \mathbb{C} \mid |z| < R\}$ with $0 < R < \infty$. Let f be holomorphic in some domain $G \supset D_R$. We want to construct algorithms for computing f(z) for $z \in G$, especially for $z \in G \setminus D_R$, where (1.1) is divergent. Usually G is assumed starlike with respect to 0, such that the analytic continuation f(z) is uniquely determined for $z \in G$.

The tools we want to use are classical methods of summability applied to the series (1.1). We confine ourselves to methods determined by a triangular matrix

(1.2) $$\mathcal{P} := (p_{n,k})_{n\geq 0,\ 0\leq k\leq n}, \quad p_{n,k} \in \mathbb{C}.$$

With $\mathcal{P}$, a summability method in "series-sequence-form" is defined by

(1.3) $$t_n(z) := \sum_{k=0}^{n} p_{n,k} u_k z^k .$$

If $U(z) := (u_0, u_1 z, u_2 z^2, \ldots)^T$ and $T(z) := (t_0(z), t_1(z), t_2(z), \ldots)^T$, then (1.3) can be written formally as

(1.3)' $\quad T(z) = \mathcal{P}\, U(z).$

As usually in the following we will identify $\mathcal{P}$ and the summability method induced by $\mathcal{P}$.

The following questions arise:

1. For what subset of G does the sequence $\{t_n(z)\}_{n \geq 0}$ converge to f(z)?
2. How can an appropriate $\mathcal{P}$ be generated?
3. How can the transform (1.3) be effectively computed, eventually without explicit generation of the matrix $\mathcal{P}$?
4. There exist possibly different methods appropriate for a certain problem; does there exist an optimal one?

In Section 2 we describe a formal connection between summability methods of type (1.3) and certain interpolating schemes for the special function $g(y) := 1/(1-y)$. Using this connection, an answer to questions 1 and 2 is given in Section 3. From the theory of interpolation a result to question 4 is derived in Section 4. Some of the computational aspects of the problem are described in Section 5.

2. Formal relations

Let $\mathbb{C}[X]$ be the vector space of formal polynomials on $\mathbb{C}$. Then for f given by the power series (1.1) we introduce the linear functional

(2.1) $\quad \Omega_f : \mathbb{C}[X] \to \mathbb{C} \quad$ by

(2.2) $\quad \Omega_f(X^i) := u_i \;;$

Ω_f is also used in [1]. It holds formally

(2.3) $\quad f(z) = \sum_{i=0}^{\infty} u_i z^i = \sum_{i=0}^{\infty} \Omega_f(X^i z^i) \sim \Omega_f(1/(1-Xz)).$

Similar to the construction of an interpolatory quadrature formula by the formal relation (2.3) we are led to an

Interpolation Method (IM)

1. Given a triangular matrix K of nodes from $\mathbb{C}$,

$$K: = \begin{bmatrix} x_o^o & & \\ x_o^{(1)} & x_o^{(1)} & \\ x_o^{(2)} & x_1^{(2)} & x_2^{(2)} \\ \vdots & \vdots & \ddots \end{bmatrix}$$

where we assume $x_o^{(n)} = 0$ and $x_i^{(n)} \neq 1$ $(n = 0,1,\ldots;\ 0 \leq i \leq n)$.

2. Generate for $n \in \mathbb{N}_o$ the unique Hermite interpolation polynomial

$$(2.4) \qquad P_n(y) := p_{n,o} + p_{n,1}y + \ldots + p_{n,n}y^n \in \mathbb{C}[y],$$

which interpolates $g(y) := 1/(1-y)$ at the nodes $x_i^{(n)}$ $(i = 0,\ldots,n)$. Since $x_o^{(n)} = 0$ we have $p_{n,o} = 1$ $(n=0,1,\ldots)$.

3. Let

$$(2.5) \qquad \Omega_n(f;z) := \Omega_f(P_n(Xz)) = p_{n,o} + p_{n,1}u_1z + \ldots + p_{n,n}u_nz^n.$$

Hopefully, the sequence $\{\Omega_n(f;z)\}_{n\geq 0}$ can be seen as an approximation for $f(z)$. It is now important for our further investigations, that formulas (1.3) and (2.5) are identical, i.e., to each IM generated by the node matrix K there corresponds a matrix summability method $\mathcal{P}$, where the rows of $\mathcal{P}$ consist of the coefficients of the interpolating polynomials P_n in (2.4).

We denote

$$(2.6) \qquad W_n(y) := \prod_{i=1}^{n} (y-x_i^{(n)})/(1-x_i^{(n)})$$

the n-th characteristic polynomial of IM, whereas

$$(2.7) \qquad w_n(y) := a_n \prod_{i=1}^{n} (y-x_i^{(n)})$$

is called a generating polynomial of the IM; evidently, for each generating polynomial w_n it holds that $W_n(y) = w_n(y)/w_n(1)$. For notational reasons the node $x_o^{(n)} = 0$ is not a zero of W_n or of w_n.

An easy calculation leads to

Lemma 1: Let w_n be a generating polynomial of an IM. Then for the interpolating polynomials P_n according to (2.4) it holds

$$(2.8) \qquad P_n(y) = \frac{1}{w_n(1)} \; \frac{w_n(1) - y w_n(y)}{1-y} \; ;$$

especially for the n-th characteristic polynomial W_n it holds that

$$(2.9) \qquad P_n(y) = (1 - yW_n(y)/(1-y) .$$

We have seen that to each IM there corresponds a summability method; to a certain extent the reverse holds, too: If a given summability method $\mathcal{P}$ transforms the series (1.1) into the sequence (1.3), then there exists a "sequence-to-sequence" matrix $\mathcal{Q} := (q_{n,k})_{n \geq 0, 0 \leq k \leq n}$, which transformes the sequence of partial sums of (1.1) to the sequence (1.3); between $\mathcal{P}$ and $\mathcal{Q}$ holds the relation $\mathcal{P} = \mathcal{Q} \cdot \Sigma$, $\mathcal{Q} = \mathcal{P} \cdot \Sigma^{-1}$ with

$$\Sigma := \begin{bmatrix} 1 & 0 & 0 & 0 & .. \\ 1 & 1 & 0 & 0 & .. \\ 1 & 1 & 1 & 0 & .. \\ . & . & . & . & \end{bmatrix} , \quad \Sigma^{-1} = \begin{bmatrix} 1 & 0 & 0 & 0 & .. \\ -1 & 1 & 0 & 0 & .. \\ 0 & -1 & 1 & 0 & .. \\ . & . & . & . & .. \end{bmatrix}$$

(see [5], p.7). If now the condition

$$(2.10) \qquad \sum_{i=0}^{n} q_{n,i} = 1 \qquad (n = 0,1...)$$

is imposed on $\mathcal{Q}$ then the rows of $\mathcal{Q}$ determine a sequence of char. polynomials $\{W_n(z)\}_{n \geq 0}$ of an IM by

$$(2.11) \qquad W_n(z) := q_{n,o} + q_{n,1} z + ... + q_{n,n} z^n \quad (n = 0,1...),$$

such that the zeros of W_n together with O correspond to the node matrix K of an IM, i.e., each summability method with "sequence-to-sequence" matrix $\mathcal{Q}$ which satisfies (2.10) corresponds to an IM.

As an example let us remember the well-known Césaro method (see [5], p.104); here we have $W_n(y) = (1+y+\ldots+y^n)/(n+1) = (1-y)^{n+1}/(1-y)$, i.e. the corresponding nodes are the (n+1)-th roots of unity without 1, together with O. The Euler-Knopp method ([5], p.130) corresponds to an IM with generating polynomial $w_n(y) = (y-a)^n$ for $a \in \mathbb{C} \setminus \{1\}$.

The following lemma is needed in the next section; the second part follows directly from (2.9) in Lemma 1.

Lemma 2: Let $\{W_n(z)\}_{n\geq 0}$ be the characteristic polynomials of an IM. Then for the "error" it holds formally

$$f(z) - \Omega_n(f;z) \sim z\Omega_f\left(X \frac{W_n(Xz)}{1-Xz}\right); \tag{2.12}$$

especially for $g(z) = 1/(1-z)$ it follows

$$g(z) - \Omega_n(g;z) = z \cdot \frac{W_n(z)}{1-z} \quad . \tag{2.13}$$

3. Convergence

Given a summability method $\mathcal{R}$, resp. the corresponding IM, what can be said about the convergence of the sequences $\{t_n(z)\}_{n\geq 0}$ resp. $\{\Omega_n(f;z)\}_{n\geq 0}$ to f(z)? We cite a classical result which gives an answer to this question for arbitrary f, if the result is known for $g(z) := 1/(1-z)$.

Let f be holomorphic in $G \subset \mathbb{C}$ with $0 \in G$, such that (1.1) holds, and let R be some open subset of G. We say: A summability method $\mathcal{R}$ sums f in $R \subset G$ iff the sequence $\{t_n(z)\}_{n\geq 0}$ according to (1.3) converges to f(z) for $z \in R$, where the convergence is uniform with respect to each compact subset of R.

Theorem 1 (Perron 1923, Okada 1925; cf [4], p.155). Let f be holomorphic in G with $0 \in G$ and let $\mathcal{P}$ sum the function $g(z) := 1/(1-z)$ for $z \in R$, $R \subset \mathbb{C} \setminus \{1\}$; then $\mathcal{P}$ sums f in

(3.1) $$F(R,G) := \bigcap_{\zeta \in \mathbb{C} \setminus G} \zeta \cdot R .$$

E.g., if f has the only singularities $\zeta_1, \zeta_2 \in \mathbb{C}$, then $F(R,G) = \zeta_1 R \cap \zeta_2 R$. Thus, given $\mathcal{P}$, the problem is to describe the region R where $\mathcal{P}$ sums the "geometric series" $g(z) = 1/(1-z)$. Here the correspondence between $\mathcal{P}$ and an IM is very useful.

Let $\{W_n(z)\}_{n\geq 0}$ be the sequence of characteristic polynomials of the IM, corresponding to $\mathcal{P}$, and generated by K. Then from (2.13) in Lemma 2 we see directly that $\{\Omega_n(g;z)\}_{n\geq 0}$ converges to g(z) iff $\{W_n(z)\}_{n\geq 0}$ converges to 0. As a meassure for the rate of convergence of the $\{W_n(z)\}_{n\geq 0}$ we introduce

(3.2) $$\sigma_K(z) := \overline{\lim_{n\to\infty}} \, |W_n(z)|^{1/n}.$$

Then, if $\sigma_K(z) < 1$, $\{\Omega_n(g;z)\}_{n\geq 0}$ converges to g(z), whereas the same sequence diverges if $\sigma_K(z) > 1$.

The following definition excludes IMs which are difficult to handle.

Definition 1: An IM generated by K with meassure σ_K according to (3.2) is called normal iff

(a) σ_K is continuous on the preimage $\sigma_K^{-1}([0,1])$ of $[0,1]$,

(b) $\sigma_K(0) < 1$,

(c) $\mathbb{C} \setminus \sigma_K^{-1}(0,r)$ is connected for $0 < r \leq 1$.

If a normal IM with node matrix K is given, the "region of continuation" of the geometric series

(3.2) $$R_K := \{z \in \mathbb{C} \mid \sigma_K(z) < 1\}$$

is well defined. For $0 < r < 1$, a closed subset of R_K is

(3.3) $\quad R_K^r := \{z \in \mathbb{C} \mid \sigma(z) \leq r\}$.

It may be difficult to describe R_K explicitly, but we will see later that there are special IMs where R_K can be easily described. Given R_K, Theorem 1 describes the region of continuation $F(R_K, G)$ for an arbitrary function f. It may be helpful to use instead of (3.1) the equivalent definition

(3.4) $\quad F(R,G) = \{z \in \mathbb{C} \mid z/\zeta \in R,\ \zeta \in \mathbb{C} \setminus G\}$.

4. Optimal methods

If we try to find summability methods resp. IMs which sum a given function in some subset of G, by Theorem 1, we can confine ourselves to the special function $g(z) = 1/(1-z)$.

Let $H \subset \mathbb{C}$ be compact with $0 \in H$, $1 \notin H$. Then we call an IM, generated by the node matrix K, admissible with respect to H, if $\sigma_{H,K} := \sup\{\sigma_K(z) \mid z \in H\} < 1$. An IM admissible with respect to H and generated by K* is called optimal with respect to H, if $\sigma_{H,K^*} \leq \sigma_{H,K}$ for all IM which are admissible with respect to H.

For constructing optimal methods we hope that known results on interpolation can be applied. Especially the concept of "equidistributed nodes" is very useful. For simplicity we make the additional assumption that H posesses at least two boundary points and that $\bar{\mathbb{C}} \setminus H$ is simply connected where $\bar{\mathbb{C}} := \mathbb{C} \cup \{\infty\}$.

By the Riemann mapping theorem there exists a conformal mapping

(4.1) $\quad \psi : \bar{\mathbb{C}} \setminus \{\omega \mid |\omega| \leq 1\} \quad \rightarrow \quad \bar{\mathbb{C}} \setminus H$

with $\psi(\infty) = \infty$, $\psi'(\infty) > 0$. ψ has the expansion

(4.2) $\quad \psi(\omega) = c\omega + c_0 + c_1\omega^{-1} + c_2\omega^{-2} + \dots$.

Let Φ be the inverse mapping and let, for $R > 1$,

$$\Gamma_R := \{z \in \mathbb{C} \mid |\Phi(z)| = R\} \quad \text{and}$$

(4.3) $\quad R_o := | \Phi(1) |$;

since $1 \notin H$ we have $R_o > 1$.

Definition 2 ([3], p.65): The nodes $x_i^{(n)} \in H$ (with $x_o^{(n)} = 0$), corresponding to the zeros of the characteristic polynomial W_n, are called equidistributed on H, if

$$(4.4) \qquad \lim_{n\to\infty} [\sup_{z\in H} |W_n(z)|^{1/n}] = c,$$

where c is defined in (4.2).

The following theorem gives a connection between c and the constant R_o in (4.3).

Theorem 2:

(a) Let K generate a normal IM admissible with respect to H such that the nodes are in H. Then

$$\sigma_{H,K} \geq 1/R_o.$$

(b) If the nodes are equidistributed on H, then $\sigma_{H,K} = 1/R_o$, i.e., the corresponding IM is optimal with respect to H.

A proof is given in [2]; it uses essentially the Theorem of Kalmâr and Walsh ([3], p.66).

Example 1: Let $H := [a,b]$ be the line segment between a and b where $a,b \in \mathbb{C}$ with $1 \notin H$, and let T_n be the n-th Tschebyscheff polynomial of the first kind with zeros in [-1,1]. If W_n is the corresponding Tschebyscheff polynomial, transformed to H, i.e.,

$$(4.5) \qquad W_n(X) = T_n\left(\frac{2}{b-a}X - \frac{a+b}{b-a}\right) / T_n\left(\frac{2-a-b}{b-a}\right) ,$$

then the zeros of W_n are equidistributed on $H = [a,b]$ and the IM generated by $\{W_n\}_{n\geq 0}$, is optimal with respect to a,b . It holds

$$(4.6) \qquad 1/R_o = |\sqrt{1-a} - \sqrt{1-b}\,|^2 / |b-a|.$$

Example 2: Let H satisfy the conditions above such that the map ψ in (4.1) is defined, and let ψ have a continuous extension to the boundary of the unit disk; this holds, e.g., if ∂H is a Jordan curve. The so-called "Fejer nodes"

$$x_j^{(n)} = \psi(e^{2\pi ij/(n+1)}), \quad (n=0,1,\ldots; \ 0 \le j \le n)$$

are equidistributed on H.

There are other sets of equidistributed nodes (see, e.g., [3], p.70).

5. Computational aspects

Given the power series (1.1) we want to compute the sequence $\{t_n(z)\}_{n\geq 0}$ (see (1.3)) without explicit use of the transformation matrix $\mathcal{R}$.

It is assumed that the corresponding characteristic polynomials (2.6) obey the recursion

$$(5.1) \quad \begin{aligned} &W_0(y) = 1; \quad W_1(y) = s_0^{(1)}y + s_1^{(1)} \\ &W_n(y) = (s_0^{(n)}y + s_1^{(n)})W_{n-1}(y) + s_2^{(n)}W_{n-2}(y), \quad (n \ge 2), \end{aligned}$$

where the $s_i^{(n)}$ are complex constants with $\sum_{i=0}^{2} s_i^{(n)} = 1$ (similar results will be obtained if a n-term recurrence relation holds).

The quantities

$$(5.2) \quad \Omega_{m,n} := \sum_{j=0}^{m-n} u_j z^j + z^{m-n+1}\Omega_f(X^{m-n+1}P_n(Xz)), \quad (m \ge n \ge 0)$$

will be placed in the following two dimensional array.

$$\begin{matrix} \Omega_{0,0} & & & \\ \Omega_{1,0} & \Omega_{1,1} & & \\ \Omega_{2,0} & \Omega_{2,1} & \ddots & \\ \vdots & \vdots & & \ddots \\ \Omega_{m,0} & \Omega_{m,1} & & \Omega_{m,m} \\ \vdots & \vdots & & \vdots \quad \ddots \end{matrix}$$

The partial sums of (1.1) form the first column of this array and the sequence $\{t_n(z)\}_{n\geq 0} = \{\Omega_n(f;z)\}_{n\geq 0}$ is the first diagonal of this scheme. The quantities (5.2) obey the relations

$$\text{(5.3)}\quad \begin{aligned} \Omega_{m,1} &= s_0^{(1)}\Omega_{m,0} + s_1^{(1)}\Omega_{m-1,0} && (m \geq 1)\\ \Omega_{m,n} &= s_0^{(n)}\Omega_{m,n-1} + s_1^{(n)}\Omega_{m-1,n-1} + s_2^{(n)}\Omega_{m-2,n-2} && (m,n \geq 2). \end{aligned}$$

The characteristic polynomials of Example 4.1 obey the assumptions (5.1) with

$$s_0^{(n)} = \frac{4}{b-a}\,\frac{T_{n-1}(\frac{2-a-b}{b-a})}{T_n(\frac{2-a-b}{b-a})}\,,\quad s_1^{(n)} = -2\,\frac{a+b}{b-a}\,\frac{T_{n-1}(\frac{2-a-b}{b-a})}{T_n(\frac{2-a-b}{b-a})}$$

and $s_2^{(n)} = 1-s_0^{(n)} - s_1^{(n)}$. The corresponding IM can therefore be computed by the relations (5.3).

Above all we will sum divergent power series and in this case, the partial sums of (1.1) usually grow rapidly in magnitude. For numerical purpose, it seems to be better to use algorithms which have the coefficients u_j of (1.1) - instead of the partial sums - as starting quantities. Such algorithms are obtained easily from (5.3) (compare [2]).

Finally we will construct an optimal IM for a compact and simply connected subset H of $\mathbb{C}$, which can be computed recursively. In (4.2) we mentioned the Fejer nodes as equidistributed on H. Unfortunately, these nodes do not allow a recursive computation of the corresponding IM because the n-th Fejer nodes cannot be obtained from the j-th nodes $(j \leq n-1)$ by recursion. We therefore choose a special subsequence of the Fejer nodes.

If $j \in \mathbb{N}$ there are uniquely determined integers k and m with $2^{k-1} \leq j < 2^k$ and $j = 2^{k-1} + m$ $(0 \leq m \leq 2^{k-1} - 1)$.

We define $y_j := \exp(2\pi i\,\frac{2m+1}{2^k})$, that means that y_j is a 2^k-th root of unity, and

(5.4)
$$x_o = 0$$
$$x_j = \psi(y_j), \text{ where } \psi \text{ is the conformal mapping described in 4.}$$

It is clear that the nodes x_i are equidistributed on H and that the IM corresponding to the node matrix

$$K^* = \begin{bmatrix} x_o & & \\ x_o & x_1 & \\ x_o & x_1 & x_2 \\ \vdots & \vdots & \vdots & \ddots \end{bmatrix}$$

can be computed recursively (the nodes in each column of K^* are equal). The assumptions (5.1) are fulfilled with

$$s_o^{(n)} = \frac{1}{1-x_n}, \quad s_1^{(n)} = \frac{-x_n}{1-x_n}, \quad s_2^{(n)} = 0.$$

References:

[1] Brezinski, C.: Padé-approximation and general orthogonal polynomials, ISNM 50. Basel-Boston-Stuttgart: Birkhäuser 1980

[2] Eiermann, M.: Numerische analytische Fortsetzung durch Interpolationsverfahren. Dissertation Universität Karlsruhe 1982.

[3] Gaier, D.: Vorlesungen über Approximation im Komplexen. Basel-Boston-Stuttgart: Birkhäuser 1980.

[4] Powell, R.E. and Shah, S.M.: Summability theory and its applications. London : Van Nostrand 1972.

[5] Zeller, K. und Beekmann, W.: Theorie der Limitierungsverfahren. Berlin-Heidelberg-New York: Springer 1970.

Institut für Praktische Mathematik
Universität Karlsruhe
Englerstr. 2
D-7500 Karlsruhe

Blending-Splines auf Dreiecksnetzen

H. Engels und A. Merschen, RWTH Aachen

1. Aufgabenstellung

Die üblichen Konstruktionen für Blending-Splines auf Dreiecksnetzen benutzen die Technik der Boole'schen Summen und sind demzufolge recht kompliziert (vgl. [1], [2], [3]), entsprechend auch Fehleranalyse und Konvergenzbeweise, ihre Realisierung auf dem Rechner ist zeitaufwendig.

Von elementaren geometrischen Überlegungen ausgehend entwickeln wir hier Blending-Splines einfachster Bauart (die Ausdehnung auf höhere Ordnungen liegt auf der Hand): Wir beziehen uns auf das Standarddreieck D (vgl. Fig. 1) und suchen einen Blending-Spline, der mit einer gegebenen Funktion f(x,y) auf dem Rand ∂D von D übereinstimmt und auch nur auf ∂D Werte von f benutzt.

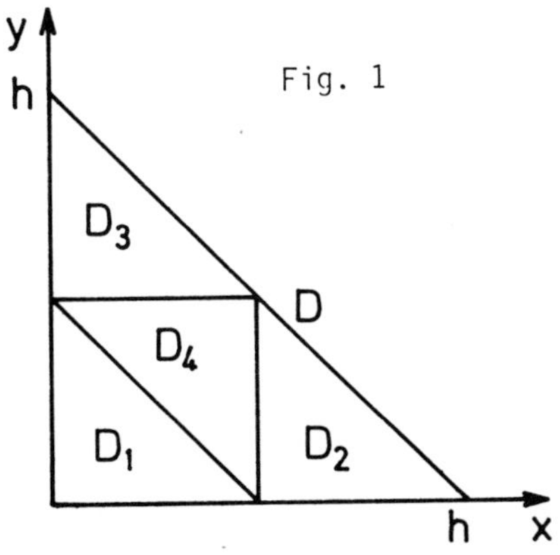

Fig. 1

D wird in vier kongruente Teildreiecke D_1, D_2, D_3 und D_4 unterteilt. In den Dreiecken D_1, D_2 und D_3 wird die gegebene Funktion f(x,y) durch Stücke einer Regelfläche (im einfachsten Falle Ebene oder hyperbolisches Paraboloid) so approximiert, daß die Näherungsfunktion S(f,x,y) auf dem Rand ∂D von D mit f(x,y) übereinstimmt. Auf dem Rand ∂D_4 von D_4 ist S(f,x,y) jeweils ein Polynom ersten Grades, so daß f(x,y) in D_4 durch eine Ebene approximiert werden kann, ohne die Stetigkeit der Gesamtkonstruktion oder die Exaktheit der Approximation auf ∂D zu verletzen.

Wir beschreiben $S(f,x,y)$ in zwei Varianten und geben im Anschluß daran einige Fehlerabschätzungen an.

2. Darstellung von $S(f,x,y)$

Wir beschreiben unsere beiden Konstruktionsvarianten exemplarisch im Dreieck D_1:

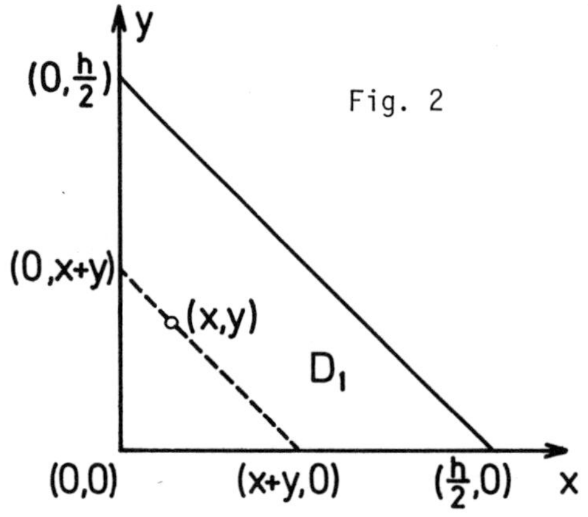

Fig. 2

a) Die Funktion f wird also in einem Punkt $(x,y) \in D_1$ linear interpoliert längs der Geraden zwischen den Punkten $(0,x+y)$ und $(x+y,0)$. Setzt man diese Konstruktion analog in D_2 bzw. D_4 fort, indem man längs vertikaler bzw. horizontaler Geraden interpoliert, und ersetzt man schließlich f in D_4 durch das Ebenenstück mit den Stützpunkten $(0,\frac{h}{2})$, $(\frac{h}{2},0)$, $(\frac{h}{2},\frac{h}{2})$, so erhält man:

$$S_1(f,x,y) = \begin{cases} \frac{1}{x+y}\,[xf(x+y,0) + yf(0,x+y)] & (x,y) \in D_1 \\ \frac{1}{h-x}\,[(h-x-y)\; f(x,0) + yf(x,h-x)] & (x,y) \in D_2 \\ \frac{1}{h-y}\,[(h-x-y)\; f(0,y) + xf(h-y,y)] & (x,y) \in D_3 \\ \frac{1}{h}\,[(h-2x)\; f(0,\frac{h}{2}) + (2x-2y-h)\; f(\frac{h}{2},\frac{h}{2}) + (h-2y)\; f(\frac{h}{2},0)] & (x,y) \in D_4 \end{cases}$$

S_1 benutzt also rationale Blending-Funktionen.

b) Bei der zweiten Konstruktion variieren wir die "Side-Vertex-Methode" (vgl. [4]):

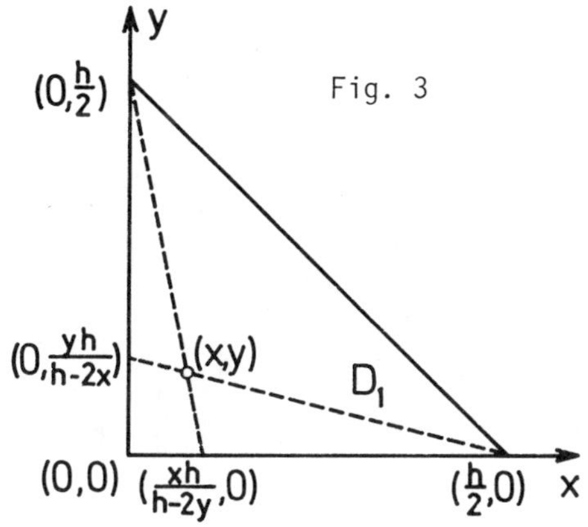

Fig. 3

Wir interpolieren linear längs der Geraden durch $(\frac{h}{2},0)$ und (x,y) bzw. $(0,\frac{h}{2})$ und (x,y) und bilden die "Quasi-Boole'sche-Summe" (vgl. [1]) dieser beiden Interpolierenden, die so vorteilhaft ausfällt, daß als Korrekturterm nur $f(0,0)$ übrigbleibt; man erhält insgesamt:

$$S_2(f,x,y) = \begin{cases} \frac{1}{h}\Big[(h-2x)\, f(0,\frac{yh}{h-2x}) + (h-2y)\, f(\frac{xh}{h-2y},0) \\ \qquad + (2x+2y-h)\, f(0,0)\Big] & (x,y) \in D_1 \\ \frac{1}{h}\Big[(h-2x)\, f(h,0) + (h-2y)\, f(\frac{h(x-y)}{h-2y},0) \\ \qquad + (2x+2y-h)\, f(\frac{h(2x+y-h)}{2x+2y-h},\frac{yh}{2x+2y-h})\Big] & (x,y) \in D_2 \\ \frac{1}{h}\Big[(h-2x)\, f(0,\frac{h(x-y)}{h-2x}) + (h-2y)\, f(0,h) \\ \qquad + (2x+2y-h)\, f(\frac{xh}{2x+2y-h},\frac{h(x+2y-h)}{2x+2y-h})\Big] & (x,y) \in D_3 \\ \frac{1}{h}\Big[(h-2x)\, f(0,\frac{h}{2}) + (h-2y)\, f(\frac{h}{2},0) \\ \qquad + (2x+2y-h)\, f(\frac{h}{2},\frac{h}{2})\Big] & (x,y) \in D_4 \end{cases}$$

S_2 benutzt polynomiale Blending-Funktionen.

Wir resumieren die Eigenschaften der beiden Spline-Konstruktionen: S_1 und S_2 werten die Funktion f nur auf dem Rand von D aus und geben f dort exakt wieder, die Splines sind stetig in ganz D, und sie interpolieren lineare Funktionen in ganz D exakt. Erhöht man die Ordnung, so sind auch innere Interpolationspunkte erforderlich.

3. Fehlerabschätzungen

Entsprechend der Eigenschaft, lineare Funktionen exakt zu interpolieren, ergibt sich eine Fehlerordnung $O(h^2)$, $h \to 0$. Es gilt etwa für $f \in C^2(D)$:

$$|f(x,y) - S_1(f,x,y)| \leq \frac{h^2}{32} \max\left\{ \|D^2f\|_\infty^1 , \|f_{yy}\|_\infty^2 , \|f_{xx}\|_\infty^3 , \|D^2f\|_\infty^4 \right\}$$

sowie

$$|f(x,y) - S_2(f,x,y)| \leq \frac{h^2}{27} \|D^2f\|_\infty$$

mit $\|D^2f\|_\infty := \|f_{xx}\|_\infty + 2\,\|f_{xy}\|_\infty + \|f_{yy}\|_\infty$. Die oberen Indizes der ersten Abschätzung beziehen sich auf die einzelnen Teildreiecke.

4. Anwendungen

Mit den beiden Splines S_1 und S_2 wurde die Funktion $f(x,y) = 0.5 \quad \cos^4(4(x^2+y-1))$ auf $[0,1] \times [0,1]$ interpoliert. Zum Vergleich wurde der lineare Spline S_L (Fehlerordnung ebenfalls $O(h^2)$) sowie der von Barnhill et. al. (vgl. [1]) angegebene Blending-Spline S_B herangezogen (Fehlerordnung $O(h^3)$). Die Figuren 4 - 7 zeigen die entsprechenden interpolierenden Flächen für $h = \frac{1}{8}$. Erwartungsgemäß liegt die Approximationsgüte von S_1 bzw. S_2 zwischen der von S_B bzw. S_L. Bei S_B sind kaum Unterschiede zu $f(x,y)$ wahrnehmbar, S_L hingegen läßt kaum die Struktur von f erkennen. Da die Rechenzeiten t_i zur Auswertung der Interpolierenden in einem beliebigen Punkt (x,y) sich verhalten wie

$$t_L : t_1 : t_2 : t_B = 1 : 1 : 1{,}4 : 2{,}6 ,$$

sind bei nicht zu hohen Genauigkeitsansprüchen die Konstruktionen S_1 bzw. S_2 zu empfehlen.

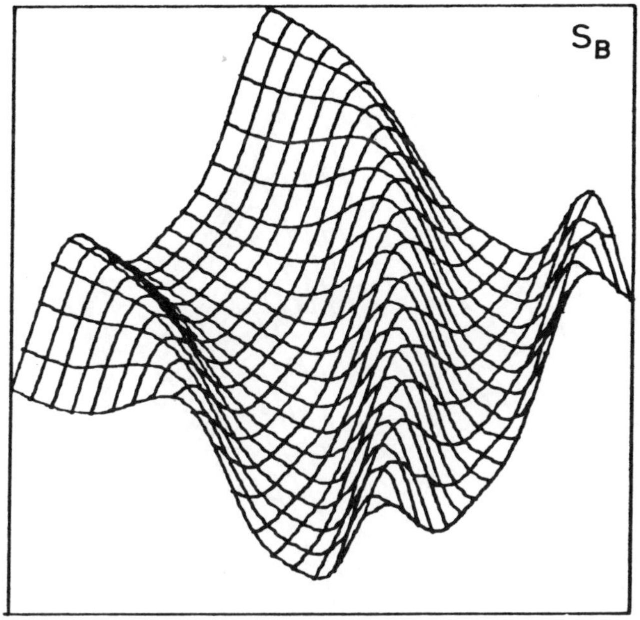

Fig. 4

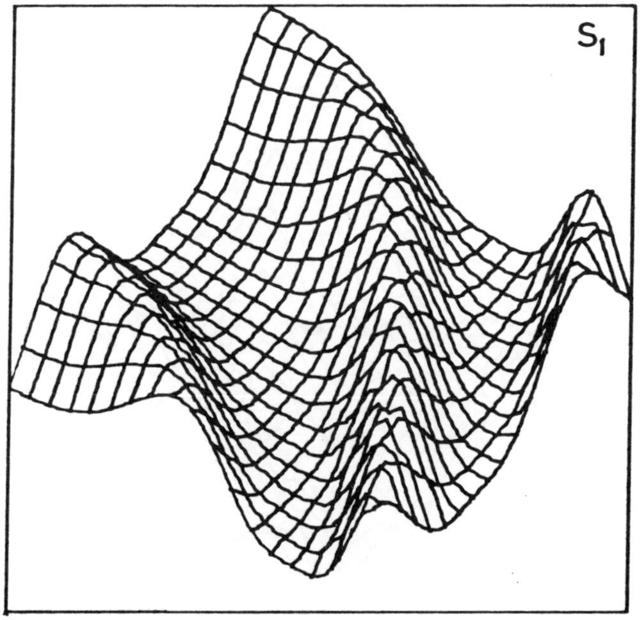

Fig. 5

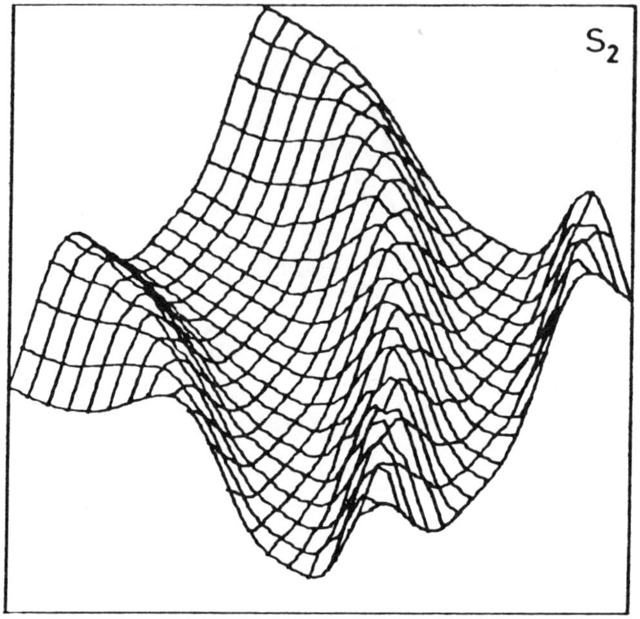

Fig. 6

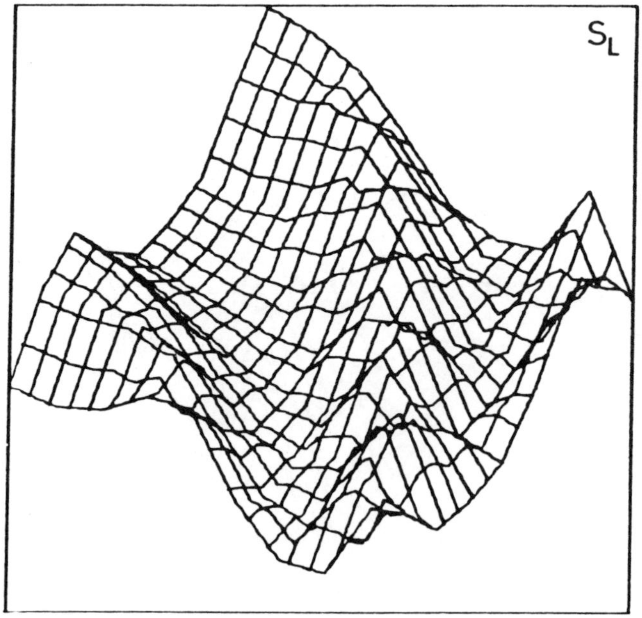

Fig. 7

Literatur:

[1] Barnhill, R. E., Birkhoff, G., Gordon, W. J. (1973)
Smooth Interpolation in Triangles, Journal of Approximation Theory, 8 , 114 - 128

[2] Barnhill, R. E., Mansfield, N. (1974)
Error Bounds for Smooth Interpolation in Triangles, Journal of Approximation Theory, 11 , 306 - 318

[3] Böhmer, K., Coman, Gh. (1977)
Blending Interpolation Schemes on Triangles with Error Bounds, in: Zeller, K., Schempp, W. (eds.): Constructive Theory of Functions of Several Variables, Lecture Notes in Mathematics 571, Berlin

[4] Nielson, G. M. (1979)
The Side-Vertex-Method for Interpolation in Triangles, Journal of Approximation Theory, 25 , 318 - 336

REMAINDER TERMS IN NUMERICAL INTEGRATION FORMULAS OF THE SPHERE

W. Freeden, R. Reuter

Rheinisch-Westfälische Technische Hochschule Aachen

The purpose of the present paper is the study of formulas for numerical computation of integrals over the (unit) sphere. The theory of Green's functions on the sphere with respect to the (Laplace-)Beltrami-operator is the main tool. General cubature formulas are considered. Estimates of the truncation error are given. Best approximations are discussed for regular figures on the sphere.

1. Definitions and Notations

Let us use $x_{(q)}, y_{(q)}, \ldots$ (or when confusion is not likely to arise $x, y, \ldots$) to represent the elements of Euclidean space $\mathbb{R}^q$. For all $x_{(q)} \in \mathbb{R}^q$, $x_{(q)} = (x_1, \ldots, x_q)$, different from the origin, we have the representation

$$x_{(q)} = r\,\xi_{(q)}\ , \quad r = |x_{(q)}| = \sqrt{x_1^2 + \ldots + x_q^2}\ , \qquad (1.1)$$

where $\xi_{(q)} = (\xi_1,\ldots,\xi_q)$ is the uniquely determined directional (unit) vector of $x_{(q)} \in \mathbb{R}^q$. The unit sphere in $\mathbb{R}^q$ will be called Ω_q. The total surface of Ω_q will be denoted by ω_q. As is well-known,

$$\omega_q = \int\limits_{\Omega_q} d\omega_q = \frac{2\,\pi^{q/2}}{\Gamma(q/2)} \qquad (d\omega : \text{volume element}) .$$

If the vectors $e^1,\ldots,e^q$ form the (canonical) orthonormal basis in $\mathbb{R}^q$, we may represent the points $\xi_{(q)}$ of Ω_q by

$$\xi_{(q)} = t\, e^q + \sqrt{1-t^2}\; \xi_{(q-1)} \;, \quad -1 \le t \le 1 \;, \tag{1.2}$$

where $\xi_{(q-1)}$ is a unit vector in the space spanned by $e^1,\ldots,e^{q-1}$. In particular, for $q = 3$, we have

$$\begin{aligned} &\xi_{(3)} = t\, e^3 + \sqrt{1-t^2}\;(e^1 \cos\varphi + e^2 \sin\varphi) \\ &-1 \le t \le 1 \;,\; 0 \le \varphi < 2\pi \;,\; t = \cos\theta \;. \end{aligned} \tag{1.3}$$

In terms of the polar coordinates (1.1) the Laplace-operator $\Delta = \nabla\nabla$ in $\mathbb{R}^q$

$$\Delta_{x_{(q)}} = \left(\frac{\partial}{\partial x_1}\right)^2 + \ldots + \left(\frac{\partial}{\partial x_q}\right)^2 \tag{1.4}$$

has the representation

$$\Delta_{x_{(q)}} = r^{1-q}\,\frac{\partial}{\partial r}\, r^{q-1}\,\frac{\partial}{\partial r} + \frac{1}{r^2}\,\Delta^*_{\xi_{(q)}} \;, \tag{1.5}$$

where $\Delta^* = \nabla^*\nabla^*$ describes the (Laplace-)Beltrami-operator of the unit sphere Ω_q

$$\begin{aligned} \Delta^*_{\xi_{(q)}} &= (1-t^2)\left(\frac{\partial}{\partial t}\right)^2 - (q-1)\,t\,\frac{\partial}{\partial t} + \frac{1}{1-t^2}\,\Delta^*_{\xi_{(q-1)}} \\ \Delta^*_{\xi_{(2)}} &= \left(\frac{\partial}{\partial\varphi}\right)^2 \;. \end{aligned} \tag{1.6}$$

2. Theory of Spherical Harmonics

The (Laplace-)spherical harmonics S_n of order n and dimension q are defined as the everywhere on the unit sphere Ω_q infinitely differentiable eigenfunctions of the Beltrami (differential) equation

$$(\Delta^*_\xi + \lambda_n)\; S_n(q;\xi) = 0 \qquad (2.1)$$

corresponding to the eigenvalues $\lambda_n = \lambda_n(q) = n(n+q-2)$. As is well-known (cf. [7]), the functions $H_n(q;x) = r^n\, S_n(q;\xi)$ are polynomials in rectangular coordinates which satisfy the Laplace equation $\Delta_x\, H_n(q;x) = 0$ and are homogeneous of degree n. Conversely, every homogeneous harmonic polynomial of degree n restricted to the unit sphere Ω_q is a spherical harmonic of order n and dimension q.

The Legendre polynomials P_n are the only everywhere on the interval [-1,1] infinitely differentiable eigenfunctions of the Legendre (differential) equation

$$\left((1-t^2)\left(\frac{d}{dt}\right)^2 - (q-1)\,t\,\frac{d}{dt} + \lambda_n\right) P_n(q;t) = 0\;, \qquad (2.2)$$

which in t = 1 satisfy $P_n(q;1) = 1$. Apart from a constant factor, the Legendre polynomials are the only spherical harmonics which are invariant under orthogonal transformations with the "north pole" e^q as fixed point.

Spherical harmonics of different order are orthogonal in the sense of the L^2-inner product

$$(S_n, S_m)_2 = \int_{\Omega_q} S_n(q;\xi)\; S_m(q;\xi)\; d\omega_q(\xi) = 0\;.$$

The linear space $H(q;n)$ of all spherical harmonics of order n and dimension q is of the dimension (cf. e. g. [7])

$$N(q;n) = \dim H(q;n) \; \begin{cases} \dfrac{(2n+q-2)\;\Gamma(n+q-2)}{\Gamma(n+1)\;\Gamma(q-1)} & , n \geq 1 \\ 1 & , n = 0 \;. \end{cases}$$

In other words, there must be $N(q;n)$ linearly independent spherical harmonics $S_{n1},\ldots,S_{n\,N(q;n)}$. We assume this system to be orthonormalized in the sense of the L^2-inner product

$$(S_{nj},S_{mk})_2 = \int_{\Omega_q} S_{nj}(q;\xi)\; S_{mk}(q;\xi)\; d\omega_q(\xi) = \delta_{nm}\,\delta_{jk} \;.$$

For any two vectors $\xi,\eta \in \Omega_q$ the sum

$$F_n(q;\xi,\eta) = \sum_{j=1}^{N(q;n)} S_{nj}(q;\xi)\; S_{nj}(q;\eta)$$

is invariant under all orthogonal transformations A, i. e. $F_n(q;A\xi,A\eta) = F_n(q;\xi,\eta)$. For fixed $\xi \in \Omega_q$, $F_n(q;\xi,\eta)$ is as function of η a spherical harmonic of order n and dimension q. $F_n(q;\xi,\eta)$ is symmetric in ξ and η and depends only on the scalar product of ξ and η. Thus we have apart from a multiplicative constant $\alpha_n = \alpha_n(q)$

$$\sum_{j=1}^{N(q;n)} S_{nj}(q;\xi)\; S_{nj}(q;\eta) = \alpha_n\; P_n(q;\xi\eta) \;.$$

In order to evaluate α_n we set $\xi = \eta$. Then we obtain

$$\sum_{j=1}^{N(q;n)} \left[S_{nj}(q;\xi)\right]^2 = \alpha_n\; P_n(q;1) = \alpha_n \;.$$

Integration over Ω_q yields $N(q;n) = \omega_q\,\alpha_n$. Therefore we find the addition theorem

$$\sum_{j=1}^{N(q;n)} S_{nj}(q;\xi)\; S_{nj}(q;\eta) = \frac{N(q;n)}{\omega_q}\; P_n(q;\xi\eta) \;. \tag{2.3}$$

Let ϕ be a function of class $L^1[-1,1]$. Then, for all

$S_n \in H(q;n)$ Hecke's formula gives

$$\int_{\Omega_q} \phi(\xi\eta) S_n(q;\eta)\, d\omega_q(\eta) = \beta_n\, S_n(q;\xi) \tag{2.4}$$

where

$$\beta_n = \omega_{q-1} \int_{-1}^{+1} \phi(t)\, P_n(q;t)\, (1-t^2)^{(q-3)/2}\, dt\ . \tag{2.5}$$

This formula establishes the close connection between the orthogonal invariance of the sphere and the addition theorem.

3. Green's Functions

We next define the Green function of the unit sphere with respect to an operator $L_1(\Delta^*) = \Delta^* + \rho$. Although the procedure to prove the main properties of Green's function is nearly standard, a short approach shall be presented here.

Definition 1: Let ρ be a complex number. A function $G_q(L_1;\xi,\eta)$ is called Green's function of the unit sphere Ω_q with respect to the operator $L_1(\Delta^*) = \Delta^* + \rho$ and the vector $\xi \in \Omega_q$, if it satisfies the following properties.

(i) For fixed $\xi \in \Omega_q$, $G_q(L_1;\xi,\eta)$ is infinitely differentiable for all $\eta \in \Omega_q$, $\eta \neq \xi$, with

$$L_1(\Delta^*_\eta)\ G_q(L_1;\xi,\eta) = \omega_q \sum_{L_1(-\lambda_n)=0} \sum_{j=1}^{N(q;n)} S_{nj}(q;\xi) S_{nj}(q;\eta).$$

(ii) In the neighbourhood of the point $\xi \in \Omega_q$ the estimates

$$\left.\begin{aligned} G_q(L_1;\xi,\eta) + \ln(1-\xi\eta) &= O(1) \\ \nabla^*_\eta\, G_q(L_1;\xi,\eta) + \nabla^*_\eta \ln(1-\xi\eta) &= O(1) \end{aligned}\right\} \quad (q=3)$$

and

$$G_q(L_1;\xi,\eta) - C_q(1-\xi\eta)^{(3-q)/2} = \begin{cases} O(\ln(1-\xi\eta)) & (q = 5) \\ O((1-\xi\eta)^{(5-q)/2}) & (\text{otherwise}) \end{cases}$$

$$\nabla^*_\eta G_q(L_1;\xi,\eta) - C_q\nabla^*_\eta(1-\xi\eta)^{(3-q)/2} = \begin{cases} O(\nabla^*_\eta\ln(1-\xi\eta)) & (q = 5) \\ O(\nabla^*(1-\xi\eta)^{(5-q)/2}) & (\text{otherwise}) \end{cases}$$

are valid $\left(C_q = \frac{\omega_q}{q-3} \cdot \frac{2^{(3-q)/2}}{\omega_{q-1}} \right)$.

(iii) For all orthogonal transformations A

$$G_q(L_1;\xi,\eta) = G_q(L_1;A\xi,A\eta) .$$

(iv) For all integers n with $L_1(-\lambda_n) = 0$ and $j = 1,\ldots,N(q;n)$

$$\int_{\Omega_q} G_q(L_1;\xi,\eta)\, S_{nj}(q;\eta)\, d\omega_q(\eta) = 0$$

uniformly with respect to $\xi \in \Omega_q$.

Remark: For an operator $L(\Delta^*)$, the symbol $\sum_{L(-\lambda_n)=0}$ means that the sum is to be taken over all non-negative integers n for which $L(-\lambda_n)=0$. In the case $L(-\lambda_n)\neq 0$ for all non-negative integers n, the sum is assumed to be zero.

By the defining properties Green's function is uniquely determined (cf. [2], [3]). Following Hilbert's approach to the theory of Green's functions [5] we prove the existence of $G_q(L_1;\xi,\eta)$ by first giving an explicit representation of Green's function to the operator Δ^*.

<u>Lemma 1:</u> (cf. [8]*). For* $L_1 = L_1(\Delta^*) = \Delta^*$

$$G_q(\Delta^*;\xi,\eta) = \frac{1}{q-2}\int_0^1 \frac{1+r^{q-2}}{r}\left[\frac{1}{(1-2r(\xi\eta)+r^2)^{(q-2)/2}} - 1\right]dr.$$

In particular,

$$G_3(\Delta^*;\xi,\eta) = -1 + \ln 2 - \ln(1-\xi\eta) \text{ , } (cf.\ [2],\ [3]).$$

In order to assure the existence of $G_q(L_1;\xi,\eta)$, $L_1(-\lambda_n) \neq 0$ for all integers n, we consider the integral equation

$$
\begin{aligned}
G_q(L_1;\xi,\eta) &= G_q(\Delta^*;\xi,\eta) \\
&+ \frac{L_1(-\lambda_n)+\lambda_n}{\omega_q} \int_{\Omega_q} G_q(L_1;\xi,\zeta)\, G_q(\Delta^*;\eta,\zeta)\, d\omega_q(\zeta) \\
&- \frac{1}{L_1(-\lambda_n)+\lambda_n} \,,
\end{aligned}
\tag{3.1}
$$

which establishes the close relation between Green's function and the resolvent of the kernel $G_q(\Delta^*;\xi,\eta)$. The factor $L_1(-\lambda_n) + \lambda_n$ is an eigenvalue of the kernel $G_q(\Delta^*;\xi,\eta)$ if and only if $L_1(-\lambda_n) + \lambda_n$ is an eigenvalue with respect to the Beltrami operator Δ^*. Thus the existence of $G_q(L_1;\xi,\eta)$ follows from standard arguments of the theory of integral equations.

In the case $L_1(-\lambda_n) = 0$ for an integer $n > 0$, we discuss the integral equation

$$
\begin{aligned}
G_q(L_1;\xi,\eta) &= G_q(\Delta^*;\xi,\eta) \\
&+ \frac{L_1(-\lambda_n)+\lambda_n}{\omega_q} \int_{\Omega_q} G_q(L_1;\xi,\zeta)\, G_q(\Delta^*;\eta,\zeta)\, d\omega_q(\zeta) \\
&- \frac{1}{L_1(-\lambda_n)+\lambda_n} - \frac{\omega_q}{\lambda_n} \sum_{j=1}^{N(q;n)} S_{nj}(q;\xi)\, S_{nj}(q,\eta) \,.
\end{aligned}
\tag{3.2}
$$

It is not difficult to see that

$$
\int_{\Omega_q} G_q(\Delta^*;\xi,\eta)\; S_n(q;\eta)\; d\omega_q(\eta)
$$

$$
= \int_{\Omega_q} \left[\frac{1}{L_1(-\lambda_n)+\lambda_n} + \frac{\omega_q}{\lambda_n} \sum_{j=1}^{N(q;n)} S_{nj}(q;\xi)\, S_{nj}(q;\eta) \right] S_n(q;\eta)\, d\omega_q(\eta)
$$

for all spherical harmonics S_n of order $n > 0$.

The integral equation (3.2) therefore has a solution which is uniquely determined by the conditions

$$\int_{\Omega_q} G_q(L_1;\xi,\eta)\; S_{nj}(q;\eta)\; d\omega_q(\eta) = 0$$

for $j = 1,\ldots,N(q;n)$.

Let f be a function on Ω_q satisfying a Hölder-condition at the point $\xi \in \Omega_q$. Then

$$L_1(\Delta^*_\eta) \int_{\Omega_q} G_q(L_1;\eta,\zeta)\; f(\zeta)\; d\omega_q(\zeta)$$

$$= -\,\omega_q f(\eta) + \omega_q \sum_{L_1(-\lambda_n)=0}\; \sum_{j=1}^{N(q;n)} S_{nj}(q;\eta) \int_{\Omega_q} f(\zeta) S_{nj}(q;\zeta) d\omega_q(\zeta). \tag{3.3}$$

The proof can be given by analogous conclusions as known in potential theory.

<u>Definition 2:</u> Let $\rho_1,\ldots,\rho_m$ be complex numbers. Suppose that L_m is an operator recursively given by $L_m(\Delta^*) = (\Delta^*+\rho_m) L_{m-1}(\Delta^*)$, $L_1(\Delta^*) = \Delta^*+\rho_1$. Let $G_q(L_m;\xi,\eta)$ be defined by the convolution

$$G_q(L_m;\xi,\eta) = \int_{\Omega_q} G_q(L_{m-1};\xi,\zeta) G_q(\Delta^*+\rho_m;\zeta,\eta) d\omega_q(\zeta)\;, \quad m = 2,3,\ldots$$

$$G_q(L_1;\xi,\eta) = G_q(\Delta^*+\rho_1;\xi,\eta).$$

Then $G_q(L_m;\xi,\eta)$ is called Green's function with respect to the operator L_m and the parameter $\xi \in \Omega_q$.

In analogy to techniques known in potential theory it can be proved that

$$G_q(L_m;\xi,\eta) = \begin{cases} O((1-\xi\eta)^{m-(q-1)/2}\ln(1-\xi\eta)) & , 2m \geq q-1 \\ & q \text{ odd} \\ O((1-\xi\eta)^{m-(q-1)/2}) & , \text{otherwise} \end{cases}$$

Hence, if $m > (q-1)/2$, $G_q(L_m;\xi,\eta)$ is continuous on the whole surface Ω_q. Furthermore, for $m > (q-1)/2$, the bilinear expansion

$$(-\omega_q)^m \sum_{L_m(-\lambda_n)\neq 0} \frac{1}{L_m(-\lambda_n)} \sum_{j=1}^{N(q;n)} S_{nj}(q;\xi)\, S_{nj}(q;\eta)$$

is absolutely and uniformly convergent both in ξ and η and uniformly in ξ and η together. Therefore we have

Lemma 2: If $m > \frac{q-1}{2}$, *then*

$$G_q(L_m;\xi,\eta) = (-\omega_q)^m \sum_{L_m(-\lambda_n)\neq 0} \frac{1}{L_m(-\lambda_n)} \sum_{j=1}^{N(q;n)} S_{nj}(q;\xi) S_{nj}(q;\eta).$$

Remark: The symbol $\sum_{L_m(-\lambda_n)\neq 0}$ means that the sum is to be extended over all non-negative integers n for which $L_m(-\lambda_n) \neq 0$.

4. Integral Formulas of the Unit Sphere

One of us has presented in [2], [3],[4] a class of integral formulas for the sphere Ω_3 in Euclidean space $\mathbb{R}^3$ which can be used in (geodetic) problems to integrate experimental data or measured values numerically. The purpose now is to transcribe these integral formulas to the general q-dimensional case and to derive some generalizations.

Suppose that f is a twice continuously differentiable function on Ω_q. Then for each sufficiently small $\varepsilon > 0$ Green's integral theorem gives (cf. [2], [6])

$$\int_{\substack{|\xi-\eta|\geq\varepsilon\\|\eta|=1}} \left\{G_q(L_1;\xi,\eta)\ L_1(\Delta_\eta^*)\ f(\eta) - f(\eta)\ L_1(\Delta_\eta^*)\ G_q(L_1;\xi,\eta)\right\} d\omega_q(\eta)$$

$$= \int_{\substack{|\xi-\eta|=\varepsilon\\|\eta|=1}} \left\{G_q(L_1;\xi,\eta)\ \frac{\partial}{\partial n_\eta} f(\eta) - f(\eta)\ \frac{\partial}{\partial n_\eta}\ G_q(L_1;\xi,\eta)\right\} dF_q(\eta)\ .$$

dF denotes the surface element of the intersection of the sphere with center ξ and radius ε and Ω_q, while n is the (unit) vector normal to $|\xi-\eta| = \varepsilon$ and tangential on Ω_q and directed into the exterior of $|\xi-\eta| \geq \varepsilon$. Inserting the differential equation of Green's function we obtain

$$\int_{\substack{|\xi-\eta|\geq\varepsilon\\|\eta|=1}} f(\eta)\ L_1(\Delta_\eta^*)\ G_q(L_1;\xi,\eta)\ d\omega_q(\eta)$$

$$= \omega_q \sum_{L_1(-\lambda_n)=0}\ \sum_{j=1}^{N(q;n)} S_{nj}(q;\xi) \int_{\substack{|\xi-\eta|\geq\varepsilon\\|\eta|=1}} f(\eta)\ S_{nj}(q;\eta)\ d\omega_q(\eta)\ .$$

Observing the characteristic singularity of Green's function we are able to prove by analogous conclusions as known in potential theory

$$\int_{\substack{|\xi-\eta|=\varepsilon\\|\eta|=1}} G_q(L_1;\xi,\eta)\ \frac{\partial}{\partial n}\ f(\eta)\ dF_q(\eta) = o(1) \qquad , \ (\varepsilon\to 0)$$

$$\int_{\substack{|\xi-\eta|=\varepsilon\\|\eta|=1}} f(\eta)\ \frac{\partial}{\partial n_\eta}\ G_q(L_1;\xi,\eta)\ dF_q(\eta) = \omega_q f(\xi) + o(1),\ (\varepsilon\to 0).$$

Summarizing our results we therefore obtain

Theorem 1: If $\xi \in \Omega_q$ *and* $f \in C^{(2)}(\Omega_q)$, *then*

$$f(\xi) = \sum_{L_1(-\lambda_n)=0} \sum_{j=1}^{N(q;n)} S_{nj}(q;\xi) \int_{\Omega_q} f(\eta) S_{nj}(q;\eta)\, d\omega_q(\eta)$$

$$- \frac{1}{\omega_q} \int_{\Omega_q} G_q(L_1;\xi,\eta)\ [L_1 f(\eta)]\ d\omega_q(\eta)\ .$$

Remark: When confusion is avoided we use the more simplified notation $L_1 f(\eta)$ instead of $L_1(\Delta^*_\eta)\ f(\eta)$.

By successive integration by parts we obtain in connection with formula (3.3) the following extension of Theorem 1:

Theorem 2: Let $\xi \in \Omega_q$ *and* $f \in C^{(2m)}(\Omega_q)$. *Then*

$$f(\xi) = \sum_{L_m(-\lambda_n)=0} \sum_{j=1}^{N(q;n)} S_{nj}(q;\xi) \int_{\Omega_q} f(\eta) S_{nj}(q;\eta)\, d\omega_q(\eta)$$

$$+ \left(\frac{1}{-\omega_q}\right)^m \int_{\Omega_q} G_q(L_m;\xi,\eta)\ [L_m f(\eta)]\ d\omega_q(\eta)\ .$$

In particular, for $L_m = (\Delta^* + \lambda_0)\dots(\Delta^* + \lambda_{m-1})$,

$$f(\xi) = \sum_{n=0}^{m-1} \sum_{j=1}^{N(q;n)} S_{nj}(q;\xi) \int_{\Omega_q} f(\eta) S_{nj}(q;\eta)\, d\omega_q(\eta)$$

$$+ \left(\frac{1}{-\omega_q}\right)^m \int_{\Omega_q} G_q(L_m;\xi,\eta)\ [L_m f(\eta)]\ d\omega_q(\eta)\ .$$

This formula compares the (m-1)-th partial sum of the orthogonal expansion of f into spherical harmonics and the functional value of f taken at the point $\xi \in \Omega_q$ respectively.

5. The Differential Equation $L_m V = W$

Theorem 2 will be used now to discuss the differential equation $L_m V = W$, $V \in C^{(2m)}(\Omega_q)$. From Green's identity it is obvious that

$$\int_{\Omega_q} [L_m V(\eta)]\; S(\eta)\; d\omega_q(\eta) = \int_{\Omega_q} V(\eta)\; [L_m\, S(\eta)]\; d\omega_q(\eta) = 0 \tag{5.1}$$

for all elements S belonging to the null space of the operator L_m. On the other hand, any function S of the null space of L_m can be added to V without changing the differential equation. However, if we require that V is orthogonal to the null space of L_m, then the differential equation is uniquely solvable.

Theorem 3: Let W *be a function of class* $C^{(0)}(\Omega_q)$ *orthogonal to the null space of the operator* L_m. *Then the function* V *given by*

$$V(\xi) = \left(\frac{1}{-\omega_q}\right)^m \int_{\Omega_q} G_q(L_m;\xi,\eta)\; W(\eta)\; d\omega_q(\eta)$$

represents the only 2m-*times continuously differentiable solution of the differential equation* $L_m V = W$ *on* Ω_q *which is orthogonal to the null space of* L_m.

6. A General Summation Formula

By $V = L_m^{-1} W$ we denote a function of class $C^{(2m)}(\Omega_q)$ satisfying $L_m V = W$ on Ω_q (and it does not matter which we choose). Integration by parts yields

$$\int_{\Omega_q} W(\eta) f(\eta) d\omega_q(\eta) = \int_{\Omega_q} L_m^{-1} W(\eta) \, [L_m f(\eta)] \, d\omega_q(\eta) \qquad (6.1)$$

for every function $f \in C^{(2m)}(\Omega_q)$. From Theorem 2, it follows that

$$\sum_{k=1}^{T} a_k f(\eta_k) = \sum_{L_m(-\lambda_n)=0} \sum_{j=1}^{N(q;n)} \sum_{k=1}^{T} a_k S_{nj}(q;\eta_k) \int_{\Omega_q} f(\eta) S_{nj}(q;\eta) d\omega_q(\eta)$$
$$+ \left(-\frac{1}{\omega_q}\right)^m \int_{\Omega_q} \sum_{k=1}^{T} a_k G_q(L_m;\eta_k,\eta) \, [L_m f(\eta)] \, d\omega_q(\eta) \qquad (6.2)$$

holds for any choice of (real) coefficients $a_1,\dots,a_T$ and nodes $\eta_1,\dots,\eta_T$ of the unit sphere Ω_q. Combining (6.1) and (6.2) we therefore obtain the following summation formula:

Theorem 4: Let W *be a weight function of class* $C^{(0)}(\Omega_q)$ *orthogonal to the null space of the operator* L_m. *For given weights* $a_1,\dots,a_T$ *and preassigned nodes* $\eta_1,\dots,\eta_T$ *let* K *designate a function of the form*

$$K_m(\eta) = \left(-\frac{1}{\omega_q}\right)^m \sum_{k=1}^{T} a_k G_q(L_m;\eta_k,\eta) - L_m^{-1} W(\eta) \; . \qquad (6.3)$$

Then the identity

$$\sum_{k=1}^{T} a_k f(\eta_k) = \int_{\Omega_q} W(\eta) \, f(\eta) \, d\omega_q(\eta)$$
$$+ \sum_{L_m(-\lambda_n)=0} \sum_{j=1}^{N(q;n)} \sum_{k=1}^{T} a_k S_{nj}(q;\eta_k) \int_{\Omega_q} f(\eta) S_{nj}(q;\eta) d\omega_q(\eta)$$
$$+ R f$$

holds for every function $f \in C^{(2m)}(\Omega_q)$, *where the remainder* Rf *is given by*

$$Rf = \int_{\Omega_q} K_m(\eta)\ [L_m f(\eta)]\ d\omega_q(\eta) \quad .$$

Let S be an element belonging to the null space of the operator L_m, i. e. a function of the form

$$S(\eta) = \sum_{L_m(-\lambda_n)=0} \ \sum_{j=1}^{N(q;n)} d_{nj}\ S_{nj}(q;\eta) \ . \qquad (6.4)$$

A system $\eta_1,\ldots,\eta_T$ of T points of the unit sphere Ω_q with $T \geq M = \sum_{L_m(-\lambda_n)=0} N(q;n)$ is called L_m-<u>fundamental</u> <u>system</u> if the rank of the (M,T)-matrix of the linear system

$$\sum_{k=1}^{T} a_k\ S_{nj}(q;\eta_k) = d_{nj} \qquad (6.5)$$

$$L_m(-\lambda_n) = 0 \ , \ j = 1,\ldots,N(q;n)$$

is equal to M.

Given a function S of the form (6.4) and a L_m-fundamental system $\eta_1,\ldots,\eta_T$. Then the function S has the representation

$$S(\eta) = \sum_{L_m(-\lambda_n)=0} \ \sum_{j=1}^{N(q;n)} \sum_{k=1}^{T} a_k S_{nj}(q;\eta_k) S_{nj}(q;\eta) \ . \qquad (6.6)$$

Therefore Theorem 4 admits the reformulation

<u>*Corollary:*</u> *Let* $\eta_1,\ldots,\eta_T$ *be a* L_m*-fundamental system. Let* S *be a function of the form* (6.4). *Furthermore, let* K_m *be a function of the form* (6.3) *with coefficients* $a_1,\ldots,a_T$ *satisfying the linear system* (6.5). *Then the identity*

$$\sum_{k=1}^{T} a_k f(\eta_k) = \int_{\Omega_q} [S(\eta) + W(\eta)] f(\eta) \, d\omega_q(\eta)$$

$$+ \int_{\Omega_q} K_m(\eta) \, [L_m f(\eta)] \, d\omega_q(\eta)$$

holds for all $f \in C^{(2m)}(\Omega_q)$.

7. Error Bounds

For integers m with $m > \frac{q-1}{2}$ the kernel K_m is continuous on Ω_q.

Corollary: For integers m *with* $m > \frac{q-1}{2}$

$$|Rf| \leq \max_{\eta \in \Omega_q} |K_m(\eta)| \cdot \int_{\Omega_q} |L_m f(\eta)| \, d\omega_q(\eta)$$

is valid for all $f \in C^{(2m)}(\Omega_q)$.

The kernel K_m has a finite L^p-norm

$$\|K_m\|_p = \left(\int_{\Omega_q} |K_m(\eta)|^p \, d\omega_q(\eta) \right)^{1/p} < \infty$$

provided that p satisfies the condition

$$1 \leq p < \frac{q-1}{2} \cdot \frac{2}{q-1-2m} \quad . \tag{7.1}$$

Therefore Hölder's inequality gives

Corollary: For given m, *let* p *satisfy*

$$1 < p < \frac{q-1}{q-1-2m}$$

Then

$$|Rf| \leq \left(\int_{\Omega_q} |L_m f(\eta)|^{p'} \, d\omega_q(\eta) \right)^{1/p'} \cdot \left(\int_{\Omega_q} |K_m(\eta)|^{p} \, d\omega_q(\eta) \right)^{1/p}$$

for $\frac{1}{p} + \frac{1}{p'} = 1$.

For all integers m with $2m > \frac{q-1}{2}$ we have, in particular

$$|Rf| \leq \left(\int_{\Omega_q} |L_m f(\eta)|^{2} \, d\omega_q(\eta) \right)^{1/2} \cdot \left(\int_{\Omega_q} |K_m(\eta)|^{2} \, d\omega_q(\eta) \right)^{1/2} .$$

The norm $||K_m||_2$ is independent of f. If we fix the L_m-fundamental system $\eta_1,\ldots,\eta_T$ a priori, then $||K_m||_2$ is a function of the weights $a_1,\ldots,a_T$. Among all T-tupels of weights $a_1,\ldots,a_T$ satisfying (6.5) there exists one and only one having the property that $||K_m||_2$ is minimal. Following Sard (cf. e. g. [9] and for a general approach [1]), we shall say that the resulting cubature formula is the best formula.

8. Best Approximation for Regular Polyhedra

We conclude the paper by giving some numerical information about best approximate cubature for the following nets generated by the regular polyhedra on the three-dimensional sphere Ω_3.

S_1: Tetrahedron, T = 4

$\alpha(-1,-1,-1)$, $\alpha(-1, 1, 1)$ $\quad \alpha = \sqrt{1/3}$
$\alpha(1,-1, 1)$, $\alpha(1, 1,-1)$

S_2: Octahedron, T = 6

$(\pm1,0,0)$, $(0,\pm1,0)$, $(0,0,\pm1)$

S_3: Cube, T = 8

$$\alpha(\pm 1,\pm 1,\pm 1)\ ,\quad \alpha = \sqrt{1/3}$$

S_4: Icosahedron, T = 12

$$\alpha(0,\pm\tau,\pm 1)\ ,\quad \alpha(\pm 1,0,\pm\tau)\ ,\quad \alpha(\pm\tau,\pm 1,0)\ ,$$
$$\tau = (1+\sqrt{5})/2\ ,\quad \alpha = 1/\sqrt{(1+\tau^2)}$$

S_5: Dodecahedron, T = 20

$$\alpha(\pm 1/\tau,\pm\tau,0)\ ,\quad \alpha(0,\pm 1/\tau,\pm\tau)\ ,$$
$$\alpha(\pm\tau,0,\pm 1/\tau)\ ,\quad \alpha(\pm 1,\pm 1,\pm 1)\ ,$$
$$\tau = (1+\sqrt{5})/2\ ,\quad \alpha = \sqrt{1/3}\ .$$

We discuss two typical examples of simple nature.

Example 1: $S(\eta) = 1$, $W(\eta) = 0$

Considered operators $L_m(\Delta^*)$: Δ^*, $\Delta^*(\Delta^*+2)$.
In order to obtain the weights of the best approximate cubature formula

$$\sum_{k=1}^{T} a_k f(\eta_k) \approx \int_{\Omega_3} f(\eta)\, d\omega_3(\eta)$$

we have to solve the quadratic optimization problem

$$\frac{1}{\omega_3^{2m}} \sum_{k=1}^{T} \sum_{i=1}^{T} a_k a_i\, G_3(L_m^2;\eta_i,\eta_k) \longrightarrow \min.$$

under the constraints (6.5). It turns out that the unique solution is given by

$$a_1 = \ldots = a_T = \omega_3/T = 4\pi/T\ .$$

The weights $a_1,\ldots,a_T$ are independent of the special operators, whereas the quantities $\|K_m\|_2$ are not :

$$||K_m||_2 = \sqrt{\frac{\omega_3^{2-2m}}{T} \sum_{i=1}^{T} G_3(L_m^2;\eta_i,\eta_k)} ,$$

for every k=1,...,T.

Table I: $||K_m||_2$

point system	Δ^*	$\Delta^*(\Delta^* + 2)$
S_1	0.71743	0.06044
S_2	0.46736	0.02283
S_3	0.36306	0.01583
S_4	0.22964	0.00509
S_5	0.15018	0.00291

Example 2: $S(\eta) = 0$, $W(\eta) = 1$

Considered operators $L_m(\Delta^*)$: $\Delta^* - 1$, $\Delta^* + 1$, $(\Delta^* - 1)(\Delta^* + 1)$. The problem of best approximation now is equivalent to the quadratic optimization problem

$$\frac{1}{\omega_3^{2m}} \sum_{i=1}^{T} \sum_{k=1}^{T} a_i a_k G_3(L_m^2;\eta_i,\eta_k) - 2 \sum_{i=1}^{T} a_i + \omega_3 \longrightarrow \min.$$

The set of constraints (6.5) is empty, because there is no integer n with $L_m(-\lambda_n) = 0$. The unique solution depends on the chosen operators:

$$a_1 = \ldots = a_T = \omega_3^{2m} \Big/ \sum_{i=1}^{T} G_3(L_m^2;\eta_i,\eta_k) , \text{ for every } k.$$

$||K_m||_2$ now reads

$$||K_m||_2 = \sqrt{\omega_3 - a_k T} = \sqrt{4\pi - a_k T} , \text{ for every } k=1,\ldots,T.$$

Table II: Weights $a_1 = \dots = a_T$

point system	$\Delta^* - 1$	$\Delta^* + 1$	$(\Delta^* - 1)(\Delta^* + 1)$
S_1	3.031897090128	3.000626363487	3.140941664452
S_2	2.061296010638	2.055526076264	2.094324150137
S_3	1.555566892327	1.553264205723	1.570770547271
S_4	1.042986892295	1.042644446705	1.047195586674
S_5	0.627228490905	0.627154872858	0.628318144430

Table III: $\|K_m\|_2$

point system	$\Delta^* - 1$	$\Delta^* + 1$	$(\Delta^* - 1)(\Delta^* + 1)$
S_1	0.66241	0.75091	0.05103
S_2	0.44564	0.48292	0.02063
S_3	0.34905	0.37451	0.01436
S_4	0.22478	0.23375	0.00486
S_5	0.14765	0.15256	0.00278

Remark: Using the addition theorem $\omega_3^{-2m} G_3(L_m^2;\eta,\xi)$ is equal to

$$\frac{1}{\omega_3} \sum_{L_m(-\lambda_n) \neq 0} \frac{2n+1}{[L_m(-\lambda_n)]^2} P_n(3;\eta\xi) \;.$$

For the computation of the quantities in the tables the series has been approximated by its partial sum of 2000 terms.

References

[1] Delvos, F.J.; Schempp, W.: An extension of Sard's method. In: Spline Functions, Proc. Int. Symp. Karlsruhe (Ed.: Böhmer, K.; Meinardus, G.; Schempp, W.), Lecture Notes in Mathematics 501, Berlin, Heidelberg, New York, Springer 1976, 80 - 92

[2] Freeden, W.: Über eine Klasse von Integralformeln der mathematischen Geodäsie, Habilitationsschrift, Veröffentlichung des Geodätischen Instituts der RWTH Aachen, Heft Nr.27 (1979)

[3] Freeden, W.: On integral formulas of the (unit) sphere and their application to numerical computation of integrals, Computing 25 (1980), 131 - 146

[4] Freeden, W.: Eine Klasse von Kubaturformeln der Einheitssphäre, Zeitschrift für Vermessungswesen (ZfV) (1981), 200 - 210

[5] Hilbert, D.: Grundzüge einer allgemeinen Theorie der linearen Integralgleichungen, Leipzig, Berlin, Teubner 1912

[6] Hlawka, E.: Gleichverteilung auf Produkten von Sphären, Journ. Reine Angew. Math. 330 (1982), 1 - 43

[7] Müller, Cl.: Spherical harmonics, Lecture Notes in Mathematics 17, Berlin, Heidelberg, New York 1966

[8] Reuter, R.: Über Integralformeln der Einheitssphäre und harmonische Splinefunktionen, Veröffentlichung des Geodätischen Instituts der RWTH Aachen, Heft Nr.33 (1982)

[9] Sard, A.: Best approximate integration formulas, Amer. Journ. Math. 71 (1949), 80 - 91

Priv.-Doz. Dr. Willi Freeden
Institut für Reine und Angewandte Mathematik
RWTH Aachen
Templergraben 55
D-5100 Aachen

Dr. Richard Reuter
Rechenzentrum
RWTH Aachen
Seffenter Weg 23
D-5100 Aachen

A GENERALIZED SYLVESTER'S IDENTITY ON DETERMINANTS AND ITS APPLICATIONS TO INTERPOLATION PROBLEMS

M. Gasca, A. López-Carmona, V. Ramírez.

Dpto. Ecuaciones Funcionales. Univ. Granada. SPAIN.

1. Introduction

In the last few years G. Mühlbach [8 - 11] has introduced some interesting recurrence interpolation formulae that generalize those of Newton and Aitken-Neville. The interpolation space was spanned by Chebyshev systems in [8 - 9] but in [10 - 11] he has extended the formulae to the general finite linear interpolation problem.

C. Brezinski [1 - 2] has obtained in a simpler manner the same formulae for complete Chebyshev systems by using an identity on determinants due to Sylvester. The results are also valid for the general interpolation problem under similar hypotheses. Simultaneously M. Gasca and A. López-Carmona [4] have derived a recurrence interpolation formula that generalizes Mühlbach's work. In that paper the formula is applied to multivariate interpolation problems. Further examples can be found in [7].

Our aim in this paper is to derive this general formula following Brezinski's method and under less restrictive hypotheses. In section 2 a generalization of Sylvester's identity is obtained which to the authors best knowledge is new. Sylvester's original expression is recovered as the simplest particular case of this identity, as it is shown in section 3 .

In section 4 the generalized identity is applied to the derivation of a recurrence interpolation formula for the solution of the general interpolation problem. The formula, which takes the form of a quotient of two determinants, is similar, but not necessarily identical, to that of [4]. However they coincide in the most usual cases, as it is shown in sections 5 and 6.

2. A generalization of Sylvester's identity on determinants.

Let

(2.1) $$A = (a_{ij}) \qquad 1 \le i,j \le p+q$$

be a matrix whose elements belong to a conmutative field K of characteristic zero. $(p \ge 1, q \ge 2)$

Let us denote $|A| =$ det A and

(2.2) $$A\begin{bmatrix} i_1, i_2, \dots, i_h \\ j_1, j_2, \dots, j_h \end{bmatrix} = \det\,(a_{i_t j_u}) \qquad 1 \le t,u \le h.$$

Let

(2.3) $$I_k = \{r_{k1}, r_{k2}, \dots, r_{kp+1}\}, \qquad k = 1, 2, \dots, q$$

be q subsets of $\{1, 2, \dots, p+q\}$ with

(2.4) $$\begin{cases} \text{card } I_k = p+1 & k = 1, 2, \dots, q \\ \text{card } (I_k \cap I_{k+1}) = p & k = 1, 2, \dots, q-1 \,. \end{cases}$$

We call

(2.5) $$S_k = I_k \cap I_{k+1} = \{s_{k1}, s_{k2}, \dots, s_{kp}\},$$

and we assume

$$(2.6)\qquad \begin{cases} r_{k1} < r_{k2} < \dots < r_{kp+1} & k = 1, 2, \dots, q \\ s_{k1} < s_{k2} < \dots < s_{kp} & k = 1, 2, \dots, q-1 . \end{cases}$$

Let $|B|$ be the determinant of the matrix whose elements are.

$$(2.7)\qquad b_{ij} = A\begin{pmatrix} 1, & 2, & \dots, & p, & p+i \\ r_{j1}, & r_{j2}, & \dots, & r_{jp}, & r_{jp+1} \end{pmatrix} \qquad 1 \leq i, j \leq q .$$

Finally, let r_{kh_k} be the element of I_k such that

$$(2.8)\qquad \begin{cases} r_{kh_k} \in I_k - I_{k-1} = I_k - S_{k-1}, & k = 2, 3, \dots, q \\ r_{1h_1} \in I_1 - I_2 \quad = I_1 - S_1 & k = 1. \end{cases}$$

With all these notations we can establish the following result.

<u>Theorem</u>

$$(2.9)\qquad |B| = C\,|A| \prod_{k=1}^{q-1} A\begin{pmatrix} 1, & 2, & \dots, & p \\ s_{k1}, & s_{k2}, & \dots, & s_{kp} \end{pmatrix}$$

where C is a constant (which does not depend on the elements a_{ij} of A) whose value is

$$(2.10)\qquad C = \begin{cases} 0 \quad \text{if} \quad \operatorname{card} \bigcup_{k=1}^{q} I_k < p+q \\ (-1)^{q(q-1)/2 + \sum_{i=1}^{q} (r_{i h_i} - h_i)} \quad \text{if } \operatorname{card} \bigcup_{k=1}^{q} I_k = p+q . \end{cases}$$

Proof.- If $|A| = 0$ it is easily seen that the rows of $|B|$ are linearly dependent. Hence, $|A|$ is a factor of $|B|$, since $|A|$ is irreducible.

It is slightly more difficult [7] to show that if

$$(2.11)\qquad A\begin{pmatrix} 1, & 2, & \dots, & p \\ s_{k1}, & s_{k2}, & \dots, & s_{kp} \end{pmatrix} = 0$$

for any $k = 1, 2, \dots, q-1$, then the columns k, $k+1$ of $|B|$ are linearly dependent, and therefore $|B| = 0$.

The irreducibility of the left hand side of (2.11) imply that it is a factor of $|B|$. Thus

$$(2.12) \qquad |A| \prod_{k=1}^{q-1} A\begin{pmatrix} 1, & 2, & \dots, & p \\ s_{k1}, & s_{k2}, & \dots, & s_{kp} \end{pmatrix}$$

is also a factor of $|B|$. Since the degree of (2.12) and $|B|$ is $pq+q$ we obtain (2.9), where C is a constant, independent of the a_{ij}'s.

Finally we prove (2.10). If

$$\operatorname{card} \bigcup_{k=1}^{q} I_k < p + q$$

$|B|$ and

$$\prod_{k=1}^{q-1} A\begin{pmatrix} 1, & 2, & \dots, & p \\ s_{k1}, & s_{k2}, & \dots, & s_{kp} \end{pmatrix}$$

do not depend on the elements of at least one column of $|A|$, say a_{it}, $i = 1, 2, \dots, p+q$. Then, it is clear in (2.9) that C must be zero.

If

$$(2.13) \qquad \operatorname{card} \bigcup_{k=1}^{q} I_k = p + q$$

we observe that there is a set of terms in the product $b_{11} b_{22} \dots b_{qq}$:

$$(2.14) \qquad a_{p+1\, r_{1h_1}} A\begin{pmatrix} 1, & 2, & \dots, & p \\ s_{11}, & s_{12}, & \dots, & s_{1p} \end{pmatrix} \prod_{k=2}^{q} \left(a_{p+k\, r_{kh_k}} A\begin{pmatrix} 1, & 2, & \dots, & p \\ s_{k-11}, & s_{k-12}, & \dots, & s_{k-1p} \end{pmatrix} \right)$$

that does not appear twice in the expansion of $|B|$. (2.14) is affected by the sign

$$(2.15) \qquad (-1)^{p+1+h_1 + p+1+h_2 + \dots + p+1+h_q} = (-1)^{q(p+1) + \sum_{i=1}^{q} h_i}.$$

In

$$|A| \prod_{k=1}^{q} A\begin{pmatrix} 1 , & 2 , & \dots , & p \\ s_{k1}, & s_{k2}, & \dots , & s_{kp} \end{pmatrix}$$

the sign of (2.14) is

$$(2.16) \quad (-1)^{p+1+r_{1h_1}+p+2+r_{2h_2}+\dots+p+q+r_{qh_q}} = (-1)^{pq+\sum_{i=1}^{q}(i+r_{ih_i})} .$$

Hence

$$C = (-1)^{q(q-1)/2+\sum_{i=1}^{q}(r_{ih_i}-h_i)}$$

3. Special cases.

For $I_k = \{1, 2, 3, \dots , p , p+k\}$, $k = 1, 2, \dots, q$, we have the Sylvester's identity ([3], page 33):

$$|B| = |A| \cdot A\begin{pmatrix} 1, 2, \dots , p \\ 1, 2, \dots , p \end{pmatrix}^{q-1}$$

If $I_k = \{k, k+1, \dots , k+p\}$, $k = 1, 2, \dots , q$ we arrive at the identity

$$|B| = |A| \prod_{k=1}^{q-1} A\begin{pmatrix} 1 , & 2 , & \dots , & p \\ k+1, & k+2, & \dots , & k+p \end{pmatrix}$$

with

$$b_{ij} = A\begin{pmatrix} 1, & 2 , & \dots , p , & p+i \\ j, & j+1, & \dots , & j+p \end{pmatrix} \qquad 1 \le i, j \le q$$

in $|B|$.

4. Application of the identity to interpolation problems

Let V be a vector space of finite dimension k over K, $L_1, L_2, \dots L_k$, k linear forms on V and $z_1, z_2, \dots , z_k$ k elements of K.

Consider the problem of finding $p \in V$ such that

$$(4.1)\qquad L_i p = z_i \qquad i = 1, 2, \ldots, k.$$

Let $f_1, f_2, \ldots, f_k$ be a basis of V and L a linear form on V. If

$$(4.2)\qquad \det L_i f_j \neq 0 \qquad 1 \leq i, j \leq k$$

then there exists a unique solution of the problem (4.1)

$$p = \sum_{j=1}^{k} a_j f_j \;.$$

Obviously one has

$$(4.3)\qquad L p = \sum_{j=1}^{k} a_j L f_j \in K,$$

where the coefficients a_j are the solution of the system

$$(4.4)\qquad \sum_{j=1}^{k} a_j L_i f_j = z_i \,, \qquad i = 1, 2, \ldots, k.$$

Let $|A|$ and $|A'|$ be the determinants of the (k+1)x(k+1) matrices

$$(4.5)\qquad A = \begin{pmatrix} L_1 f_1 & L_2 f_1 & \ldots & L_k f_1 & L f_1 \\ L_1 f_2 & L_2 f_2 & \ldots & L_k f_2 & L f_2 \\ \ldots\ldots & \ldots & \ldots\ldots & & \\ L_1 f_k & L_2 f_k & \ldots & L_k f_k & L f_k \\ -z_1 & -z_2 & \ldots & -z_k & 0 \end{pmatrix} \qquad A' = \begin{pmatrix} L_1 f_1 & L_2 f_1 & \ldots & L_k f_1 & L f_1 \\ L_1 f_2 & L_2 f_2 & \ldots & L_k f_2 & L f_2 \\ \ldots\ldots & \ldots & \ldots\ldots & & \\ L_1 f_k & L_2 f_k & \ldots & L_k f_k & L f_k \\ 0 & 0 & \ldots & 0 & 1 \end{pmatrix}.$$

Then, we have

$$(4.6)\qquad L p = \frac{|A|}{|A'|} \;.$$

Let now $k = n+m$ (m, n natural numbers) and assume (4.2).

Let W be the vector space generated by $\{f_1, f_2, \ldots, f_n\}$ and

$$(4.7)\qquad \mathcal{L}_{n+m} = \{L_1, L_2, \ldots, L_{n+m}\} \;, \qquad z = (z_1, z_2, \ldots, z_{n+m}) \;.$$

We shall denote

$$(4.8)\qquad p_{n+m} = p_{n+m}[z]$$

the element of V that verifies

$$(4.9)\qquad L_i\, p_{n+m} = z_i \qquad i = 1,2,\ldots,n+m .$$

Let J_i, $i = 1,2,\ldots,s$ $(\leq m+1)$ subsets of $\{1,2,\ldots,n+m\}$ such that

$$(4.10)\qquad \begin{cases} \operatorname{card} J_i = n \\ \operatorname{card} (J_i \cap J_{i+1}) = n-1, \quad i = 1,2,\ldots,s-1 \text{ if } s > 1 \\ \operatorname{card} \bigcup_{i=1}^{s} J_i = n+s-1 . \end{cases}$$

If

$$(4.11)\qquad J_i = \{ j_1, j_2, \ldots, j_n \}$$

where $j_1 < j_2 < \ldots < j_n$ we shall denote

$$(4.12)\qquad \mathcal{L}_{n,i} = \{ L_{j_1}, L_{j_2}, \ldots, L_{j_n} \}, \qquad z_{n,i} = (z_{j_1}, z_{j_2}, \ldots, z_{j_n}) .$$

Assume that

$$(4.13)\qquad \det \begin{bmatrix} f_1, f_2, \ldots, f_n \\ \mathcal{L}_{n,i} \end{bmatrix} = \det (L_j f_k)_{L_j \in \mathcal{L}_{n,i},\ 1 \leq k \leq n} \neq 0, \quad 1 \leq i \leq s .$$

Then the problem (4.1) for $L_{j_1}, L_{j_2}, \ldots, L_{j_n}$ and $z_{j_1}, z_{j_2}, \ldots, z_{j_n}$ has a unique solution in W that we call

$$(4.14)\qquad p_{n,i} = p_{n,i}[z] \quad i = 1,2,\ldots,s .$$

Analogously, we denote

$$(4.15)\qquad p_{n,i}^{n+k}$$

the element of W that verifies

$$(4.16)\qquad L_j\, p_{n,i}^{n+k} = L_j\, f_{n+k},\ L_j \in \mathcal{L}_{n,i} .$$

If $s > 1$ we also assume that

$$(4.17)\qquad \det \begin{bmatrix} f_1, f_2, \ldots, f_n \\ \mathcal{L}_{n,i} \cap \mathcal{L}_{n,i+1} \cup \{L\} \end{bmatrix} \neq 0 \qquad i = 1,2,\ldots,s-1 .$$

Our aim is to construct (4.8) from (4.14) $(i = 1,2,\ldots,s)$.

Let

$$(4.18)\qquad I_i = J_i \cup \{n+m+1\} \qquad i = 1,2,\ldots,s .$$

If $s < m+1$, we define

$$I_i = J_s \cup \{t_i\}, \quad i = s+1, \dots, m+1 \tag{4.19}$$

where

$$\{t_{s+1}, t_{s+2}, \dots, t_{m+1}\} = \{1, 2, \dots, n+m\} - \bigcup_{i=1}^{s} J_i \tag{4.20}$$

with

$$t_{s+1} < t_{s+2} < \dots < t_{m+1} .$$

Let us take A and A' as in (4.5), for $k = n+m$ and apply (2.9) to them with $p = n$, $q = m+1$ and I_i defined by (4.18) (4.19). We can divide the resulting identities because from (4.13) (4.17) it follows that

$$A\begin{bmatrix} 1, 2, \dots, n \\ s_{k1}, s_{k2}, \dots, s_{kn} \end{bmatrix} = A'\begin{bmatrix} 1, 2, \dots, n \\ s_{k1}, s_{k2}, \dots, s_{kn} \end{bmatrix} \neq 0 \tag{4.21}$$

The constant C being the same for $|A|$ and $|A'|$, we have

$$\frac{|B|}{|B'|} = \frac{|A|}{|A'|} = Lp_{n+m} . \tag{4.22}$$

The elements of $|B|$ are

$$b_{ij} = A\begin{bmatrix} 1, 2, \dots, n, n+i \\ I_j \end{bmatrix} \tag{4.23}$$

where I_j is ordered with increasing indices, and analogously

$$b'_{ij} = A'\begin{bmatrix} 1, 2, \dots, n, n+i \\ I_j \end{bmatrix} . \tag{4.24}$$

Taking into account that

$$A\begin{bmatrix} 1, 2, \dots, n \\ J_i \end{bmatrix} \neq 0 \qquad j = 1, 2, \dots, s \tag{4.25}$$

we can divide the column j of $|B|$ and $|B'|$ by (4.25) for $j=1, 2, \dots, s$, and by

$$A\begin{bmatrix} 1, 2, \dots, n \\ J_s \end{bmatrix}$$

for $j = s+1, \dots, m+1$ if $s < m+1$.

For $i = 1, 2, \dots, m$ and $j = 1, 2, \dots, s$ one easily gets

$$(4.26)\qquad \frac{b_{ij}}{A\begin{bmatrix}1,2,\ldots,n\\ J_j\end{bmatrix}} = \frac{b'_{ij}}{A\begin{bmatrix}1,2,\ldots,n\\ J_j\end{bmatrix}} = L f_{n+i} - L p^{n+i}_{n,j} \;.$$

Similarly, if $s < m+1$, for $i = 1, 2, \ldots, m$ and $j = s+1, \ldots, m+1$

$$(4.27)\qquad \frac{b_{ij}}{A\begin{bmatrix}1,2,\ldots,n\\ J_s\end{bmatrix}} = \frac{b'_{ij}}{A\begin{bmatrix}1,2,\ldots,n\\ J_s\end{bmatrix}} = L_{t_j} f_{n+i} - L_{t_j} p^{n+i}_{n,s} \;,$$

and for $i = m+1$, $j = 1, 2, \ldots, s$

$$(4.28)\qquad \frac{b_{m+1j}}{A\begin{bmatrix}1,2,\ldots,n\\ J_j\end{bmatrix}} = \frac{A\begin{bmatrix}1,2,\ldots,n,n+m+1\\ I_j\end{bmatrix}}{A\begin{bmatrix}1,2,\ldots,n\\ J_j\end{bmatrix}} = L p_{n,j}$$

$$(4.29)\qquad \frac{b'_{m+1j}}{A\begin{bmatrix}1,2,\ldots,n\\ J_j\end{bmatrix}} = 1.$$

If $s < m+1$, for $i = m+1$ and $j = s+1, \ldots, m+1$ we have

$$(4.30)\qquad \frac{b_{m+1j}}{A\begin{bmatrix}1,2,\ldots,n\\ J_s\end{bmatrix}} = -z_{t_j} + L_{t_j} p_{n,s} \,, \quad \frac{b'_{m+1j}}{A\begin{bmatrix}1,2,\ldots,n\\ J_s\end{bmatrix}} = 0 \;.$$

Finally, if in order to simplify the formulae we denote

$$(4.31)\qquad r^{n+i}_{n,j} = f_{n+i} - p^{n+i}_{n,j}$$

then, for $s < m+1$, one has

(4.32)

$$
L\,p_{n+m} = \frac{\begin{vmatrix}
L\,r^{n+1}_{n,1} & \cdots & L\,r^{n+1}_{n,s}, & L_{t_{s+1}}\,r^{n+1}_{n,s} & \cdots\cdots & L_{t_{m+1}}\,r^{n+1}_{n,s} \\
L\,r^{n+2}_{n,1} & \cdots & L\,r^{n+2}_{n,s}, & L_{t_{s+1}}\,r^{n+2}_{n,s} & \cdots\cdots & L_{t_{m+1}}\,r^{n+2}_{n,s} \\
\cdots\cdots & & \cdots\cdots & \cdots\cdots & \cdots\cdots & \cdots\cdots \\
\cdots\cdots & & \cdots\cdots & \cdots\cdots & \cdots\cdots & \cdots\cdots \\
L\,r^{n+m}_{n,1} & \cdots & L\,r^{n+m}_{n,s}, & L_{t_{s+1}}\,r^{n+m}_{n,s} & \cdots\cdots & L_{t_{m+1}}\,r^{n+m}_{n,s} \\
L\,p_{n,1} & \cdots & L\,p_{n,s}, & -z_{t_{s+1}} + L_{t_{s+1}}\,p_{n,s} & & -z_{t_{m+1}} + L_{t_{m+1}}\,p_{n,s}
\end{vmatrix}}{\begin{vmatrix}
L\,r^{n+1}_{n,1} & \cdots & L\,r^{n+1}_{n,s}, & L_{t_{s+1}}\,r^{n+1}_{n,s} & \cdots\cdots & L_{t_{m+1}}\,r^{n+1}_{n,s} \\
L\,r^{n+2}_{n,1} & \cdots & L\,r^{n+2}_{n,s}, & L_{t_{s+1}}\,r^{n+2}_{n,s} & \cdots\cdots & L_{t_{m+1}}\,r^{n+2}_{n,s} \\
\cdots\cdots & & \cdots\cdots & \cdots\cdots & \cdots\cdots & \cdots\cdots \\
\cdots\cdots & & \cdots\cdots & \cdots\cdots & \cdots\cdots & \cdots\cdots \\
L\,r^{n+m}_{n,1} & \cdots & L\,r^{n+m}_{n,s}, & L_{t_{s+1}}\,r^{n+m}_{n,s} & \cdots\cdots & L_{t_{m+1}}\,r^{n+m}_{n,s} \\
1 & \cdots & 1\;, & 0 & \cdots\cdots & 0
\end{vmatrix}},
$$

For $s = m+1$ all the columns of both determinants are of the kind of the s first columns of (4.32).

5. Some particular cases.

We have already remarked at the end of § 4 the particular case $s = m+1$. In this case the most frequent choices of J_i are

(5.1) $$J_i = \{1, 2, \ldots, n-1, n+i-1\} \qquad i = 1, 2, \ldots, m+1$$

and

(5.2) $$J_i = \{i, i+1, i+2, \ldots, i+n-1\} \qquad i = 1, 2, \ldots, m+1 .$$

(5.1) leads to an Aitken-like algorithm and (5.2) to a

Neville-like algorithm.

Another interesting particular case is given by $s = 1$. Then there is no condition (4.17) on L and (4.32) can be written

(5.3)

$$L p_{n+m} = L p_n + (-1)^m \frac{\begin{vmatrix} L r^{n+1}_{n,1} & L_{t_2} r^{n+1}_{n,1} & \cdots & L_{t_{m+1}} r^{n+1}_{n,1} \\ L r^{n+2}_{n,1} & L_{t_2} r^{n+2}_{n,1} & \cdots & L_{t_{m+1}} r^{n+2}_{n,1} \\ \cdots\cdots & & \cdots & \cdots\cdots \\ L r^{n+m}_{n,1} & L_{t_2} r^{n+m}_{n,1} & \cdots & L_{t_{m+1}} r^{n+m}_{n,1} \\ 0 & -z_{t_2} + L_{t_2} p_n & \cdots & -z_{t_{m+1}} + L_{t_{m+1}} p_n \end{vmatrix}}{\begin{vmatrix} L_{t_2} r^{n+1}_{n,1} & \cdots\cdots & L_{t_{m+1}} r^{n+1}_{n,1} \\ L_{t_2} r^{n+2}_{n,1} & \cdots\cdots & L_{t_{m+1}} r^{n+2}_{n,1} \\ \cdots & \cdots\cdots & \cdots \\ L_{t_2} r^{n+m}_{n,1} & \cdots\cdots & L_{t_{m+1}} r^{n+m}_{n,1} \end{vmatrix}}$$

Since this is true for all L, we can write the same formula without it. This is a generalized Newton's formula in the sense of Mühlbach [9].

In many applications V is a subspace of finite dimension of a linear space F of functions of one or several variables and $z_j = L_j f$ are function values or directional derivative values of $f \in F$ at points of the domain X of f. The linear form is usually given by

$$L f = f(x)$$

where x is a point of X.

6. Relations with other known results.

In § 4 we have derived the formula (4.32) which can be also proved in a very simple way.

As both sides of the equality (4.32) are the value of a linear form on K^{n+m} at z, it is sufficient to show they coincide for a basis of K^{n+m}, for example

$$(6.1)\left\{\mathcal{L}_{n+m}f_i,\ i=1,2,\dots,n+m\right\}=\left\{(L_1 f_i, L_2 f_i,\dots,L_{n+m} f_i)\ i=1,2,\dots,n+m\right\}.$$

This allows us to conclude that (4.32) is true even in conditions less restrictive than those of § 4, namely in the case of nonzero denominator.

In [4] we have proved the following result for $s>1$: if (4.2), ($k=n+m$), (4.10), (4.13) and (4.17) hold and

$$\det\begin{bmatrix} f_1, f_2, \dots, f_{n+s-1} \\ \bigcup_{i=1}^{s}\mathcal{L}_{n,i}\end{bmatrix}\neq 0,$$

then

$$(6.2)\qquad L\,p_{n+m}=\sum_{i=1}^{s}\lambda_i\ L\,p_{n,i}+\sum_{i=1}^{s}\ \sum_{k=n+s}^{n+m}\lambda_i\alpha^k_{n+m}\ L(f_k-p^k_{n,i})$$

where $(\lambda_1, \lambda_2, \dots, \lambda_s)$ is the unique solution of the system

$$(6.3)\qquad \begin{cases}\displaystyle\sum_{i=1}^{s}\lambda_i=1\\[2ex] \displaystyle\sum_{i=1}^{s}\lambda_i\ L\,p^j_{n,i}=L\,f_j\ ,\quad j=n+1,\dots,n+s-1\end{cases}$$

and

$$\alpha^k_{n+m}$$

is the coefficient of f_k in (4.8).

Also, for $s=1$, if (4.2) and (4.13) hold

$$(6.4)\qquad p_{n+m}=p_{n,1}+\sum_{k=n+1}^{n+m}\alpha^k_{n+m}\,(f_k-p^k_{n,1}).$$

If we expand the numerator of (4.33) by the last row and then divide we have

$$(6.5) \qquad L\,p_{n+m} = \sum_{i=1}^{s} \mu_i \; L\,p_{n,i} + R\,.$$

In general this formula is not the same as (6.3). From (4.33) it is clear that

$$(6.6) \qquad \sum_{i=1}^{s} \mu_i = 1$$

but generally

$$\sum_{i=1}^{s} \mu_i \; L\,p_{n,i}^{n+1} \neq L\,f_{n+1}$$

and therefore $(\mu_1, \ldots, \mu_s) \neq (\lambda_1, \ldots, \lambda_s)$, as it can be seen in simple examples. Nevertheless, in most interesting particular cases ($s = m+1$, $s = 1$) (6.5) and (6.2) coincide.

The particular case of $s = m+1$ which corresponds to the choice (5.2) for J_i, was obtained by C. Brezinski [2]. He begins with $m = 1$ and then applies his E-algorithm and the original Sylvester's identity [3] to reach the general formula. In the proofs it is implicitely assumed that, in the case of the usual Lagrange's interpolation problem for one variable, the functions $\{f_i\}_{i \geq 0}$ are a complete Chebyshev system. Similarly, in the case of the general problem the hypothesis

$$\det L_i\, f_j \neq 0 \qquad 1 \leq i,j \leq r \qquad r = 1,2, \ldots$$

is required.

Our identity (2.10) enables us to abandon these assumptions.

Also include in [2] is the particular case $m=1$ of (5.1).

References

[1] Brezinski, C. (1980). A general extrapolation algorithm. Num. Math. 35, 175-187.

[2] Brezinski, C. (1980). The Mühlbach-Neville-Aitken algorithm and some extensions. Pub. A.N.O. 19, Université de Lille.

[3] Gantmacher, F.R. (1966). Théorie des matrices. Vol. I, Dunod, París.

[4] Gasca, M., López-Carmona, A. (1982). A general recurrence interpolation formula and its applications to multivariate interpolation. Jour. Approx. Th. in press.

[5] Havie, T. (1979). Generalized Neville type extrapolation schemes. B.I.T. 19, 204-213.

[6] Havie, T. (1980). Remarks on the Mühlbach-Neville-Aitken algorithm. Dept. of Mathem. Univ. of Trondheim, Norway (preprint).

[7] López-Carmona, A. (1981). Interpolación por recurrencia (in Spanish). Tesis. Univ. Granada, Spain.

[8] Mühlbach, G. (1979). The general Neville-Aitken algorithm and some applications. Num. Math. 31, 97-110.

[9] Mühlbach, G. (1979). The general recurrence relation for divided differences and the general Newton-interpolation algorithm with applications to trigonometric interpolation. Num. Math. 32, 393-408.

[10] Mühlbach, G. (1980). An algorithmic approach to finite linear interpolation. (Approximation Theory III, E.W. Cheney editor) Academic Press, 655-660.

[11] Mühlbach, G. (1980). On two general algorithms for extrapolation with applications to numerical differentiation and integration. Procee. Confer. on "Rational approximation, theory and applications". Amsterdam, Oc. 29-31 (preprint).

M. Gasca , A. López-Carmona , V. Ramírez

Departamento de Ecuaciones Funcionales
Facultad de Ciencias.
Universidad de Granada.
Granada. Spain.

BIORTHOGONALITY IN APPROXIMATION

Werner Haußmann, Eberhard Luik and Karl Zeller

Department of Mathematics
University of Duisburg
D-4100 Duisburg

Department of Mathematics
University of Tübingen
D-7400 Tübingen

1. Introduction

Numerical approximation can be carried out by ascent or descent methods - or in a more explicit way by expansion methods (like truncation, telescoping, pre-iteration). We use general biorthogonal systems (BOGS) to describe procedures of the latter type. This setting leads easily to useful results and provides good insight. The basic task is to improve or shorten a given expression by changing the coefficients. Thereby one employs information comprised in the elements and functionals of the BOGS. More specifically we consider expansions of Fourier (Chebyshev) type in the univariate and bivariate case.

2. Elements

Assume we are given a normed vector space X (with dual space X') and a finite normalized biorthogonal system (BOGS)

$$L_j(g_i) = \delta_{ij}; \quad g_i \in X,\ L_j \in X'; \quad i,j = 0,\dots,s\,.$$

In practice the L_j can be given by sums or integrals (e.g. Fourier transforms), in some cases with $\|L_j\|$ near to 1, $\|g_i\| = 1$ (cf. [2]). $P_k f$ denotes a proximum to $f \in X$ with respect to $\mathrm{span}(g_0,\dots,g_k)$, and $E_k(f)$ the corresponding degree of approximation.

Lemma. Let $q = c_0 g_0 + \dots + c_s g_s$, and $f \in X$. Then

$$|c_j| \le \|L_j\| (\|f - q\| + E_k(f)) \quad \text{for } 0 \le k < j \le s.$$

Indeed, this follows by applying L_j to $q-f + f-P_k f$.
The inequality is useful in both directions: upper bounds for the coefficients, lower bounds for the degree of approximation. These give corresponding information about the approximations appearing below. Of course it is helpful to have $\|f-q\|$ small (i.e. q a good approximation to f). Then one will try to reduce the number of terms in q, for instance as follows:

$$p := c_0 g_0 + \dots + c_k g_k + c_{k+1} P_k g_{k+1} + \dots + c_s P_k g_s$$

(a similar method proceeds recursively; see also below). Our Lemma yields the

Corollary. In the given setting we have

$$\|p - q\| \le \sum_{j=k+1}^{s} \|L_j\| (\|f-q\| + E_{j-1}(f)) \cdot E_k(g_j)\,.$$

p will be advantageous if the $c_j E_k(g_j)$ are small enough. If the $E_k(g_j)$ are large (as for Chebyshev polynomials), one will try approximations of binary (ternary,...) combinations (for example of $g_{k+1} + c g_{k+2}$), as we explain now.

3. Chebyshev Polynomials

We put $X := C[-1,1]$ (continuous functions, max-norm) and $g_j := T_j$, the Chebyshev polynomials of the first kind. In this setting we approximate

$$q := T_n + c_{n+1}T_{n+1} + \ldots + c_{2n}T_{2n} \quad \text{by} \quad p := a_{n-1}T_{n-1} + \ldots + a_0 T_0 .$$

Thus in the notation above we have $s = 2n$, $k = n-1$; further we have reduced q modulo P_{n-1}, and then normalized to $c_n = 1$ (assuming that originally $c_n \neq 0$). The (main) error term T_n alone would not allow an improvement. But for the full error term we have several possibilities. Our first choice is $a_{n-j} := c_{n+j}$ ($j = 1,\ldots,n$). The error without T_n is described by

$$\sum_{j=1}^{n} c_{n+j}(T_{n+j} - T_{n-j})(x) = -2 \sin nt \sum_{j=1}^{n} c_{n+j} \sin jt$$

(where $x := \cos t$). For the latter sum several estimates are known from the theory of Fourier series. But we have also to watch the interplay with T_n, which is most easily done in the case $q := T_n + cT_{n+1}$, $p := a\,T_{n-1}$, $a := c$.

<u>Lemma.</u> In this case the following inequality holds:

$$\|p - q\| \leq \sqrt{1 + 4c^2} .$$

This bound (obtained by trigonometry) is rather sharp (and almost sharp for large n). The approximation error is very close to 1 for c small. If we allow larger $|c|$ (near 1), then it is better to apply attenuation factors, e.g.

$$a \approx \frac{c}{1 + 2|c|} .$$

And somewhat better approximations can be obtained by introducing T_{n-2} or more terms, ending with polynomials p of Zolotarev type. Similar considerations are possible for functions q containing more terms (tabulation of near best approximation, construction of generalized Zolotarev polynomials).

4. Bivariate Approximation

We use the space $C(J^2)$, where $J := [-1,1]$. Suppose

$$q(x,y) := \sum_{k=0}^{m} \sum_{l=0}^{n} c_{kl} x^k y^l \qquad (\text{given } m,n \geq 1).$$

Then we replace

$$x^m \text{ by } x^m - 2^{-m+1} T_m(x) \quad \text{and} \quad y^n \text{ by } y^n - 2^{-n+1} T_n(y),$$

thus reducing the number of terms. The process could be repeated and it can be applied to other sets of indices (k,l). For coefficient estimates as in Section 2 one will restrict f to a suitable subset of $C(J^2)$. Next we replace

$$q := \sum_{k=0}^{m+1} \sum_{l=0}^{n+1} c_{kl} T_k T_l \quad \text{by} \quad p := \sum_{k=0}^{m-1} \sum_{l=0}^{n-1} a_{kl} T_k T_l$$

(where $T_k T_l$ means $T_k(x) T_l(y)$). For the determination of the a_{kl} we use intermediate values b_{kl} (corresponding to approximation in x):

$$b_{kl} := c_{kl} + c_{k+2,l} \ (k=m-1), \quad b_{kl} := c_{kl} \ (k<m-1), \ l=0,\dots,n+1,$$

$$a_{kl} := b_{kl} + b_{k,l+2} \ (l=n-1), \quad a_{kl} := b_{kl} \ (l<n-1), \ k=0,\dots,m-1.$$

Theorem. The given polynomials p and q satisfy

$$\| p - q \| \leq \sum_{l=0}^{n+1} \sqrt{c_{ml}^2 + 4c_{m+1,l}^2} + \sum_{k=0}^{m-1} \sqrt{b_{kn}^2 + 4b_{k,n+1}^2} .$$

The proof rests upon the Lemma in Section 3. We remark that $b_{ij} = c_{ij}$ in the second sum except for the last term. Our approximation and estimate is quite good in many cases, e.g. if the c_{kl} decrease geometrically in k and in l. For finer investigations it can be better to collect terms and write

$$q(x,y) = \sum_{k=0}^{m+1} c_k(y) \cdot T_k(x) = \sum_{l=0}^{n+1} c_l^*(x) \cdot T_l(y) .$$

Using point functionals we recognize $\|p-q\| \geq \|c_m\|$ (also $\geq \|c_{m+1}\|$, $\|c_n^*\|$, $\|c_{n+1}^*\|$) for any p (i.e. arbitrary a_{kl}) of the type above (with $\|.\| :=$ max-norm).

5. Remarks

For numerical approximation in general we refer to the books [6] and [9]. Approximation in connection with expansions is treated in [3], [4], [5] and [6]. The papers [1] and [8] deal with "best vs. good" approximations. Fourier coefficients can be determined by discrete point functionals (in connection with error representations); if one uses fast Fourier transforms (see [7]) then the main computing work consists (in general) in the evaluation of the function. It is also possible to use implicit information about the coefficients (contained for instance in the error functions of initial approximations). The methods described here could be also useful for improving approximation operators for theoretical purposes. Finally one should use not only polynomials but also other (adapted) function systems for approximation.

References

[1] de Boor, C. (1978) The approximation of functions and linear functionals: best vs. good approximation. Proc. Symp. Appl. Math., Vol. 22, 53-70.

[2] Haußmann, W., Luik, E., Zeller, K. (1982) Cubature remainder and biorthogonal systems. This volume.

[3] Hollenhorst, M. (1981) Improved lower and upper bounds in polynomial Chebyshev approximation based on a pre-iteration formula. J. Approx. Theory 32, 170-188.

[4] Hornecker, G. (1958) Evaluation approchée de la meilleure approximation polynômiale d'ordre n de f(x) sur un segment fini [a,b]. Chiffres 1, 157-169.

[5] Mason, J.C. (1980) Near-best multivariate approximation by Fourier series, Chebyshev series and Chebyshev interpolation. J. Approx. Theory 28, 349-358.

[6] Meinardus, G. (1964) Approximation von Funktionen und ihre numerische Behandlung (Springer, Berlin).

[7] Nussbaumer, H.J. (1981) Fast Fourier transform and convolution algorithms (Springer, Berlin).

[8] Scherer, R., Zeller, K. (1982) Floppy vs. fussy approximation. Internat. Ser. Numer. Math., Vol. 59, 171-178.

[9] Watson, G.A. (1980) Approximation theory and numerical methods (John Wiley & Sons, Chichester).

Prof. Dr. Werner Haußmann
Department of Mathematics
University of Duisburg
Lotharstraße 65
D-4100 Duisburg

Prof. Dr. Karl Zeller
Dipl.-Math. Eberhard Luik
Department of Mathematics
University of Tübingen
Auf der Morgenstelle 10
D-7400 Tübingen

CUBATURE REMAINDER AND BIORTHOGONAL SYSTEMS

Werner Haußmann, Eberhard Luik and Karl Zeller

Department of Mathematics
University of Duisburg
D-4100 Duisburg

Department of Mathematics
University of Tübingen
D-7400 Tübingen

1. Introduction

Error estimates for cubature formulas are usually given in terms of higher derivatives (Peano-Sard) or in terms of analyticity properties (Davis-Hämmerlin). The approximation method has found little attention. We present the latter method in a generalized and refined form, based on biorthogonal systems (BOGS). The degrees of approximation and coefficient estimates connected with a BOGS lead to rather good and versatile inequalities for the error. More specifically, we consider Chebyshev polynomials, Clenshaw-Curtis procedures and product formulas. Estimates for the employed degrees of approximation are available in the theory of approximation, and these estimates can be supported or refined by numerical computation.

2. Notation

Assume we are given a normed vector space X with norm $\|.\|$ and the dual space X', further a normalized biorthogonal system (BOGS) in (X,X'):

$$L_j(g_i) = \delta_{ij}; \qquad g_i \in X, \; L_j \in X'; \; i,j=0,\ldots,s .$$

The basic functions g_i determine degrees of approximation:

$$E_k(f) := \text{Min} \, \|f-p\| \qquad (\text{over all } p = c_0 g_0 + \ldots + c_k g_k) .$$

A proximum (in general not uniquely determined) will be denoted by p_k^*, i.e. we have $\|f - p_k^*\| = E_k(f)$.
The following estimate is basic for the considerations below.

Lemma. Given $f \in X$, suppose $p_k^* = c_0^* g_0 + \ldots + c_k^* g_k$ is a proximum to f. Then the inequalities

$$|c_j^*| \le \|L_j\| \, (E_{j-1}(f) + E_k(f)) \qquad (j=0,\ldots,k)$$

hold true (with $E_{-1}(f) := \|f\|$).

Indeed, we have

$$|c_j^*| = |L_j(p_k^*)| \le |L_j(f-p_k^*)| + |L_j(f)| .$$

The first term is estimated by $\|L_j\| \cdot E_k(f)$, for the second one we employ the biorthogonality and get

$$|L_j(f)| = |L_j(f-p_{j-1}^*)| \le \|L_j\| \cdot E_{j-1}(f) .$$

The latter estimate could be refined by using approximations of the type $c_0 g_0 + \ldots + c_s g_s$ with $c_j = 0$. This is especially useful in considerations with multiindices (where one could also modify the assumption of the Lemma by considering an approximation of other type). In the applications the g_j will often be orthogonal polynomials, in particular Chebyshev polynomials (see below).

3. BOGS Estimates

We consider an arbitrary continuous linear functional R on the normed vector space X (later R will be the remainder of a cubature formula). The estimate (in case $R(g_j) = 0,\ j=0,\dots,k$)

$$|R(f)| \leq \|R\| \cdot E_k(f)$$

can be refined in the following way.

Theorem. Suppose R is a continuous linear functional on the normed vector space X, and the degrees of approximation $E_j(f)$ are determined by the basic functions $g_0,\dots,g_j$ of a normalized BOGS $(g_j, L_j)_{j=0,\dots,s}$. Then for any $f \in X$ and any $n = 0,\dots,s$ the following estimate holds (with $E_{-1}(f) := \|f\|$):

$$|R(f)| \leq \|R\| \cdot E_n(f) + \sum_{j=0}^{n} \|L_j\| \cdot (E_{j-1}(f) + E_n(f)) \cdot |R(g_j)| .$$

Proof. Inserting a proximum p_n^* we decompose

$$R(f) = R(f - p_n^*) + R(p_n^*) .$$

The first term is covered by the first term in the inequality. As to the second one, we have (by the Lemma in Section 2)

$$|R(p_n^*)| \leq \sum_{j=0}^{n} |c_j^*| |R(g_j)| \text{ with } |c_j^*| \leq \|L_j\| (E_{j-1}(f) + E_n(f)).$$

In many cases some $R(g_j)$ will vanish. For easier reference we state the

Corollary. The estimate of the Theorem reduces to

$$|R(f)| \leq \|R\| \cdot E_n(f) + \sum_{j=k+1}^{n} \|L_j\| \cdot (E_{j-1}(f) + E_n(f)) \cdot |R(g_j)|$$

if we have $R(g_j) = 0$ $(j=0,\dots,k)$ for a certain $k < s$.

Further it can happen in practice that $R(g_j) = 0$ for all even resp. odd j, reducing the estimate correspondingly (cf. below).

4. Cubature

Now we use the space $X = C(B)$ of all continuous functions $f : B \to \mathbb{R}$, where B is a nonvoid compact set (in a Hausdorff space, e.g. $\mathbb{R}^d$). The set $C(B)$ will be endowed with the usual algebraic structure and with the maximum-norm

$$\|f\| := \operatorname{Max}_{x \in B} |f(x)|$$

such that it is a Banach space (even a Banach algebra). On $C(B)$ we consider two continuous linear functionals

$$J, Q := C(B) \to \mathbb{R}$$

and the remainder (difference)

$$R := J - Q .$$

In the applications J will be an integral and Q a quadrature or cubature formula of the following type:

$$Q(f) := \sum_{i=1}^{n} a_i \cdot f(x_i) \qquad (a_i \in \mathbb{R},\ x_i \in B).$$

We investigate R by employing the BOGS functions, thus the values $R(g_j)$ as well as the norms $\|L_j\|$ will be important.

In constructing a quadrature (cubature) formula, one can follow different policies with regard to the basic remainders $R(g_j)$: Make many of them vanish, make them in the average small, make some of them vanish and others small (compare the types Gauß, Wilf, Clenshaw-Curtis). Besides this also the choice of the g_j is important: nice analytical properties (e.g. orthogonality over B), good approximation properties (cf. the notion "width"). Also one will try to have the $\|L_j\|$ close to 1.

In the following we shall show the influence of $R(g_j)$ and $E_k(f)$, thus giving some information concerning proper choices of BOGS and quadrature (cubature) rules. In particular, we shall deal with Clenshaw-Curtis procedures as well as with tensor product formulas in some detail. Therefore we consider as a possible choice of BOGS functions the Chebyshev polynomials T_j.

5. Chebyshev Polynomials

We consider $B := [-1,1]$ and the Chebyshev polynomials

$$T_k(x) \quad := \quad \cos(k \arccos x) .$$

Our BOGS functions will be such polynomials:

$$g_j \quad := \quad T_j \qquad\qquad (j = 0,1,\dots,s) .$$

For the L_j we have several choices, in first place the coefficient functionals

$$L_j(f) := \frac{2}{\pi} \int_{-1}^{1} f(x) T_j(x) \cdot (1-x^2)^{-1/2} dx \quad (j = 0^*,1,\dots,s)$$

(0^*: for $j = 0$ replace the factor $\frac{2}{\pi}$ by $\frac{1}{\pi}$) . We note

$$\|L_0\| = 1; \quad \|L_j\| = \frac{4}{\pi} \qquad\qquad (j = 1,2,\dots,s) .$$

Since we have a finite system only, we can replace the integral by discrete point functionals (see Fox-Parker [3], Rivlin [8] pp. 40, 49, 149, 151). Especially important for us is the possibility to replace many of the L_j given above by functionals of norm 1 :

$$\|L_k^*\| = 1 \qquad\qquad (\tfrac{s}{3} < k \le s) ,$$

with the definition (cf. Meinardus [7] p. 74)

$$L_k^*(f) \quad := \quad \frac{1}{k} \sum_{i=0}^{k}{}'' \ (-1)^i f(y_i)$$

where the y_i are the extremal points of T_k, i.e.

$$y_i \quad := \quad \cos \frac{i}{k} \pi \qquad\qquad (i = 0,1,\dots,k),$$

and Σ'' means that we have a "trapezoidal sum" (insert the factor 1/2 for i=0 and for i=k). We mention that in our setting the norm of any admissible L_j is ≥ 1.

These preliminaries are important for quadrature in general, and especially for rules of the types Pólya, Filippi, and Clenshaw-Curtis (where the T_k enter directly, cf. Braß [1] pp. 116, 117).

6. Clenshaw-Curtis Procedures

The basic rule (for the ordinary integral over [-1,1])

$$Q(f) \quad := \quad Q_n(f) \quad := \quad \sum_{j=1}^{n} a_j f(x_j) \qquad (n \geq 3)$$

has as nodes the extremal points of T_{n-1} : $x_j := -\cos \frac{j-1}{n-1}\pi$ $(j = 1,\ldots,n)$ and (positive) weights a_j determined by

$$R_n(T_k) \quad = \quad 0 \qquad (k = 0,\ldots,n-1) .$$

By skew symmetry we have further:

$$R_n(T_{2k+1}) \quad = \quad 0 \qquad (k = 0,1,2,\ldots) .$$

For simplicity we now consider only odd n (slight modifications for even n). We put

$$d_{2k} \quad := \quad \frac{2}{4k^2-1}$$

and note (cf. Braß [1] p. 145)

$$J(T_{2k}) \quad := \quad \int_{-1}^{1} T_{2k}(x)\,dx \quad = \quad -d_{2k} \,;$$

$$Q_n(T_{2k}) \quad = \quad -d_{2r}, \text{ where } r \equiv k \pmod{n-1}, \; -\frac{n-1}{2} < r \leq \frac{n-1}{2} .$$

We apply the Corollary in Section 3 (with j only even, L_k^* of Section 5, change of indices) and obtain the following

Theorem. The Clenshaw-Curtis remainder $R_n := J - Q_n$ (where $n \geq 3$, n odd) satisfies for any non-negative $m \leq n-1$

$$|R_n(f)| \;\leq\; 4\, E_{n+2m}(f) \;+\; \sum_{k=0}^{m-1} (E_{n+2k}(f)+E_{n+2m}(f))\,|d_{n-3-2k}-d_{n+1+2k}| \,.$$

Remark. The latter condition yields $n-1-2m \geq -n+1$ and thus the expression for the basic remainders (not employing r). Further it assures that the L_k^* are feasible (exact bound $m \leq n+1$). Using the ordinary L_j would introduce a factor $4/\pi$ (cf. [6]). Similar results are available for other quadrature and cubature procedures connected with orthogonal polynomials.

7. Tensor Product Formulas

Here we take $B := [-1,1]^2$,

$$J(f) := \int_{-1}^{1}\int_{-1}^{1} f(x,y)\,dxdy\,,$$

$$Q(f) := Q_{mn}(f) := \sum_{i=1}^{m}\sum_{j=1}^{n} a_i b_j \cdot f(x_i,y_j)$$

(with nodes in B). The latter cubature formula is based on the two univariate formulas

$$Q_m(g) := Q_m^I(g) := \sum_{i=1}^{m} a_i g(x_i)\ ; \quad Q_n(h) := Q_n^{II}(h) := \sum_{j=1}^{n} b_j h(y_j)$$

(we mostly omit the superscripts I and II, respectively). The relation of these formulas can be described by a tensor product

$$Q_{mn} = Q_m \hat{\otimes} Q_n\,, \qquad J = J^I \hat{\otimes} J^{II}$$

(cf. Treves [11], Haußmann-Pottinger [5]) or, more directly,

$$Q_{mn}(f) = Q_m(Q_n(f)) = Q_n(Q_m(f)),\ J(f) = J^I(J^{II}(f)) = J^{II}(J^I(f)).$$

The inner operators are applied to a family of functions, inner and outer operators commute. The remainders $R := J - Q$ satisfy

$$R_{mn}(f) = J^I(R_n(f)) + Q_n(R_m(f)) = Q_m(R_n(f)) + J^{II}(R_m(f)).$$

In particular, the following estimate holds (see the literature given in Stroud [10] p. 138, especially Stroud-Secrest p. 72):

<u>Lemma.</u> In the given setting one has

$$|R_{mn}(f)| \le \|J^I\|\cdot\|R_n(f)\| + \|Q_n\|\cdot\|R_m(f)\|\,,$$

$$|R_{mn}(f)| \le \|Q_m\|\cdot\|R_n(f)\| + \|J^{II}\|\cdot\|R_m(f)\|\,.$$

<u>Remarks.</u> Since $R_n(f)$ and $R_m(f)$ are continuous functions of the parameter x resp. y, the maximum-norm is defined. In our case we have $\|J^I\| = \|J^{II}\| = 2$. But $\|Q_m\|$ and $\|Q_n\|$ can be large; the value 2 is attained in case of positive coefficients with sum 2, i.e. for positive formulas which are exact for constant functions.

8. Approximation

In the space C(B), where $B := [-1,1]^2$, we use different sets of approximating functions and obtain corresponding degrees of approximation:

$$E_{mn}(f) \quad \text{by} \quad \sum_{i=0}^{m} \sum_{j=0}^{n} \alpha_{ij} x^i y^j ,$$

$$E_{m\infty}(f) \quad \text{by} \quad \sum_{i=0}^{m} \beta_i(y)\, x^i , \quad E_{\infty n}(f) \quad \text{by} \quad \sum_{j=0}^{n} \gamma_j(x)\, y^j .$$

The $\beta_i(y)$ resp. $\gamma_j(x)$ are arbitrary functions (we could also demand continuity); a proximum is obtained if we take for every y resp. x the best (admissible) polynomial.

The relations between such bivariate and univariate approximations have been investigated by Bernstein, Temljakov, Scherer-Zeller [9]. We note

$$E_{mn}(f) \le (E_{m\infty}(f) + E_{\infty n}(f))\cdot\Lambda_n + E_{\infty n}(f),$$

where Λ_n are the ordinary Lebesgue constants. The inequality is almost sharp (there are examples where a logarithmic factor like Λ_n is realistic).

The Lemma in Section 7 shows that the situation is more favourable in the realm of cubature if we keep the norms of the quadrature operators bounded. We formulate a special case:

Theorem. Suppose that a product formula $Q_{mn} = Q_m \hat{\otimes} Q_n$ satisfies $\|Q_m\| = \|Q_n\| = 2$ and

$$R_m(x^i) = 0 \quad (i=0,\ldots,k), \quad R_n(y^j) = 0 \quad (j=0,\ldots,l).$$

Then these estimates hold:

$$|R_{mn}(f)| \le 8\, E_{kl}(f) ,$$

$$|R_{mn}(f)| \le 8\, (E_{k\infty}(f) + E_{\infty l}(f)) .$$

Thus it has certain advantages to use the second estimate.

9. Biorthogonal Systems

Now we refine the approximation method by introducing biorthogonal systems. This can be done in several ways. We formulate a result somewhat in the middle between abstraction and specialization, using general product formulas in connection with Chebyshev polynomials.

Theorem. Suppose that a product formula $Q_{mn} = Q_m \hat{\otimes} Q_n$ (see Section 7) satisfies $\|Q_m\| = \|Q_n\| = 2$ and

$$R_m(x^i) = 0 \quad (i=0,\ldots,k), \quad R_n(y^j) = 0 \quad (j=0,\ldots,l)$$

(for the ordinary integral over $[-1,1]^2$). Then the estimate

$$|R_{mn}(f)| \le 8\, E_{k+1,\infty}(f) + 2(E_{k,\infty}(f) + E_{k+1,\infty}(f))\, |R_m(T_{k+1})|$$

$$+ \; 8\, E_{\infty,l+1}(f) + 2(E_{\infty,l}(f) + E_{\infty,l+1}(f))\, |R_n(T_{l+1})|$$

holds.

For the proof we apply the Lemma in Section 7 (slightly specialized) and a bivariate BOGS estimate corresponding to the Corollary in Section 3.

The estimate is good if the new (higher order) degrees of approximation are known to be small and the introduced remainders corresponding to Chebyshev polynomials are also small. As in Section 6 we can employ more terms. Our result corresponds to the second part of the Theorem in Section 8; there are several possibilities to extend the first part of that Theorem in a similar way. A first specialization of the result above would refer to Clenshaw-Curtis cubature.

For integration over domains other than a square one can employ product formulas together with transformations of the variables (see Stroud [10], Engels [2]). Or one can use direct formulas together with biorthogonal systems adapted to the domain (by harmonic analysis or by the definition of the cubature formula).

10. Remarks

We mention some literature useful to the reader. The books by Braß [1], Engels [2], and Stroud [10] treat quadrature and cubature (and give further references). Braß discusses the approximation method for error estimates. Much information about Chebyshev polynomials is contained in the books by Fox-Parker [3], Meinardus [7], and Rivlin [8]. It seems that the use of degrees of approximation in connection with coefficient functionals is rather new here. Companion publications to the present paper are [4] and [6]. For tensor products in our context see Haußmann-Pottinger [5], and Treves [11].

References

[1] Braß, H. (1977) Quadraturverfahren (Vandenhoeck & Ruprecht, Göttingen).

[2] Engels, H. (1980) Numerical quadrature and cubature (Academic Press, London).

[3] Fox, L., Parker, I. B. (1968) Chebyshev polynomials in numerical analysis (Oxford University Press, London).

[4] Haußmann, W., Luik, E., Zeller, K. (1982) Biorthogonality in approximation. This volume.

[5] Haußmann, W., Pottinger, P. (1977) On the construction and convergence of multivariate interpolation operators. J. Approx. Theory 19, 205-221.

[6] Haußmann, W., Zeller, K. (1982) Quadraturrest, Approximation und Chebyshev-Polynome. To appear in: Internat. Ser. Numer. Math., Vol. 57.

[7] Meinardus, G. (1964) Approximation von Funktionen und ihre numerische Behandlung (Springer, Berlin).

[8] Rivlin, Th. J. (1974) The Chebyshev polynomials (John Wiley & Sons, New York).

[9] Scherer, R., Zeller, K. (1981) Bivariate polynomial approximation. Proc. of the Conf. on Approximation and Function Spaces (Gdańsk, 1979), 621-628.

[10] Stroud, A. H. (1971) Approximate calculation of multiple integrals (Prentice-Hall, Englewood Cliffs, N. J.).

[11] Treves, F. (1967) Topological vector spaces, distributions and kernels (Academic Press, London).

Prof. Dr. Werner Haußmann
Department of Mathematics
University of Duisburg
Lotharstraße 65
D-4100 Duisburg

Prof. Dr. Karl Zeller
Dipl.-Math. Eberhard Luik
Department of Mathematics
University of Tübingen
Auf der Morgenstelle 10
D-7400 Tübingen

PRODUCT APPROXIMATION: ERROR ESTIMATES

Myron S. Henry
Department of Mathematics
Central Michigan University
Mount Pleasant, MI, USA

1. Introduction

A number of recent papers [3,4,5,7,8] (also see the references of [5]) have considered various computational and theoretical aspects of uniform product approximation. In the present paper an error estimate for uniform product approximation is reviewed, and error estimates are established for certain kinds of discrete uniform product approximations. These error estimates are compared with their counterparts for linear product approximation methods, (Powell, [6]).

Let $D = IxJ$, where $I=J=[-1,1]$, and suppose that $F \varepsilon C(D)$. Next let

$$R_n = \langle r_0(x), r_1(x), \ldots, r_n(x) \rangle, \ x \varepsilon I, \tag{1.1}$$

where r_i is a polynomial of degree exactly i, i=o,1,...,n. Similarly, let

$$S_m = \langle s_0(y), s_1(y), \ldots, s_m(y) \rangle, \ y \varepsilon J, \tag{1.2}$$

where s_j is a polynomial of degree exactly j, j=o, 1,...,m. Then

$$\text{span } R_n S_m = \text{span } \langle r_i s_j \rangle_{i=o,}^{n} \ {}_{j=o}^{m} \tag{1.3}$$

is a finite dimensional subspace contained in $C(D)$. For $F \in C(D)$, the classical uniform best approximation to F from (1.3) is defined to be

$$P_{n,m}(F,J)(x,y) = \sum_{i=0}^{n} \sum_{j=0}^{m} b_{ij} r_i(x) s_j(y), \tag{1.4}$$

where $P_{n,m}(F,J)$ satisfies

$$\inf_{c_{ij}} || F - \sum_{i=0}^{n} \sum_{j=0}^{m} c_{ij} r_i s_j ||_{IxJ} = || F - P_{n,m}(F,J) ||_D = E_{n,m}(F,J). \tag{1.5}$$

The notation employed in (1.4) and (1.5) displays J and not I because only J will be discretized in the subsequent analysis. It is well known that the classical best uniform approximation problem described in (1.5) lacks an elegant theory to parallel the univariate theory of best approximation, [8]. Efficient algorithms to calculate (1.4) are also few in number.However, much is known about certain near best approximations [2,6], and these near best approximations often yield errors that are acceptably close to the error $E_{n,m}(F,J)$ given by (1.5). In the next section a class of linear product approximation methods that often produce near best approximations is discussed.

2. Linear Product Approximation

Let $L:C(I) \to \text{span } R_n$ and $M:C(J) \to \text{span } S_m$ be linear operators. For $F \in C(D)$, define $F_y \in C(I)$ by $F_y(x) = F(x,y)$, $(x,y) \in IxJ$. Then LF_y may be written as

$$(LF_y)(x) = \sum_{i=0}^{n} b_i(y) r_i(x), \tag{2.1}$$

where $b_i \in C(J)$. Next apply M to b_i to obtain

$$(Mb_i)(y) = \sum_{j=0}^{m} b'_{ij} s_j(y). \tag{2.2}$$

Equations (2.1) and (2.2) define $N:C(D) \to \text{span } R_n S_m$ by

$$(NF)(x,y) = [MoL](F)(x,y)$$

$$= \sum_{i=0}^{n} (Mb_i)(y)r_i(x)$$

$$= \sum_{i=0}^{n} \sum_{j=0}^{m} b'_{ij}s_j(y)r_i(x). \qquad (2.3)$$

Powell [6] calls the methods described by (2.1), (2.2), and (2.3) linear product methods (also see [7]). Product interpolation and product least squares are product methods in the Powell sense [6]. As an example, we focus on product interpolation in two variables. Let $T_k = \langle t_i \rangle_{i\ 0}^{k}$ be the set of zeros of the Chebyshev polynomial of degree k+1. If $\langle \ell_i \rangle_{i=0}^{k}$ is the set of Lagrange polynomials of a single variable determined by T_k, then linear operators L:C(I)--> span R_n and M:C(J)--> span S_m are defined by

$$(LF_y)(x) = \sum_{i=0}^{n} F(t_i,y)\ \ell_i(x), \qquad (2.4)$$

and

$$M(F(t_i,\cdot))(y) = \sum_{j=0}^{m} F(t_i,t_j)\ \ell_j(y), \qquad (2.5)$$

respectively. Following (2.1), (2.2), and (2.3), we use (2.4) and (2.5) to define the product interpolation operator N:

$$(NF)(x,y) = \sum_{i=0}^{n} \sum_{j=0}^{m} F(t_i,t_j)\ \ell_j(y)\ \ell_i(x). \qquad (2.6)$$

For (2.6) it is well known [2] that

$$||F-NF||_D \leq K \log(n+1) \log(m+1)\ E_{n,m}(F,J), \qquad (2.7)$$

and that (2.7) is sharp in an asymptotic sense. Since log(n+1)log(m+1) grows very slowly in comparison with the rate at which $E_{n,m}(F,J)$ generally converges to zero, (2.6) can normally be characterized as a near best approximation to F on D, [2].

In the next section we consider a slightly nonlinear companion to (2.3) and compare subsequent error estimates to the error estimate given in (2.7).

3. Uniform Product Approximation

The construction of a best uniform product approximation parallels the construction described by (2.1), (2.2), and (2.3). In particular, let

$$(B_n F_y)(x) = \sum_{i=0}^{n} f_i(y) r_i(x) \tag{3.1}$$

be the best uniform approximation from the span of R_n to F_y on I. Next let

$$\beta_m(f_i, J)(y) = \sum_{j=0}^{m} a_{ij} s_j(y) \tag{3.2}$$

be the best uniform approximation from the span of S_m to f_i on J, i=o,...,n. The slightly nonlinear counterpart to (2.3) is then defined by

$$\overline{P}_{n,m}(F,J)(x,y) = \sum_{i=0}^{n} \beta_m(f_i,J)(y)\; r_i(x)$$

$$= \sum_{i=0}^{n} \sum_{j=0}^{m} a_{ij} s_j(y)\; r_i(x). \tag{3.3}$$

The polynomial $\overline{P}_{n,m}$ is defined to be the best uniform product approximation from the span of $R_n S_m$ to F on D = I x J.

If Y is a discretization of J with cardinality $|Y| \geq m+2$, then a discrete uniform product approximation can be constructed by appropriately modifying (3.2) and (3.3). Let

$$\overline{P}_{n,m}\,(F,Y)(x,y) = \sum_{i=0}^{n} \beta_m\,(f_i,Y)(y)\; r_i(x)$$

$$= \sum_{i=0}^{n} \sum_{j=0}^{m} a^*_{ij} s_j(y) r_i(x) \tag{3.4}$$

be the discretized best uniform product approximation from the span of $R_n S_m$ to F on D* = IxY.

In contrast to (1.4), the discrete uniform product approximation (3.4) is generally efficiently calculatable [3] and can be described as a near best approximation to F on D. We will shortly compare certain error estimates for (3.3) and (3.4) with (2.7).

The following error estimate is one of several that appear in [5].

Theorem 1. *Let* $< \ell_i(x) >_{i=0}^{n}$ *be the set of Lagrange polynomials determined by* $T_n = < t_i >_{i=0}^{n}$. *Suppose that* $R_n = < \ell_i(x) >_{i=0}^{n}$, *and that* $S_m = < s_j(y) >_{j=0}^{m}$, *where each* s_j *is a polynomial of degree exactly* j. *If* $\overline{P}_{n,m}(F,J)$ *is given by* (3.3), *then*

$$||F - \overline{P}_{n,m}(F,J)||_D \leq [3+(4/\pi) \ln(n+1)]E_{n,m}(F,J). \qquad (3.5)$$

We note that (3.5) is theoretically superior to (2.7).

4. Discretization

In actual computations involving uniform product approximation, (3.4) is calculated. Specifically, let Y be a finite subset of J. If $|Y| = \mu \geq m+2$, then as described in (3.1), μ univariate best uniform approximation problems on I are solved. This process determines a_i on Y, $i=0,1,\ldots,n$. Then n+1 mini-max problems are solved on Y to obtain the n+1 best approximations given in (3.2) (where J is replaced by Y). Finally, (3.4) is constructed. Thus it is of interest to investigate the properties of (3.4) as a function of Y. To this end, let the density of Y in J be given by

$$d(Y,J) = \sup_{y \in J} \inf_{y^* \in Y} |y^*-y|.$$

The result of the next theorem is expected and desirable. The proof is omitted.

Theorem 2. *Let* $\overline{P}(F,Y)$ *and* $\overline{P}(F,J)$ *be given by* (3.3) *and* (3.4), *respectively. Then*

$$\lim_{d(Y,J)\to 0} ||\overline{P}_{n,m}(F,Y)-F||_{D^*} = ||\overline{P}_{n,m}(F,J)-F||_D.$$

In what follows, $E_{n,m}(F,Y)$ is the discrete analog to (1.5). The next two theorems are the main theorems of this paper.

<u>Theorem</u> 3. <u>Let</u> R_n <u>and</u> S_m <u>be as described in</u> (1.1) <u>and</u> (1.2), <u>respectively. If</u> $|Y| = m+2$, <u>then</u>

$$||F - \overline{P}_{n,m}(F,Y)||_{D^*} \leq 5\, E_{n,m}(F,Y). \tag{4.1}$$

Proof: On Y, the best approximation operator β_m defined in (3.2) with Y instead of J is linear. Therefore from (3.4) we have that

$$\overline{P}_{n,m}(F,Y)(x,y) = \beta_m(\sum_{i=0}^{n} f_i r_i(x), Y)(y). \tag{4.2}$$

Similarly, if

$$P_{n,m}(F,Y)(x,y) = \sum_{i=0}^{n} \sum_{j=0}^{m} b^*_{ij} s_j(y) r_i(x)$$

is the classical best uniform approximation from the span of $R_n S_m$ to F on D*((1.4) with J replaced by Y), then

$$P_{n,m}(F,Y)(x,y) = \beta_m(\sum_{i=0}^{n} \sum_{j=0}^{m} b^*_{ij} s_j r_i(x), Y)(y). \tag{4.3}$$

By subtracting (4.2) and (4.3) we obtain

$$|\overline{P}_{n,m}(F,Y)(x,y)-P_{n,m}(F,Y)(x,y)|$$

$$\leq ||\beta_m||\,||\sum_{i=0}^{n} f_i r_i(x) - \sum_{i=0}^{n} \sum_{j=0}^{m} b^*_{ij} s_j r_i(x)||_Y$$

$$\leq 2(||\sum_{i=0}^{n} f_i r_i - F||_{D^*} + ||F - \sum_{i=0}^{n} \sum_{j=0}^{m} b^*_{ij} s_j r_i||_{D^*}$$

$$\leq 4\, E_{n,m}(F,Y),$$

which basically implies (4.1).

For the last theorem, let

$$Y = (y_0, y_1, \ldots, y_{m+1}), \tag{4.4}$$

where $\langle y_j \rangle_{j=0}^{m+1}$ are the extreme points of the Chebyshev polynomial of degree m+1.

<u>Theorem</u> 5. <u>Let</u> Y <u>be given by</u> (4.4), <u>and assume that</u> $R_n = \langle \ell_i \rangle_{i=0}^{n}$ <u>and</u> S_m <u>are as in Theorem</u> 1.
<u>Then</u>

$$||F - \overline{P}(F,Y)||_D \leq K \log[(m+1)(n+1)] E_{n,m}(F,J). \tag{4.5}$$

Proof: By (3.4) we have that

$$\overline{P}(F,Y)(x,y) = \sum_{i=0}^{n} \beta_m(f_i,Y)(y)\ell_i(x). \tag{4.6}$$

From [1, p. 76],

$$\beta_m(f_i,Y)(y) = P_{m+1}(f_i)(y) - \lambda_i Q_{m+1}(y), \tag{4.7}$$

where

$$|\lambda_i| = \max_{y \in Y} |f_i(y) - \beta_m(f_i,Y)(y)| = ||e_m(f_i,Y)||_Y$$

$$\leq ||f_i - \beta_m(f_i,J)||_J = ||e_m(f_i,J)||_J, \tag{4.8}$$

where

$$P_{m+1}(f_i)(y_j) = f_i(y_j), \; j=0,1,\ldots,m+1, \tag{4.9}$$

and where

$$Q_{m+1}(y_j) = (-1)^j, \; j=0,1,\ldots,m+1. \tag{4.10}$$

Thus (4.6) implies that

$$\overline{P}(F,Y)(x,y) = \sum_{i=0}^{n} P_{m+1}(f_i)(y)\ell_i(x) - Q_{m+1}(y)\sum_{i=0}^{n} \lambda_i\ell_i(x)$$

$$= \sum_{j=0}^{m+1} F(x,y_j)\ell_j^{m+1}(y) - \sum_{j=0}^{m+1} F(x,y_j)\ell_j^{m+1}(y)$$

$$+ \sum_{i=0}^{n}\sum_{j=0}^{m+1} f_i(y_j)\ell_j^{m+1}(y)\ell_i(x) - Q_{m+1}(y)\sum_{i=0}^{n}\lambda_i\ell_i(x). \quad (4.11)$$

Here $\langle \ell_j^{m+1}(y)\rangle_{j=0}^{m+1}$ are the Lagrange polynomials of degree m+1 determined by (4.4). Subtracting F from (4.11) and utilizing the triangle inequality yields

$$|F(x,y) - \overline{P}(F,Y)(x,y)| \leq ||F(x,\cdot) - \sum_{j=0}^{m+1} F(x,y_j)\ell_j^{m+1}||_J$$

$$+ |\sum_{j=0}^{m+1} [F(x,y_j) - \sum_{i=0}^{n} f_i(y_j)\ell_i(x)]\ell_j^{m+1}(y)|$$

$$+ ||Q_{m+1}||_J|\sum_{i=0}^{n}\lambda_i\ell_i(x)|. \quad (4.12)$$

But

$$||B_nF_{y_j} - F_{y_j}||_I \leq E_{n,m}(F,J), \quad (4.13)$$

and (4.8) and [5, p. 10] imply that

$$|\lambda_i| \leq 2\ E_{n,m}(F,J). \quad (4.14)$$

Considering x as a fixed parameter we also have that

$$||\ e_m\ (F_x,J)||_J \leq E_{n,m}(F,J). \quad (4.15)$$

Applying (4.13), (4.14), and (4.15) to (4.12) yields

$$|F(x,y) - \overline{P}(F,Y)(x,y)| \leq K_1 \log(m+1)E_{n,m}(F,J)$$

$$+ K_2 \log(n+1)E_{n,m}(F,J),$$

$(x,y) \in I \times J$, which in turn implies (4.5).

In concluding we note that (4.5) is theoretically sharper than (2.7) and is basically the same as (3.5) for n=m. The results of Theorem 5 appear to be closely related to error estimates obtained in [7,p. 286] for appolation (best approximation, then interpolation). In fact, for a favorable choice of nodes in appolation, the proof of Theorem 5 yields an error estimate that essentially has the same asymptotic order as the appolation error estimate given in [7].

References

1. Cheney, E.W., "Introduction to Approximation Theory", McGraw-Hill, New York, 1966.

2. deBoor, C., A comment on "Numerical Comparisons of Algorithms for Polynomial and Rational Multivariate Approximations", SIAM J. Numer. Anal., 15 (1978), 1208-1211.

3. Henry, J.N., M.S. Henry, and D. Schmidt, Numerical comparisons of algorithms for polynomial and rational multivariate approximations, SIAM J. Numer. Anal., 15 (1978), 1197-1207.

4. Henry, M.S., and D. Schmidt, Continuity theorems for the product approximation operator, in "Theory of Approximation with Applications", A.G. Law and B.N. Sahney, editors, Academic Press, New York, 1976, 24-42.

5. Henry, M.S., and D. Schmidt, Error bounds for polynomial product approximation, J. Approx. Theory, 31 (1981), 6-21.

6. Powell, M.J.D., Numerical methods for fitting functions of two variables, in "The State of the Art in Numerical Analysis", Academic Press, London, 1977, 563-604.

7. Scherer, R., and K. Zeller, Stepwise approximation in two variables, in "Numerical Methods of Approximation Theory", L. Collatz, G. Meinardus, and H. Werner, editors, BirkhäuserVerlag, Stuttgart, 1980, 282-288.

8. Weinstein, S.E., Approximation of functions of several variables: product Chebyshev approximations I, J. Approximation Theory, 2 (1969), 433-447.

Myron S. Henry
Department of Mathematics
Central Michigan University
Mount Pleasant, Michigan 48859
U.S.A.

A LAGRANGIAN METHOD FOR MULTIVARIATE CONTINUOUS CHEBYSHEV APPROXIMATION PROBLEMS

Kristján Jónasson and G. Alistair Watson

1. Introduction

Let $X \subset R^N$ be a Cartesian product of closed intervals, and let $f : X \times R^n \to R$ be a twice continuously differentiable function of its parameters. We are concerned here with the numerical solution of the problem

$$\text{find } a \in R^n \text{ to minimize } \| f(\cdot,a) \| \qquad (1.1)$$

where the norm is the Chebyshev norm on X. This problem, particularly in the case when f is an affine function of a, has received a great deal of attention, and a number of algorithms have been proposed for its solution. The ascent algorithms of Remes are perhaps the best known, and variants of these for linear problems have been proposed by a number of authors, see e.g.[2] [3] [4], [13]. Methods of descent type have also been suggested (e.g.[11],[12]), but arguably the best currently available methods for both general linear and non-linear problems are those of two-phase type: the second phase is a fast locally convergent method, usually Newton's method applied to the first order necessary conditions for a local solution; the first phase supplies a good initial approximation, for example through the solution of a discretized problem or by the application of a robust slowly convergent method. Methods of this type have been given, for example, in [1], [6], [7], [8], [14].

The main disadvantages of two-phase methods relate to the interface: it is not easy to decide in advance how accurate the first phase solution should be (for example, how fine a grid should be used, if a discretization of the problem is being solved), and there is no guarantee that Newton's method will converge. Another possible drawback is that two separate methods have to be analysed (and implemented). The purpose of this paper is to develop an alternative approach which combines many of the merits of previously used methods: we propose a descent method based on the use of a Lagrangian function which applies to both linear and nonlinear problems, which is globally convergent and which usually has a second order rate of convergence. The assumptions which are required, and a development of the method, are given in the next section, together with a basic convergence result. In section 3 we consider the rate of convergence, and also show the connection of the method with the second algorithm of Remes for linear Chebyshev set problems.

2. A Lagrangian method

For given ρ , $0 \le \rho < 1$ and given $a \in R^n$, let $E(a,\rho)$ denote the set of local maxima of $|f(x,a)|$ satisfying $|f(x,a)| \ge \rho \| f(\cdot,a)\|$. Although ρ is permitted to vary, its role is a fairly minor one, and we will subsequently suppress the explicit dependence of quantities on ρ . It is however convenient to make the assumption that in a neighbourhood of a point a^* satisfying first order necessary conditions for a solution to (1.1), ρ is sufficiently close to 1 that $E(a^*)$ is the set of points where the norm is attained. For any $x \in X$, let $\kappa_1,\ldots,\kappa_\ell$ be the indices of the components of x on the boundary of X . Also let $\nabla_1 f$ denote the vector in R^ℓ whose components are the partial derivatives of f with respect to the components of x in these positions (evaluated at x,a), let $\nabla_2 f$ denote the vector in $R^{N-\ell}$ whose components are the partial derivatives of f with respect to the remaining components of x , and let $\nabla_2^2 f$ denote the corresponding $(N-\ell) \times (N-\ell)$ matrix of second partial derivatives of f . Then we define $Z \subset R^n$ as the (open) set of points such that

(1) there is a finite number of points $x \in X$ such that $\nabla_2 f = 0$,

(2) at each point of (1) $\nabla_2^2 f$ is nonsingular, and each component of $\nabla_1 f$ is nonzero.

Let $a' \in Z$ and let $E(a') = \{x_1,x_2,\ldots,x_{t'}\}$ where t' denotes $t(a')$. Then in a neighbourhood of a', $E(a)$ will contain $\le t'$ points, with equality holding if $|f(x_i,a')| \ne \rho \| f(\cdot,a')\|$, $i = 1,2,\ldots,t'$. By the implicit function theorem, there exist continuously differentiable functions

$x_1, x_2, \ldots, x_{t'}$ in a neighbourhood $N(a')$ of a', such that

$$\nabla_2 f(x_i(a), a) = 0, \quad i = 1,2,\ldots,t' , \tag{2.1}$$

where components at a boundary of X remain fixed. (Note that for $a \in N(a')$, $E(a) \subset \{x_1(a), x_2(a), \ldots, x_{t'}(a)\}$.) It follows that

$$\| f(\cdot,a) \| = \max_{1\le i\le t'} |g_i(a)| \tag{2.2}$$

where $\quad g_i(a) = f(x_i(a),a), \quad i = 1,2,\ldots,t',$

for all $a \in N(a')$ (cf. Theorem 5 of [9]). Thus if $a^* \in Z$ solves (1.1), it is clear that locally (1.1) is equivalent to the problem

$$\text{find } a \in R^n \text{ to minimize } \max_{1\le i\le t^*} |g_i(a)| . \tag{2.3}$$

For arbitrary $a \in Z$, $h \in R$, let $\bar{a} \in R^{n+1}$ denote the vector

$$\bar{a}^T = (a^T, h) , \tag{2.4}$$

let $E(a) = \{x_1, x_2, \ldots, x_t\}$, and let

$$r_i(\bar{a}) = \sigma_i g_i(a) - h , \quad i = 1,2,\ldots,t , \tag{2.5}$$

where $\quad \nabla_2 f(x_i(a), a) = 0 , \quad i = 1,2,\ldots,t ,$ (2.6)

and $\quad \sigma_i = \text{sign } (f(x_i,a)), \quad i = 1,2,\ldots,t .$

Then we may define in a neighbourhood of a the <u>Lagrangian function</u>

$$\begin{aligned}\mathcal{L}(\bar{a},\lambda) &= e_{n+1}^T \bar{a} + \sum_{i=1}^{t} \lambda_i r_i(\bar{a}) \\ &= h(1 - \sum_{i=1}^{t} \lambda_i) + \sum_{i=1}^{t} \lambda_i \sigma_i g_i(a) .\end{aligned} \tag{2.7}$$

The significance of this function is evident if we rewrite (2.3) as

$$\begin{aligned}&\text{find } \bar{a} \in R^{n+1} \text{ to minimize } e_{n+1}^T \bar{a} \\ &\text{subject to } r_i(\bar{a}) \le 0 , \quad i = 1,2,\ldots,t^* .\end{aligned} \tag{2.8}$$

In terms of $\mathcal{L}$, we can obtain by standard optimization techniques applied to (2.8) the well known

<u>Theorem 1</u> (First order necessary conditions). Let $a^* \in Z$ solve (1.1). Then there exist nonnegative numbers λ_i^*, $i = 1,\ldots,t^*$ such that

$$r_i(\bar{a}^*) \le 0 , \quad i = 1,2,\ldots,t^*, \tag{2.9a}$$

$$\nabla_{\bar{a}} \mathcal{L}(\bar{a}^*,\lambda^*) = 0 , \tag{2.9b}$$

$$\lambda_i^* r_i(\bar{a}^*) = 0, \quad i = 1,2,\ldots,t^* \tag{2.9c}$$

where $\bar{a}^{*T} = (a^{*T}, h^*)$, and $h^* = \| f(\cdot, a^*) \|$.
(Notice that (2.9a,c) are actually trivial by our assumption on ρ.)

<u>Definition 1</u> A point a^* satisfying the conditions of the theorem will be called a <u>stationary point</u>.

<u>Definition 2</u> If $\lambda_i^* > 0$, $i = 1,2,\ldots,t^*$, then we say that <u>strict complementarity</u> holds (at a^*, λ^*) .

We now proceed to develop a method for (1.1) based on the application of a variant of the SOLVER method (for example [5]) to (2.8). Normally the globalization of the method requires the introduction of an exact penalty function. However, for (2.8) we will exploit the special structure of the problem so that this is unnecessary: in particular, feasibility with respect to (2.8) is maintained throughout, so that descent is achieved with respect to the norm. The approach is similar in this respect to that used in [10] for discrete Chebyshev approximation problems.

Let $a \in Z$ be an estimate of a stationary point a^* of (1.1) and let

$$h = \| f(\cdot, a) \|$$

with $\bar{a}$ given by (2.4). Then we can define the quadratic programming problem

$$\begin{aligned} &\text{find } \bar{d} \in R^{n+1} \text{ to minimize } \tfrac{1}{2}\bar{d}^T \bar{H} \bar{d} + e_{n+1}^T \bar{d} \\ &\text{subject to } r_i(\bar{a}) + \bar{d}^T \nabla_{\bar{a}} r_i(\bar{a}) \le 0, \quad i = 1,2,\ldots,t \end{aligned} \tag{2.10}$$

where $\bar{H}$ is a given $(n+1) \times (n+1)$ matrix. We will in fact assume that $\bar{H}$ has the form $\begin{bmatrix} H & 0 \\ 0 & 0 \end{bmatrix}$ where H is $n \times n$. Then letting $\bar{d}^T = (d^T, p)$ where $d \in R^n$, $p \in R$, $r \in R^t$ have i^{th} component $r_i(\bar{a})$, and C be the $n \times t$ matrix with i^{th} column $\nabla_a r_i(\bar{a})$ $(= \sigma_i \nabla_a g_i(a))$ the problem may be rewritten

$$\begin{aligned} &\text{find } d \in R^n,\ p \in R \text{ to minimize } \tfrac{1}{2}d^T H d + p \\ &\text{subject to } r + C^T d - pe \le 0 \end{aligned} \tag{2.11}$$

where $e = (1,1,\ldots,1)^T \in R^t$. (Notice that $r \le 0$ by the way h is defined, and so the feasible region is always nonempty.) The significance of (2.11) is explained by the following theorem.

<u>Theorem 2</u> Let H be positive semidefinite, and $\bar{d}^T = (d^T, p)$ solve (2.11) at $a \in Z$. Then (i) if $p < 0$, d is a descent direction for $\| f(\cdot,a) \|$ at a, (ii) if $p = 0$, a is a stationary point of (1.1).

<u>Proof</u> Clearly we must have $p \leq 0$. For any i, $1 \leq i \leq t$, $\gamma > 0$ small enough we have

$$g_i(a + \gamma d) = g_i(a) + \gamma d^T \nabla g_i(a) + O(\gamma^2)$$

and since by the constraints of (2.11) we have

$\sigma_i d^T \nabla g_i(a) \leq p$ for $r_i(a) = 0$, we get

$$\| f(\cdot,a + \gamma d) \| \leq \| f(\cdot,a) \| + \gamma p + O(\gamma^2) \tag{2.12}$$

and (i) is proved. Now the Kuhn-Tucker conditions for (2.11) give the existence of a nonnegative vector $\mu \in R^t$ such that

$$Hd + C\mu = 0 \tag{2.13a}$$

$$e^T \mu = 1 \tag{2.13b}$$

$$\mu^T (r + C^T d - pe) = 0 . \tag{2.13c}$$

Let $p = 0$. Then

$$d^T H d = -\mu^T C^T d = \mu^T r \leq 0 .$$

Thus $d^T H d = 0$ and so a (possibly different) solution to (2.11) is given by $\bar{d} = 0$. Thus there exists a (possibly different) $\mu \geq 0$ such that $C\mu = 0$, $e^T \mu = 1$. The conditions of Theorem 1 are therefore satisfied and the proof is completed. □

Let d^k, p^k solve (2.11) at $a^k \in Z$, and let γ^k be the value of γ satisfying

$$\gamma = \max \{1, \tfrac{1}{2}, \tfrac{1}{4}, \ldots\} \text{ such that } T(a^k, \gamma) \geq \tau , \tag{2.14}$$

where $0 < \tau < 1$ and τ is constant (independent of k) and

$$T(a^k, \gamma) = \frac{\| f(\cdot,a^k + \gamma d^k) \| - \| f(\cdot,a^k) \|}{\gamma p^k} . \tag{2.15}$$

Clearly by (2.12) there exists γ^k satisfying (2.14) if $p^k \neq 0$; that γ^k is a suitable steplength in the direction d^k is shown by the next theorem.

<u>Remark</u> When r, C and other quantities are used with superscripts k or * it will mean that they are evaluated at a^k or a^*, and therefore have the appropriate dimension.

Theorem 3 Let (2.11) be applied at a sequence of points $\{a^k\} \subset Z$, with $H = H^k$ positive semidefinite and $a^{k+1} = a^k + \gamma^k d^k$ where $d^k \neq 0$ solves (2.11) and γ^k satisfies (2.14). If $\{d^k\}$, $\{H^k\}$ are bounded, then $p^k \to 0$ as $k \to \infty$. In addition the limit points of $\{a^k\}$ in Z are stationary points of (1.1).

Proof The proof of $p^k \to 0$ may be obtained by the application of standard techniques available from optimization theory, and will not be given here.

Let there exist a subsequence (which we do not rename) $a^k \to a^* \in Z$. From (2.13)

$$0 \le d^{k^T} H^k d^k = -d^{k^T} C^k \mu^k = \mu^{k^T} r^k - p^k ,$$

so that $\mu^{k^T} r^k \to 0$, and also $d^{k^T} H^k d^k \to 0$ as $k \to \infty$. Let k be large enough that a^k lies in a neighbourhood of a^* with $t^k \le t^*$. Restrict consideration to a further subsequence so that $E(a^k) = \{x_{\theta_1}(a^k), \ldots, x_{\theta_m}(a^k)\}$ for some constant m with $x_{\theta_i}(a^k) \to x_{\theta_i}(a^*) \in E(a^*)$, $i = 1, \ldots, m$, as $k \to \infty$. Let $r^k \to \hat{r}^*$, $C^k \to \hat{C}^*$ and (choosing another subsequence if necessary) $d^k \to d^0$, $H^k \to H^0$, $\mu^k \to \mu^0$ as $k \to \infty$. Then $\bar{d}^T = (d^{0^T}, 0)$ solves the problem

$$\begin{aligned} &\text{minimize} \quad \tfrac{1}{2} d^T H^0 d + p \\ &\text{subject to} \quad \hat{r}^* + \hat{C}^{*T} d - pe \le 0 \end{aligned} \qquad (2.16)$$

with objective function value zero. If $m = t^*$, then (2.16) is just (2.11) at a^* with $H = H^0$. Otherwise, the latter problem is (2.16) with additional constraints, and so the optimal value of the objective function cannot be negative. Thus $\bar{d} = 0$ is a solution to (2.11) at a^* and the conclusion that a^* is a stationary point follows as in Theorem 2. □

3. Rate of Convergence

The efficiency of the method described in the previous section depends primarily on the choice of the matrix H at each step, and in common with analogous methods for finite problems, close to a stationary point we look for H to be related to the Hessian matrix of the Lagrangian function. We have

$$\nabla_a^2 \mathcal{L}(\bar{a},\lambda) = \begin{bmatrix} G(a,\lambda) & 0 \\ 0 & 0 \end{bmatrix} \tag{3.1}$$

where $G(a,\lambda) = \sum_{i=1}^{t} \lambda_i \sigma_i \nabla^2 g_i(a)$. (3.2)

Thus if $G(a,\lambda)$ is positive semi-definite at the current point (with suitable choice of λ), we may take $H = G$, otherwise we could choose a positive semi-definite approximation to G , or $G + \nu I$, where $\nu > 0$ is large enough. Before making clear the significance of the choice $H = G$, we require a preliminary lemma and some notation. For given $a \in Z$, define the $(n + t + 1) \times (n + t + 1)$ matrix

$$S = \begin{bmatrix} G & 0 & C \\ 0 & 0 & -e^T \\ C^T & -e & 0 \end{bmatrix}$$

where G and C are assumed evaluated at a,λ .

<u>Lemma 1</u> Let $a^* \in Z$ be a stationary point of (1.1), let strict complementarity hold at a^*,λ^* and let S^* be nonsingular. Then if a^k is close enough to a^* and if (2.11) is applied at a^k with $H = G^k$, (2.11) has t^* constraints which hold with equality at the (unique) solution.

<u>Proof</u> At a^*, λ^* we have

$$S^* \begin{bmatrix} 0 \\ 0 \\ \lambda^* \end{bmatrix} = \begin{bmatrix} 0 \\ -1 \\ 0 \end{bmatrix} , \qquad \lambda^* > 0 .$$

Let $a^k \in N(a^*)$, and consider (2.11) at a^k with $H = G^k$. Then $t^k = t^*$ by our assumption on ρ . If it is assumed that all t^* constraints are active, Kuhn-Tucker conditions give

$$S^k \begin{bmatrix} \hat{d}^k \\ \hat{p}^k \\ \hat{\mu}^k \end{bmatrix} = \begin{bmatrix} 0 \\ -1 \\ -r^k \end{bmatrix} .$$

Since $r^k \to 0$ and $S^k \to S^*$ as $k \to \infty$, we must eventually have $\hat{\mu}^k > 0$, so that (2.11) is solved. The result is proved. □

Remark A sufficient condition for S^* to be nonsingular is that $[C^{*T} - e]$ has full rank, and second order sufficient conditions for a^* to solve (2.8) hold [5].

Theorem 4 Let $a^* \in Z$ be a stationary point of (1.1), let strict complementarity hold at a^*, λ^*, let S^* be nonsingular and let the elements of G satisfy Lipschitz conditions. Let (2.11) be applied at a^k with $H^k = G^k$, where λ^k satisfies $\|\lambda^k - \lambda^*\| = O(\|a^k - a^*\|)$. Then if a^k is close enough to a^*, and $\gamma^k = 1$ satisfies (2.14), $a^k \to a^*$ at a second order rate.

Proof Define

$$\delta^k = a^k - a^*, \quad \varepsilon^k = h^k - h^* .$$

Let a^k be close enough to a^* that $t^k = t^*$, S^k is nonsingular and equality holds in the constraints of (2.11) at the solution (using Lemma 1). Now $r^* = 0$, so that Taylor expansion gives

$$r^k = -\varepsilon^k e + C^{k^T}\delta^k + O(\|\delta^k\|^2) .$$

Using the Kuhn-Tucker conditions for (2.11),

$$S^k \begin{bmatrix} \delta^{k+1} \\ p^k + \varepsilon^k \\ \mu^k - \lambda^* \end{bmatrix} = \begin{bmatrix} 0 \\ -1 \\ -r^k \end{bmatrix} + S^k \begin{bmatrix} \delta^k \\ \varepsilon^k \\ -\lambda^* \end{bmatrix}$$

$$= \begin{bmatrix} G^k\delta^k - C^k\lambda^* \\ e^T\lambda^* - 1 \\ C^{k^T}\delta^k - \varepsilon^k e - r^k \end{bmatrix} = \begin{bmatrix} G^k\delta^k - C^k\lambda^* \\ 0 \\ O(\|\delta^k\|^2) \end{bmatrix} .$$

Also by Taylor expansion

$$G^k\delta^k - C^k\lambda^* = \sum_{i=1}^{t^*} (\lambda_i^k - \lambda_i^*)\, \nabla_a^2 r_i(\bar{a}^k)\delta^k - C^*\lambda^* + O(\|\delta^k\|^2)$$

and the result follows. □

Remark A satisfactory choice for λ^k satisfying the condition of Theorem 4 is given by the solution of the least squares problem

$$\text{minimize} \quad \left\| \begin{bmatrix} C^k \\ e^T \end{bmatrix} \lambda - e_{n+1} \right\|_2^2 ,$$

as may readily be shown.

Theorem 5 Let $N = 1$, $f(x,a) = a^T\phi(x) - \phi_0(x)$, where $\{\phi_1(x),\ldots,\phi_n(x)\}$ form a Chebyshev set on X . Then if at the unique solution a^* to (1.1) there are exactly (n+1) extrema, the method based on (2.11) is equivalent locally to the second algorithm of Remes, if $\gamma^k = 1$ eventually satisfies (2.14).

Proof Strict complementarity must hold, so assume that a^k is close enough to a^* that the active set of (2.11) is an alternating set of (n+1) points. Then without loss of generality

$$r_i(\bar{a}^k) = (-1)^i({a^k}^T\phi(x_i) - \phi_0(x_i)) - h^k \quad i = 1,2,\ldots,n+1 ,$$

and d^k, p^k solving (2.11) satisfy

$${C^k}^T d^k - ep^k = -r^k .$$

It is readily seen that this is equivalent to

$$\begin{bmatrix} \phi(x_1)^T \\ \vdots \\ \phi(x_{n+1})^T \end{bmatrix} (a^k + d^k) - \begin{bmatrix} \phi_0(x_1) \\ \vdots \\ \phi_0(x_{n+1}) \end{bmatrix} = \begin{bmatrix} -1 \\ 1 \\ \vdots \\ (-1)^{n+1} \end{bmatrix} (p^k + h^k) ,$$

the (nonsingular) system of equations which arises in a step of the second algorithm of Remes. □

Remark This result may be generalised to the case where there are more than (n+1) extrema, provided that the (n+1) maxima chosen in the Remes method correspond to those which are associated with non-zero multipliers at a solution to (2.11).

It is not necessary for G^* to be positive semi-definite for second order convergence, merely that S^* be nonsingular. However, for an accompanying guarantee of descent, positive semi-definiteness of G^* is required. For linear problems, the matrix G has the form $G = A^TDA$, where D is a block diagonal matrix arising from the use of the implicit function theorem, positive definite by assumption. Thus G in this case is always (at least) positive semi-definite.

4. Concluding remarks

The purpose of this paper has been to present the theory of a new method for solving (1.1). Any implementation of the method obviously requires a prescription for fixing ρ, and also for choosing H^k at each step. We have seen that the role of H^k is an extremely important one: close to the solution it effectively controls the rate of convergence (except in special cases when there are (n+1) extrema), and we would like if possible to choose $H^k = G^k$; far from the solution, we merely want to ensure a reasonable descent step, preferably with $\gamma^k = 1$ satisfying (2.14). The formula $H^k = G^k + \nu^k I$, where $\nu^k \geq 0$ is a suitably chosen parameter, has been effective in similar situations [15]. We do not attempt here to go into greater detail about more practical matters relevant to the implementation of the method. These will require extensive numerical experimentation, and it is intended to report on this at a later date.

References

1. Andreassen, D.O. and G.A. Watson. Linear Chebyshev approximation without Chebyshev sets, B.I.T. 16 (1976), 349-362.

2. Blatt, H-P. A simultaneous exchange algorithm for semi-infinite linear programming, paper given at International Symposium on Semi-Infinite Programming and Applications 1981, Austin, Texas.

3. Carasso, C. and P.J. Laurent. Un algorithme général pour l'approximation au sens de Tchebycheff de fonctions bornées sur un ensemble quelconque, in Approximation Theory, Lecture Notes in Math. 556, eds. R. Schaback and K. Scherer, Springer Verlag, Berlin (1976).

4. Defert, Ph. and J.P. Thiran. Chebyshev approximation by multivariate polynomials, University of Namur Report 80/10 (1980).

5. Fletcher, R. Practical methods of optimization, Vol. II Constrained optimization, Wiley, Chichester (1981).

6. Gustafson, S.-A. Nonlinear systems in semi-infinite programming, in Solutions of nonlinear algebraic equations, eds. G.B. Byrnes and C.A. Hall, Academic Press (1973).

7. Gustafson, S.-A. and K.O. Kortanek. Numerical solution of a class of semi-infinite programming problems, NRLQ 20 (1973), 477-504.

8. Hettich, R. A Newton method for nonlinear Chebyshev approximation, in Approximation Theory, Lecture Notes in Math. 556, eds. R. Schaback and K. Scherer, Springer Verlag, Berlin (1976).

9. Hettich, R. and H. Th. Jongen. On the local continuity of the Chebyshev operator, J. Approx. Theory 33 (1981), 296-307.

10. Murray, W. and M.L. Overton. A projected Lagrangian algorithm for nonlinear minimax optimization, SIAM J. Sci. Stat. Comput. 1 (1980), 345-390.

11. Opfer, G. An algorithm for the construction of best approximations based on Kolmogorov's criterion, J. Approx. Theory 23 (1978), 299-317.

12. Scott, P.D. and J.S. Thorp. A descent algorithm for linear continuous Chebyshev approximation, J. Approx. Theory 6 (1972), 231-241.

13. Watson, G.A. A multiple exchange algorithm for multivariate Chebyshev approximation, SIAM J. Num. Anal. 12 (1975), 46-52.

14. Watson, G.A. A method for calculating best nonlinear Chebyshev approximations, JIMA 18 (1976), 241-250.

15. Watson, G.A. Numerical experiments with globally convergent methods for semi-infinite programming problems, University of Dundee Report NA/49 (1981) (to appear in the Proceedings of the International Symposium on Semi-Infinite Programming and Applications 1981, Austin, Texas).

Dr G. Alistair Watson, Department of Mathematical Sciences, University of Dundee, Dundee DD1 4HN, Scotland.

CONSTRUCTION OF SURFACES OF CLASS $\mathcal{C}^k$ ON A DOMAIN $\Omega \subset \mathbb{R}^2$, AFTER TRIANGULATION

Alain LE MEHAUTE

Abstract : An algorithm is given for the construction of Hermite triangular finite elements of class $\mathcal{C}^k$ in $\mathbb{R}^2$. This allowed us to generalize some classical elements (for exemple : Zienkiewicz's, Argyris' , Bell's éléments...). These elements are used to construct surfaces of class $\mathcal{C}^k$ on a domain $\Omega \subset \mathbb{R}^2$, interpolating a given function and its derivatives at some randomly selected points in $\overline{\Omega}$, after an automatic triangulation of Ω.

Résumé : On propose une méthode générale de construction d'éléments finis triangulaires d'Hermite de classe $\mathcal{C}^k$ dans $\mathbb{R}^2$, permettant une généralisation à l'ordre k des éléments classiques (par exemple ceux de Zienkiewicz, d'Argyris ou de Bell). Ces éléments sont utilisés pour construire une surface de classe $\mathcal{C}^k$ sur un domaine Ω de $\mathbb{R}^2$, interpolant une fonction et ses dérivées en un nombre quelconque de points distribués de façon aléatoire dans $\overline{\Omega}$, après triangulation automatique du domaine Ω .

INTRODUCTION :

Given a domain $\Omega \subset \mathbb{R}^2$, where we choose some randomly selected points t_i, $i = 1,2,\dots,N$, we want to construct a surface that interpolates a given function f and its derivatives at the t_i.

After an automatic triangulation τ of Ω using the t_i, we construct a surface S which is of class $\mathcal{C}^k$ on Ω, and is polynomial on each triangle T of τ. To do this, we have to construct, on each triangle, an approximation of f using what we call, following CIARLET [2], an Hermite finite element of class $\mathcal{C}^k$. In this way, we obtain a generalization of some classical triangular elements, such as Zienkiewicz's, Argyris', or Bell's elements.

Finally, we present some numerical experiments.

Notation :

* for $\alpha = (\alpha_1,\alpha_2,\dots,\alpha_p) \in \mathbb{N}^p$,

$$|\alpha| = \alpha_1 + \alpha_2 + \dots + \alpha_p, \qquad \alpha! = \alpha_1!\,\alpha_2!\,\dots\,\alpha_p!\,,$$

$$\binom{\alpha}{\beta} = \binom{\alpha_1}{\beta_1}\binom{\alpha_2}{\beta_2}\cdots\binom{\alpha_p}{\beta_p}$$

$$\alpha + \beta = (\alpha_1 + \beta_1,\ \alpha_2 + \beta_2,\ \dots,\ \alpha_p + \beta_p)$$

$$\alpha \leqslant \beta \text{ iff } \alpha_j \leqslant \beta_j \text{ for } j = 1,2,\dots,p$$

* $$D^\alpha f(M) = \frac{\partial^{|\alpha|} f}{\partial x_1^{\alpha_1}\,\partial x_2^{\alpha_2} \cdots \partial x_p^{\alpha_p}}(M)$$

* If A and B are two points in $\mathbb{R}^p$, $A = (a_1,a_2,\dots,a_p)$, $B = (b_1,b_2,\dots,b_p)$

$$AB = ((a_1-b_1)^2 + \dots + (a_p-b_p)^2)^{1/2}$$

$$AB^\alpha = (a_1-b_1)^{\alpha_1}\,(a_2-b_2)^{\alpha_2} \cdots (a_p-b_p)^{\alpha_p}$$

* If $A_1, A_2, \dots, A_{p+1}$ are the vertices of a simplexe in $\mathbb{R}^p$, for $M \in \mathbb{R}^p$, for $i = 1,2,\dots,p+1$, $\lambda_i = \lambda_i(M)$ denotes the barycentric coordinate of M relatively to A_i.

* $T_A^k f$ is the Taylor polynomial expansion of order k of f at A, i.e :

$$T_A^k f : M \longrightarrow T_A^k f(M) = \sum_{|\alpha| \leq k} \frac{1}{\alpha!} D^\alpha f(A) \quad AM^\alpha$$

Definition : A polynomial p realizes Taylor interpolation of f (or interpolation of the Taylor field of f) of order k at a point A if

$$D^\alpha p(A) = D^\alpha f(A) \quad \textit{for all } \alpha \ , \ |\alpha| \leq k \ .$$

Following ŽENIŠEK [9], we know that, if p is the restriction on a triangle $T \in \tau$ of a function f of class $\mathcal{C}^k$ on Ω, if p is a polynomial that interpolates the Taylor field of f of order m at a point A, vertex of T, then $m \geq 2k$ and the simplest polynomial p is, on the triangle T, uniquely determined by the values :

$$D^\alpha p(P_i) \qquad i = 1,2,3 \quad ; \quad |\alpha| \leq 2k$$

$$D^\alpha p(P_o) \qquad |\alpha| \leq k-2$$

$$\frac{\partial^\ell p}{\partial \nu_i^\ell}(Q_{ri}^\ell) \qquad \ell=1,2,\dots,k \quad ; \quad r=1,2,\dots,\ell \quad ; \ i=1,2,3$$

where P_1, P_2, P_3 are the vertices of the triangle, P_o its centroïd (or another interior point), Q_{ri}^ℓ the points dividing the side $P_{i+1}P_{i-1}$ into $(\ell+1)$ equal parts and $\frac{\partial p}{\partial \nu_i}$ the normal derivative along $P_{i+1}P_{i-1}$.

I. TAYLOR INTERPOLATION OF ORDER k AT THE VERTICES OF A SIMPLEX IN $\mathbb{R}^p$:

Problem 1 : Given (p+1) points $A_1, A_2, \dots, A_{p+1}$, vertices of a simplex (non degenerate) in $\mathbb{R}^p$, construct a polynomial p, of degree 2k+1 that realizes Taylor interpolation of order k of f at each vertex A_i, $i=1,2,\dots,p+1$: i.e : $D^\alpha p(A_i) = D^\alpha f(A_i) \quad |\alpha| \leq k$

Proposition : The polynomial $P_{2k+1}[f]$*, explicitly defined by*

(1) $$P_{2k+1}[f](M) = \sum_{i=1}^{p+1} \lambda_i^{k+1} \sum_{j=0}^{k} \frac{(k+j)!}{k!\,j!} (1-\lambda_i)^j \, T_{A_i}^{k-j} f(M)$$

is a solution of problem 1.

Proof : $*\ \lambda_i(A_k) = \delta_{ik} = 0$ if $i \neq k$

$= 1$ if $i = k$

$*\ P(A_j) = \lambda_j^{k+1}(A_j)\, T_{A_j}^{k} f(A_j) = f(A_j)$

* for the derivatives, if

$$\nabla_i = \begin{pmatrix} \dfrac{\partial \lambda_i}{\partial x_1} \\ \vdots \\ \dfrac{\partial \lambda_i}{\partial x_p} \end{pmatrix} \qquad \nabla_i^{\beta} = \left(\frac{\partial \lambda_i}{\partial x_1}\right)^{\beta_1} \cdots \left(\frac{\partial \lambda_i}{\partial x_p}\right)^{\beta_p}$$

We obtain explicitly, for $|\alpha| \leqslant k$

(2) $$D^{\alpha}P_{2k+1}[f](M) = \sum_{i=1}^{p+1} \Big\{ \sum_{\beta \leqslant \alpha} \binom{\alpha}{\beta} \frac{(k+1)!}{(k+1-|\beta|)!} \nabla_i^{\beta} \lambda_i^{k+1-|\beta|} \times \sum_{j=0}^{k} \frac{(k+j)!}{k!\,j!} (-1)^j \sum_{\substack{\gamma \leqslant \alpha-\beta \\ |\gamma| \leqslant j}} \binom{\alpha-\beta}{\gamma} \frac{j!}{(j-|\gamma|)!} \nabla_i^{\gamma} (\lambda_i - 1)^{j-|\gamma|} \times T_{A_i}^{k-j-|\alpha-\beta-\gamma|} D^{\alpha-\beta-\gamma} f(M) \Big\}$$

* If $M = A_j$, we obtain $D^{\alpha}P(A_j) = D^{\alpha}f(A_j)$, $|\alpha| \leqslant k$

Remarks : There is (2k+1) unisolvency only in two cases :

1. in $\mathbb{R}$, for all k
2. in $\mathbb{R}^p$, only if k=1 (this is the case of Lagrange interpolation)

Examples (in $\mathbb{R}^2$) :

→ k=0

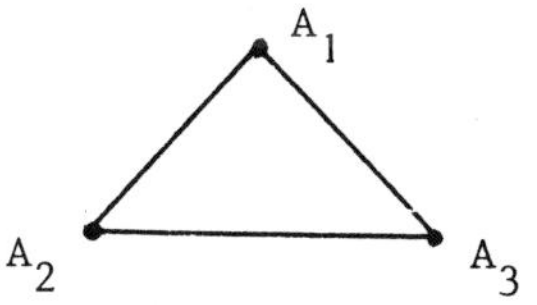

$$P_1(M) = \sum_{i=1}^{3} \lambda_i \; f(A_i)$$

→ k=1

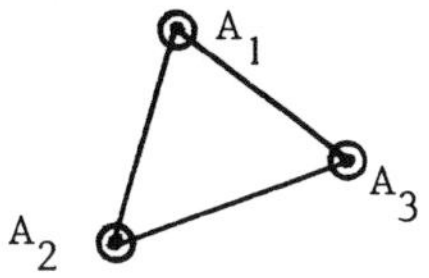

$$P_3(M) = \sum_{i=1}^{3} \lambda_i^2 \; \{T^1_{A_i} f(M) + 2(1-\lambda_i) f(A_i)\}$$

alternatively

$$P_3(M) = \sum_{i=1}^{3} \lambda_i^2 \{\lambda_{i-1} \; T^1_{A_i} \; f(A_{i-1}) + \lambda_{i+1} \; T^1_{A_i} f(A_{i+1}) + 2(1-\lambda_i) f(A_i)\}$$

→ k=2

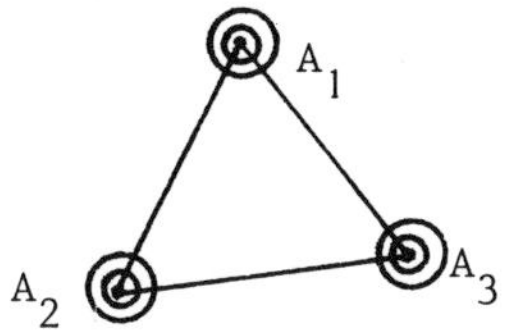

$$P_5(M) = \sum_{i=1}^{3} \lambda_i^3 \; \{T^2_{A_i} f(M) + 3(1-\lambda_i) T^1_{A_i} f(M) + 6(1-\lambda_i)^2 f(A_i)\}$$

Remark : One may obtain a formula where M appears only through its barycentric coordinates, using the fact that in each Taylor expansion of (1), one may express

$$A_iM^{\alpha} = \sum_{\beta \leqslant \alpha} \binom{\alpha}{\beta} \quad \lambda_{i+1}^{|\beta|} \; \lambda_{i-1}^{|\alpha|-|\beta|} \quad A_iA_{i+1}^{\beta} \; A_iA_{i-1}^{\alpha-\beta}$$

II. HERMITE TRIANGULAR FINITE ELEMENTS OF CLASS $\mathscr{C}^k$:

Two technical results allowed us to obtain relatively simple formulas in practical use :

1) If we rotate the axis of an angle α, writing $s = \sin \alpha$, $c = \cos \alpha$,

$$\frac{\partial^p}{\partial \nu^p} = \sum_{k=0}^{p} \binom{p}{k} (-1)^k s^k c^{p-k} \quad \frac{\partial^p}{\partial x^k \partial y^{p-k}}$$

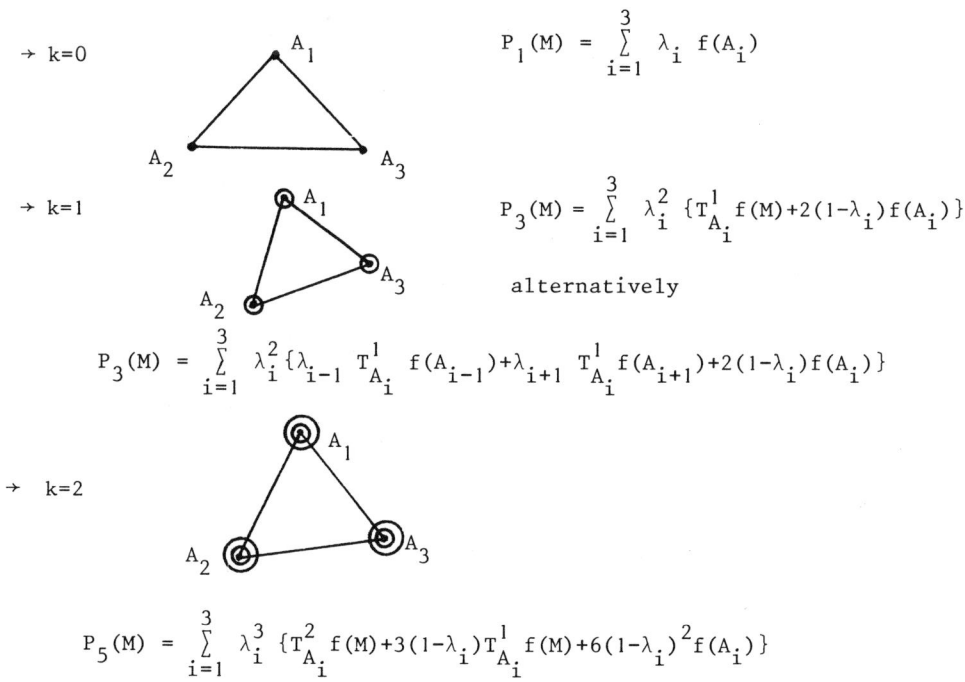

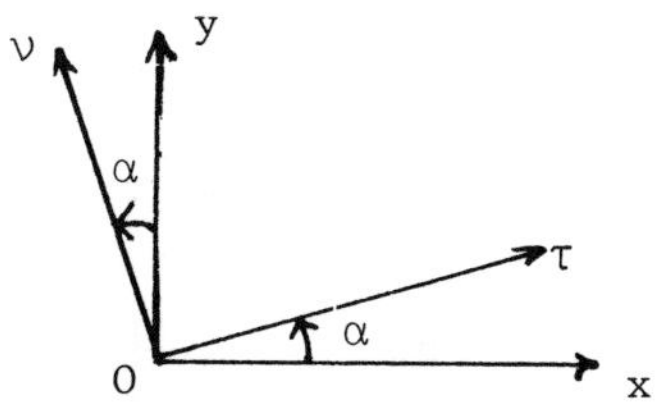

$$\frac{\partial^p}{\partial\tau^p} = \sum_{k=0}^{p} \binom{p}{k} c^k s^{p-k} \frac{\partial^p}{\partial x^k \partial y^{p-k}}$$

2) If M is on the side $A_i A_j$, then

$$(3) \quad T^n_{A_i} f(M) = \lambda_j^n T^n_{A_i} f(A_j) + (1-\lambda_j) \sum_{k=0}^{n-1} \lambda_j^k T^k_{A_i} f(A_j)$$

a) <u>triangles of type A</u> : (generalize Argyris' element)

<u>Problem 2</u> : If A_1, A_2, A_3 are the vertices of a triangle T, A_0 a strictly interior point, and Q^k_{ri}, $k=1,2,\ldots,n$; $r = 1,2,\ldots,k$; $i = 1,2,3$, points which divide $A_{i+1}\, A_{i-1}$ in (k+1) equal parts, construct a polynomial p such that :

$$(4) \quad \begin{array}{lll} \text{(i)} & D^\alpha p(A_i) = D^\alpha f(A_i) & i=1,2,3 \;;\; |\alpha| \leqslant 2n \\ \text{(ii)} & D^\beta p(A_0) = D^\beta f(A_0) & |\beta| \leqslant n-2 \\ \text{(iii)} & \dfrac{\partial^k p}{\partial \nu_i^k}(Q^k_{ri}) = \dfrac{\partial^k f}{\partial \nu_i^k}(Q^k_{ri}) & \text{all } Q^k_{ri} \end{array}$$

<u>Theorem (Zenisek [9])</u> :

1) Problem 2 is (4n+1)-unisolvent

2) Over a triangulation of Ω, the interpolant is $\mathcal{C}^n(\Omega)$.

Let $Z_{4n+1}[f]$ be the unique solution of problem 2 and $P_{4n+1}[f]$ the solution of problem 2(i) given by (1). We may express

$$(5) \quad Z_{4n+1}[f] = P_{4n+1}[f] + R_{4n+1}[f]$$

Let $R_{4n+1}[f]$ defined by ($\alpha_{pq} \in \mathbb{R}$)

(6) $$R_{4n+1}[f](M) = \lambda_1\lambda_2\lambda_3 \sum_{p+q+r=2n-1} \alpha_{pq}\{(\lambda_1\lambda_2)^p(\lambda_2\lambda_3)^q(\lambda_3\lambda_1)^r\}$$

Proposition : *a) R_{4n+1} is a polynomial of degree 4n+1, and is 2n-flat on each point A_1, A_2, A_3 .*

b) There exists n(2n+1) real constants α_{pq} such that $Z_{4n+1}[f]$ is the unique solution of problem 2.

Proof : a) obvious

b) the α_{pq} are solutions of a linear system :

(7) $$\begin{cases} \sum_{p+q=0}^{2n-1} \alpha_{pq} B^{\alpha}_{pq}(A_o) = D^{\alpha}f(A_o) - D^{\alpha}P_{4n+1}[f](A_o) , & |\alpha| \leqslant n-2 \\ \sum_{p+q=0}^{2n-1} \alpha_{pq} B^{k,\nu i}_{pq}(Q^{(k)}_{r,i}) = \frac{\partial^k f}{\partial \nu_i^k}(Q^{(k)}_{r,i}) - \frac{\partial^k}{\partial \nu_i^k} P_{4n+1}[f](Q^{(k)}_{r,i}) & \end{cases}$$

$k=1,2,\ldots,n \; ; \; r=1,2,\ldots,k \; ; \; i=1,2,3$

where explicitly :

(8) $$\frac{\partial^k}{\partial \nu^k} P_{4n+1}[f](M) = \sum_{i=1}^{3}\sum_{j=0}^{2n} \frac{(2n+j)!}{(2n)!j!} \sum_{p\leqslant k}\sum_{q\leqslant k-p} \binom{k}{p}\binom{k-p}{q} \frac{(2n+1)!}{(2n+1-p)!} \frac{j!(-1)^j}{(j-q)!} \times$$
$$\times \lambda_i^{2n+1-j} (\lambda_i-1)^{j-q} \left(\frac{\partial\lambda_i}{\partial\nu}\right)^{p+q} T^{2n-j-(k-p-q)}_{A_i} \frac{\partial^{k-p-q} f}{\partial\nu^{k-p-q}}(M)$$

$$B^{k,\nu}_{pq}(M) = \sum_{i\leqslant k}\sum_{j\leqslant i} \binom{k}{i}\binom{i}{j} \frac{(p+r+1)!\,(q+p+1)!\,(q+r+1)!}{(p+r+1-k+i)!(q+p+1-i+j)!(q+r+1-j)!} \times$$
$$\times\left(\frac{\partial\lambda_1}{\partial\nu}\right)^{k-i} \left(\frac{\partial\lambda_2}{\partial\nu}\right)^{i-j} \left(\frac{\partial\lambda_3}{\partial\nu}\right)^{j} \lambda_1^{p+r+1-k+i}\,\lambda_2^{p+q+1-i+j}\,\lambda_3^{q+r+1-j}$$

$$B^{\alpha}_{pq}(M) = \sum_{\beta \leqslant \alpha} \sum_{\gamma \leqslant \beta} \binom{\alpha}{\beta}\binom{\beta}{\gamma} \frac{(p+r+1)!(q+p+1)!(q+r+1)!}{(p+r+1-|\alpha-\beta|)!(q+p+1-|\beta-\gamma|)!(q+r+1-|\gamma|)!} \times$$

$$\times \nabla_1^{\alpha-\beta} \nabla_2^{\beta-\gamma} \nabla_3^{\gamma} \lambda_1^{p+r+1-|\alpha-\beta|} \lambda_2^{q+p+1-|\beta-\gamma|} \lambda_3^{q+r+1-|\gamma|}$$

* (7) is a linear system of n(2n+1) equations for n(2n+1) unknows
* the matrix is sparse : many of the points (all the Q^k_{ri}) lies on the boundary of the triangle, and the λ_i equal zero, hence the corresponding $B_{pq}(Q^k_{ri})$. For example :

→ n=1 : 3 constants α_{00}, α_{10}, α_{01} ; the matrix is :

$$\frac{1}{16} \begin{bmatrix} 0 & 0 & \frac{\partial\lambda_1}{\partial\eta_1} \\ \frac{\partial\lambda_2}{\partial\eta_2} & 0 & 0 \\ 0 & \frac{\partial\lambda_3}{\partial\eta_3} & 0 \end{bmatrix}$$

(Z_{4n+1} is a polynomial of degree 5)

→ n=2 : 10 constants ; polynomial Z_{4n+1} of degree 9 .

matrix of the form :

$$\begin{bmatrix} 0&0&0&0&0&0&0&0&0&\times \\ \times&0&0&0&0&0&0&0&0&0 \\ 0&0&0&0&0&0&\times&0&0&0 \\ 0&0&0&0&0&\times&0&0&\times&\times \\ 0&0&0&0&0&\times&0&0&\times&\times \\ \times&\times&\times&0&0&0&0&0&0&0 \\ \times&\times&\times&0&0&0&0&0&0&0 \\ 0&0&0&\times&0&0&\times&\times&0&0 \\ 0&0&0&\times&0&0&\times&\times&0&0 \\ \times&\times&\times&\times&\times&\times&\times&\times&\times&\times \end{bmatrix}$$

(each × is a non zero coefficient).

<u>Bounds for error interpolation</u> :

Let h be the diameter of T, ρ the diameter of the inscribed sphere

$$|f|_k = \sup_{|\beta|=k} ||D^{\alpha} f||_{\infty} = \sup_{|\beta|=k} \max_{M \in T} |D^{\alpha} f(M)|$$

Using a result of Meinguet [6] and expliciting the solution of (7) by Cramer's rule, we obtain :

Proposition : *If* $\frac{h}{\rho}$ < *constant, if* $|\alpha| \leqslant 2n$, *there exists a constant* $K(T,\alpha,n)$ *such that, for all* $f \in \mathcal{C}^{4n+1}(T)$

$$||D^{\alpha}f - D^{\alpha}Z_{4n+1}[f]||_{\infty} \leqslant h^{4n+1-|\alpha|} K(T,\alpha,n) \; |f|_{4n+1} \quad \textit{for all } f \in H^{4n+1}$$

and a constant $K'(T,\alpha,n)$ *such that*

$$||D^{\alpha}f - D^{\alpha}Z_{4n+1}[f]||_{L^2(T)} \leqslant h^{4n+1-|\alpha|} K'(T,\alpha,n) \; |f|_{4n+1}$$

<u>Examples</u> :

→ n=1

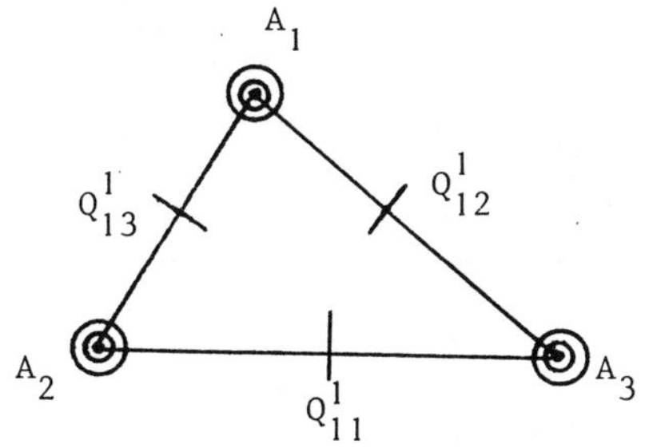

Argyris triangle polynomial of degree 5

Class $\mathcal{C}^1$

→ n=2

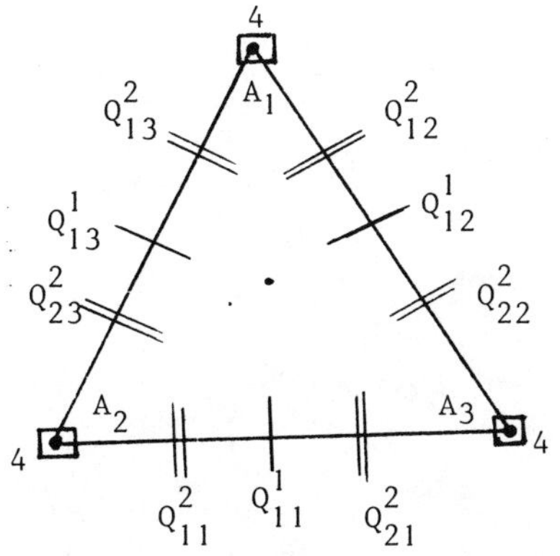

Polynomial of degree 9

Class $\mathcal{C}^2$

→ n=3

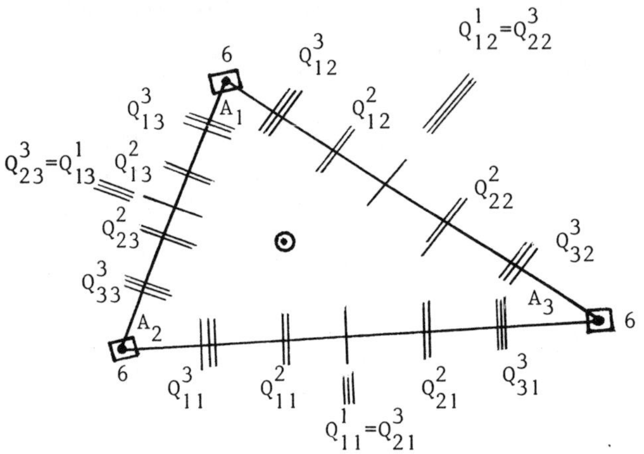

Polynomial of degree 13

Class $\mathcal{C}^3$

Remark : It is interesting to compare the number of degrees of freedom of the element and the order of the system (7) we have in fact to solve

n	degree of Z	degrees of freedom	order of Z
1	5	21	3
2	9	55	10
3	13	105	21

b) triangles of type B : (generalize Bell's element) [1]

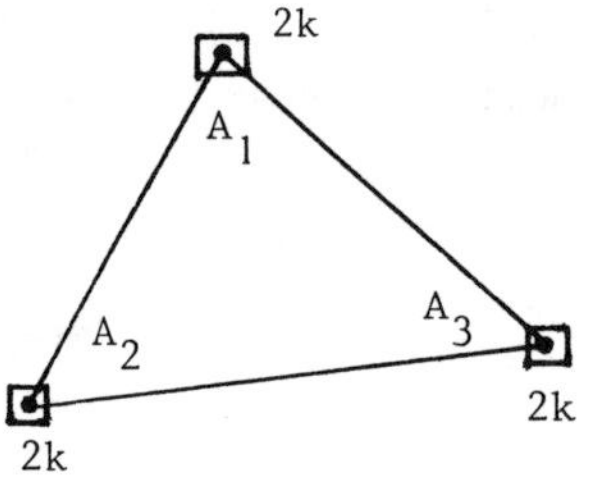

Problem 3 : Construct a reduced element, using only the values on A_1, A_2, A_3 .

The problem is not (4n+1) unisolvent but we have :

Proposition : The interpolation problem

$$D^\alpha p(A_i) = D^\alpha f(A_i) \qquad |\alpha| \leqslant 2n \; ; \; i=1,2,3 \quad \textit{is } \Pi'_{4n+1}(T)\textit{-unisol-}$$

vent, where $\Pi'_{4n+1}(T)$ is defined as the space of all polynomials of degree

4n+1 *on* T *such that :*

(i) $$\frac{\partial^q}{\partial \nu_k^q} p(M) = \sum_{\{\substack{i=k+1 \\ i=k-1}} \lambda_i^{2n-q+1} \sum_{j=0}^{2n-q} \frac{(2n-q+j)!}{(2n-q)!j!} (1-\lambda_i)^j \times T_{A_i}^{2n-j-q} \frac{\partial^q p}{\partial \nu_k^q}(M)$$

for all $M \in A_{k+1} A_{k-1}$; $k=1,2,3$; $q \leqslant n$

(ii) $D^\alpha p(A_o) = D^\alpha P_{4n+1}(A_o)$ $|\alpha| \leqslant n-2$

(i) means that $\frac{\partial^k p}{\partial \nu^k}$ is a polynomial of degree 4n+1-2k (in one variable) on each side of the triangle T.

c) triangles of type C : (generalize Zienkiewicz's element) (or Hermite triangle of type (2p+1)).
This element is only of class $\mathscr{C}^o$.

Problem 4 : Construct a polynomial p that realizes the following conditions :

(i) $D^\alpha p(A_i) = D^\alpha f(A_i)$ $|\alpha| \leqslant k$; $i=1,2,3$.

(ii) $D^\beta p(A_o) = D^\beta f(A_o)$ $|\alpha| \leqslant k-1$

where A_o *is a strictly interior (or strictly exterior) point of T.*

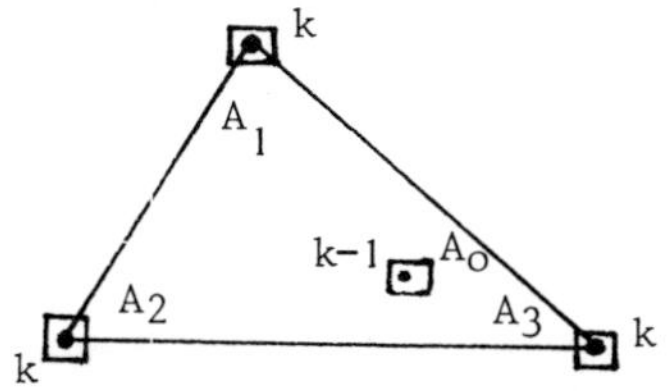

Proposition [5] *Problem 4 is (2k+1)-unisolvent.*

III. TRIANGULATION OF THE DOMAIN Ω : (CORREC [3])

Problem 5 : Given a set of randomly selected points in Ω , t_i, $i=1,\ldots,N$, *generate a triangulation* τ *that cover* Ω *such that the* t_i*'s are the vectices of the triangle T of* τ .

The idea of the algorithm originates from Saw and Smith [7] : startting from the boundary of Ω, where we number the points t_i, we choose for a "running basis" the side (j_i) where j is the last point of the list and i the first. One search $k \in \{1,\ldots,N\}$, $k \neq i$, $k \neq j$, such that the triangle (ijk) is as regular as possible and do not introduce any trouble for the future. When (ijk) is created, the list of the points on the boundary is set up, adding or substracting one point. One stop when the boundary and the free domain inside is reduced to a segment.

Optimisation of the triangulation : It is necessary to eliminate those triangles that are too flat. On each quadrilateral, we use the circle criterion (LAWSON). This criterion may be included in the creation of the triangulation, examinating then the quadrilateral having as a diagonal "the running basis".

IV. EXAMPLES :

To illustrate the behaviour of the method, we have chosen two examples : the first come from Lawson [4], the second from Utreras [8].

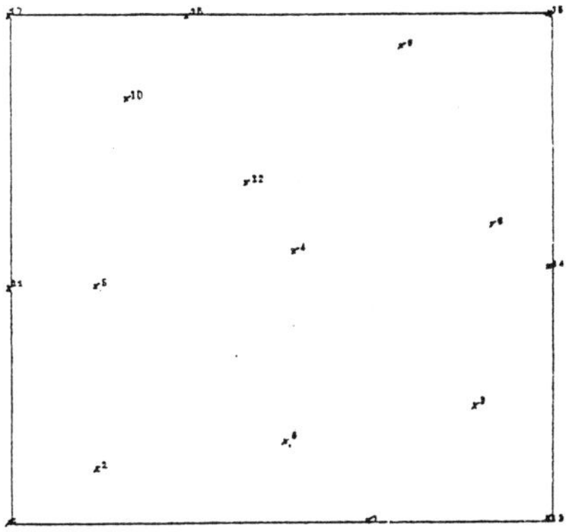

Figure 1.

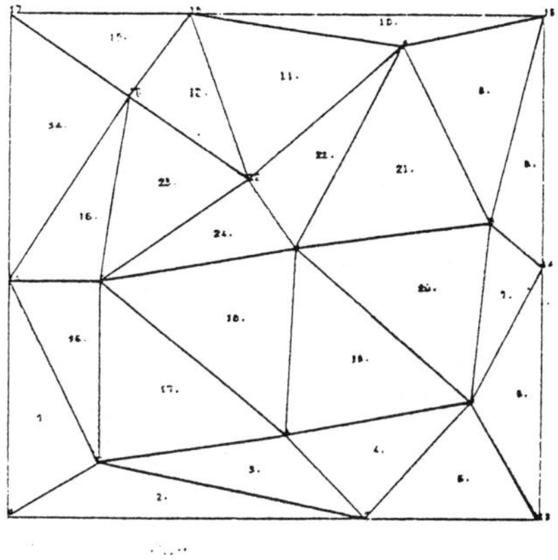

Figure 2.

Figure 1 show a set S consisting of 17 points in the plane (included the vertices of the square window where the surface is build).

Figure 2 is the result of an automatic triangulation constructed on S.

Figures 3.1, 3.2, 4.1, 4.2 show the results of interpolation in some cases, in perpective plot.

The run-time is about $\frac{N}{20}$ seconds, where N is the number of points in S.

On the drawings, one cannot see any difference when constructing triangles of type A or B, for $\mathcal{C}^k$ continuity, $k > 1$.

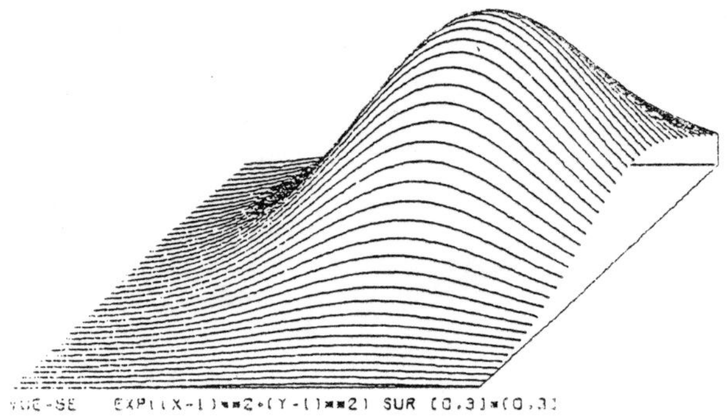

a) the test function
$f(x,y) = \exp((x-1)^2+(y-1)^2)$

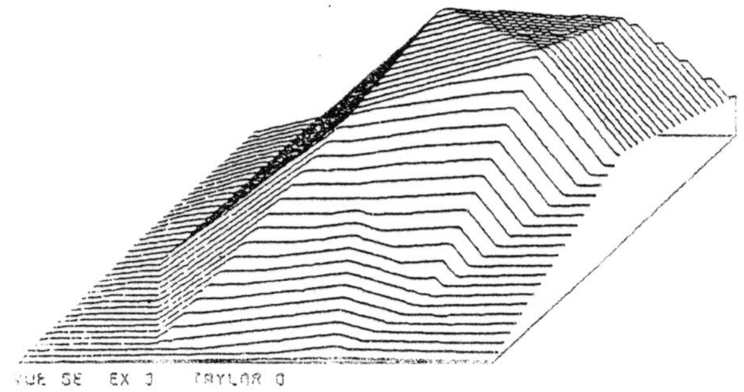

b) Lagrange interpolation
(Class $\mathcal{C}^0$)

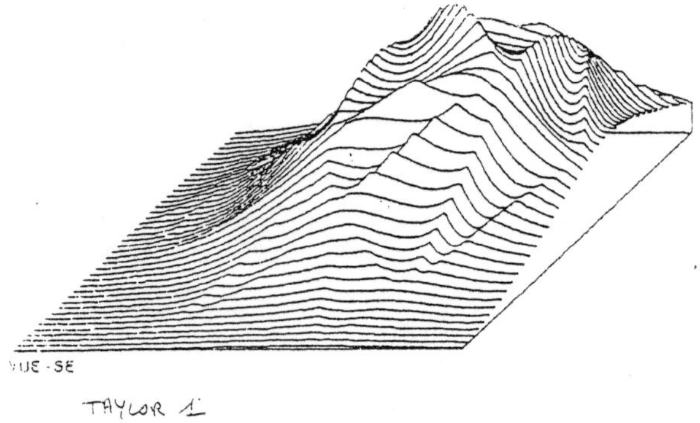

c) Taylor interpolation
of order 1 on S.

Figure 3.1.

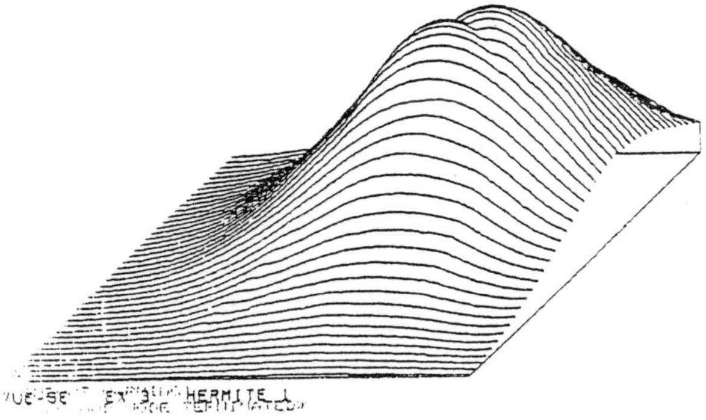

d) triangles of type C
Class $\mathcal{C}^0$

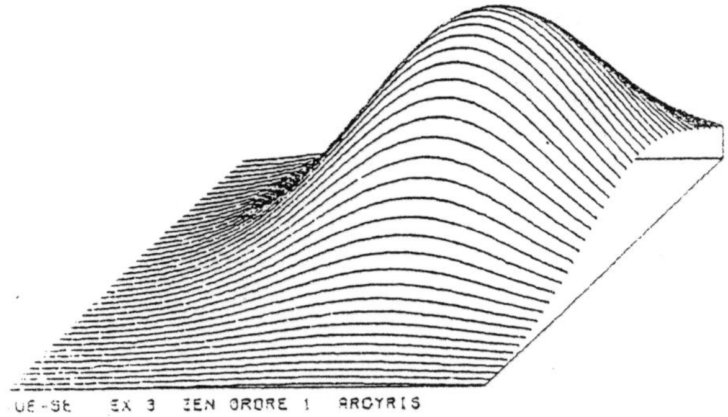

e) Triangles of type A
Class $\mathcal{C}^1$

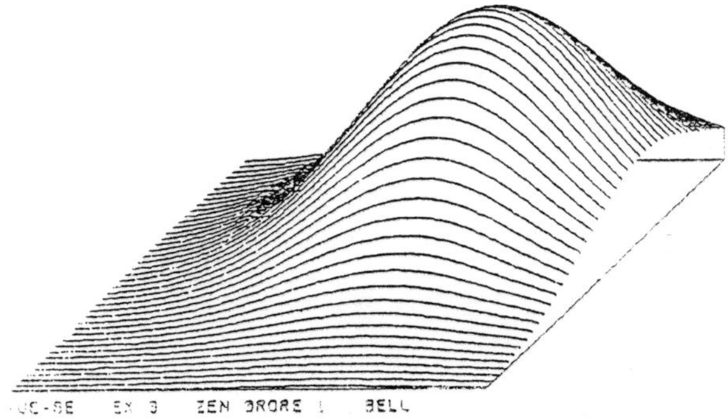

f) triangles of type B
class $\mathcal{C}^1$

Figure 3.2.

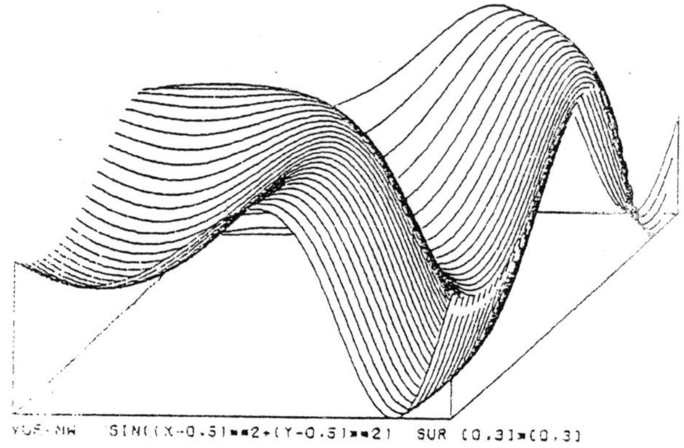

a) the test function

$$f(x,y)=\sin((x-\frac{1}{2})^2+(y-\frac{1}{2})^2)$$

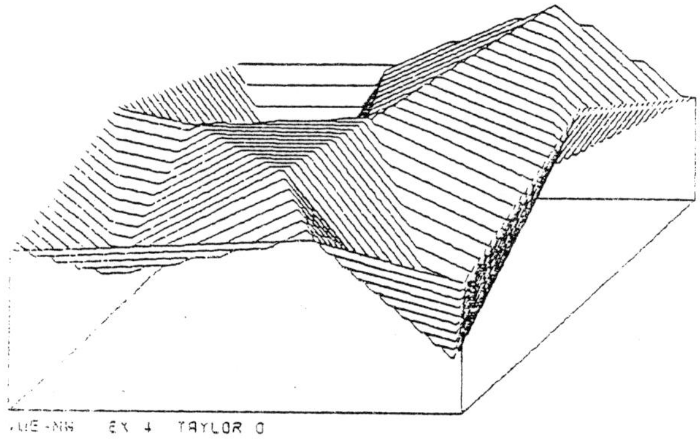

b) Lagrange interpolation, on S.

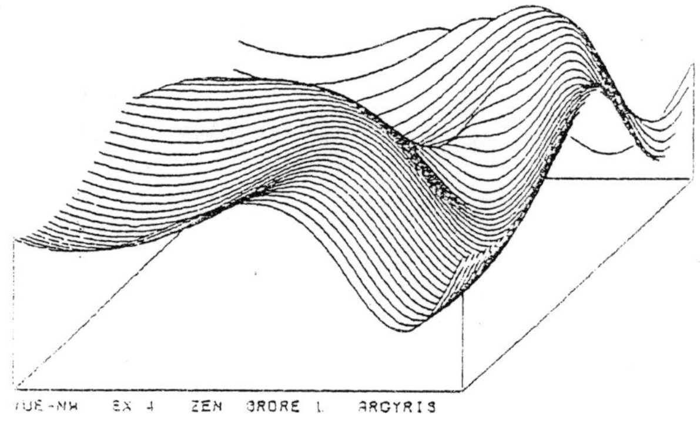

c) triangles of type A, on S

Figure 4.1.

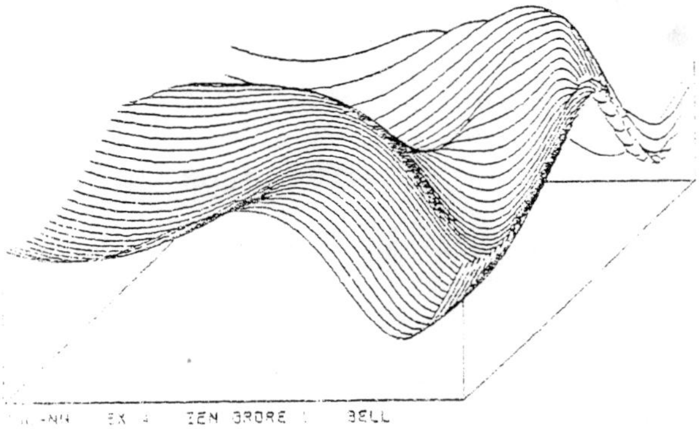

d) triangles of type B, on S.

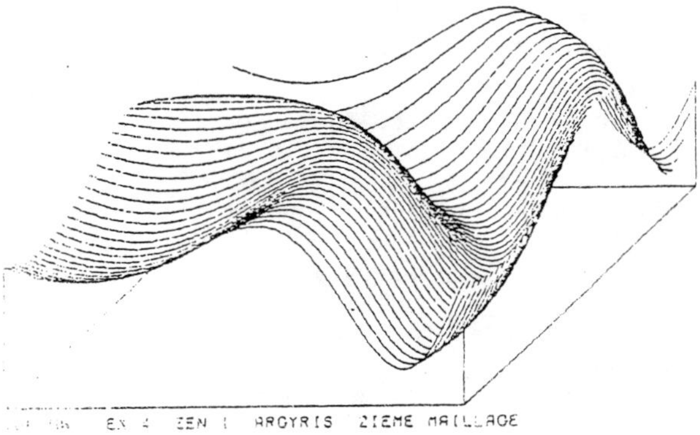

e) triangles of type A, on Σ a new set of 45 points in the square window

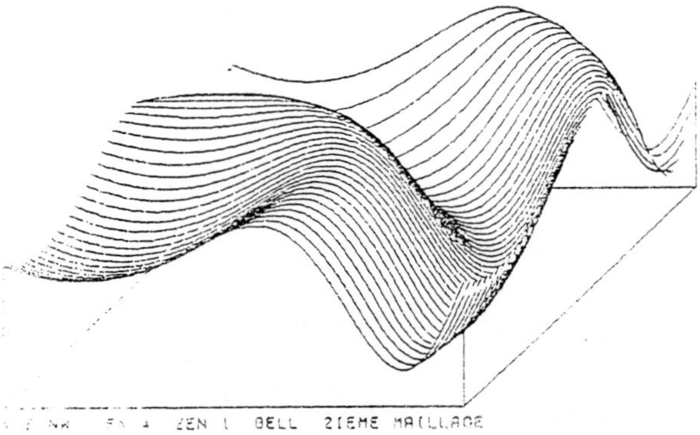

f) triangles of type B on Σ

Figure 4.2.

REFERENCES

[1] K.BELL A refined triangular plate bending element
Int. J. Num. Math. Eng 1 (1969)

[2] P.G. CIARLET The finite element method for elliptic problems
North Holland (1978)

[3] Y. CORREC Génération automatique d'une triangulation sur des points donnés
Note d'étude L.A.N. 49. CCSA. CELAR (1980)

[4] C.L. LAWSON Software for C^1 interpolation
in : Mathematical Software.
J.RICE Ed. Academic Press (1977)

[5] A. LE MEHAUTE Taylor interpolation of order n at the vertices of a triangle
in : Approximation theory and applications,
Z.ZIEGLER ed. Academic Press (1981)

[6] J. MEINGUET Structure et estimations de coefficients d'erreurs
Journées éléments finis. Université de Rennes 1977
et RAIRO, Anal. Num. Vol 11, p.355-368 (1977)

[7] C.B. SAW, R.G. SMITH Automatic Nodal triangulation for finite elements.
Computer Aided design. Vol 5, n°1 (1973)

[8] F. UTRERAS Cross validation techniques for smoothing spline functions in one or two dimensions.
in : Smoothing techniques for a curve estimation.
TH. GASSER, M. ROSENBLATT ed, Lecture notes in Math, 757,
Springer Verlag (1979)

[9] A. ŽENIŠEK Interpolation polynomials on the triangle
Numer. Math. 15 (1970)

Alain LE MEHAUTE
Laboratoire d'Analyse Numérique
Institut National des Sciences Appliquées
20, avenue des Buttes de Coësmes
35043 RENNES CEDEX - France

MINIMAL PROJECTIONS AND NEAR-BEST APPROXIMATIONS BY MULTIVARIATE POLYNOMIAL EXPANSION AND INTERPOLATION

J.C. Mason

Mathematics Branch, Royal Military College of Science,
Shrivenham, Swindon, Wilts., England.

1. Introduction

If a multivariate function f in a space X is approximated by a function f^N in a subspace Y, then f^N is near-best within a relative distance ρ according to the definition of MASON [1] if

$$\|f - f^N\| \leqslant (1 + \rho).\|f - f^B\| \tag{1}$$

where f^B is a best approximation and $\| \cdot \|$ is a chosen norm on X. Now for any projection P of f in X to Pf in Y (i.e. a bounded, linear, idempotent mapping of X into Y), it follows (see [2]) that

$$\|f - Pf\| \leqslant (1 + \|P\|).\|f - f^B\|. \tag{2}$$

Hence Pf is automatically a near-best approximation, and $\|P\|$ measures the relative distance from a best approximation. A projection for which $\|P\|$ is smallest is termed a "minimal projection". Note that "near-best" is not a relevant term unless ρ in (1) or $\|P\|$ in (2) is suitably small.

The present paper has two aspects. Firstly, a survey is given of near-best approximations and minimal projections (in both real and complex variables) based on classical series and related interpolation criteria. Secondly, relevant results are generalised to multivariate functions.

The discussion covers Fourier, Taylor, Laurent, and Chebyshev series; Chebyshev (L_∞) and L_1 norms; algebraic and trigonometric polynomials; and hypercubes (in real variables), poly-discs, poly-rings, and poly-ellipses (in complex variables). One theorem only is proved in each section of the paper, since proofs of the remaining theorems may easily be constructed by similar generalisations.

The following vector notation is used for multivariate functions. A set of N independent real or complex variables is denoted by

$$\underline{x} = (x_1, x_2, \ldots, x_N) \text{ or } \underline{z} = (z_1, z_2, \ldots, z_N), \text{ respectively, and}$$

integer vectors

$$\underline{k} = (k_1, k_2, \ldots, k_N),\ \underline{m} = (m_1, m_2, \ldots, m_N),\ \underline{n} = (n_1, n_2, \ldots, n_N)$$

are used for an index, degree negative, and degree positive, respectively. This permits the following shorthand notation

$$\sum_{\underline{k}=-\underline{m}}^{\underline{n}} \equiv \sum_{k_1=-m_1}^{n_1} \cdots \sum_{k_N=-m_N}^{n_N} \; ; \quad \prod_{\underline{k}=-\underline{m}}^{\underline{n}} \equiv \prod_{k_1=-m_1}^{n_1} \cdots \prod_{k_N=-m_N}^{n_N} .$$

Terms of the dashed sum $\sum_{\underline{k}=\underline{0}}^{\underline{n}}{}'$ include a factor $\frac{1}{2}$ for <u>every</u> zero k_j.

Define $\underline{1} = (1, 1, \ldots, 1)$ and $\underline{z}^{-1} = (z_1^{-1}, \ldots, z_N^{-1})$,

and write multiple differentials as

$$\underline{dx} \equiv dx_1\, dx_2 \ldots dx_N; \quad \underline{dz} \equiv dz_1\, dz_2 \ldots dz_N.$$

2. Spaces

2.1. Real Variables

Define hypercubes: $X = \{-1 \leqslant x_j \leqslant 1,\ (j = 1, \ldots, N)\}$, $X^0 = \{0 \leqslant \theta_j \leqslant 2\pi\}$.

Define spaces:

- $C(X)$: continuous functions on X
- $L_1(X)$: L_1-integrable functions on X
- $C^0(X^0)$: continuous functions 2π-periodic in θ_j
- $L_1^0(X^0)$: L_1-integrable functions 2π-periodic in θ_j
- $\Pi_{\underline{n}}$: polynomials of degree $\underline{n}$ in $\underline{x}$
- $\Pi_{\underline{n}}^0$: trigonometric polynomials of order $\underline{n}$ in $\underline{\theta}$.

Define norms $\|f\|_\infty = \max_X |f|$, $\|f\|_1 = \int_X |f|\, \underline{dx}$ or $\int_{X^0} |f|\, \underline{d\theta}$

2.2. Complex Variables

Define contours: poly-circle C_ρ: $\{|z_j| = \rho_j > 1\}$ $(j = 1, \ldots, N)$

poly-ellipse ξ_ρ: $\{|z_j + \sqrt{(z_j^2 - 1)}| = \rho_j > 1\}$ $(j = 1, \ldots, N)$

and, denoting an interior by I(.),

define domains: poly-disc $D = I(C_\rho)$,

poly-ellipse $E = I(\xi_\rho)$,

poly-ring $R = \{\rho_j^{-1} < |z_j| < \rho_j\}$

Define spaces: A(D), A(E), A(R): functions continuous on and analytic in D,E,R, respectively,

$\Pi_{\underline{n}}^{+}$: polynomials of degree $\underline{n}$ in $\underline{z}$,

$\Pi_{\underline{m},\underline{n}}^{+}$: polynomials of degree $\underline{m}$ in $\underline{z}^{-1}$ and $\underline{n}$ in $\underline{z}$.

Define norms: $\|f\|_\infty = \max_D |f| = \max_{C_\rho} |f|$

$$\|f\|_1^{C_\rho} = \int_{C_\rho} |f|\ |\underline{dz}| \quad \text{(contour norm)}$$

$$\|f\|_1^{D} = \int_{D} W(\underline{z})\ |f|\ \underline{dS} \quad \text{(domain norm)}$$

- weight W

(Norms are here defined on D, and analogous definitions apply on E, R). The contour C_r: $\{|z_j| = r_j\}$ is used to denote a general poly-circle within D or R, and ξ_r: $\{|z_j + \sqrt{(z_j^2 - 1)}| = r_j\}$ is used similarly in E. For a discussion of multivariate analytic functions, see [3]. (C_1 denotes $\{|z_j| = 1$, all $j\}$.)

3. Projections from Classical Series

Formulae for partial sums of classical series are summarised below. Full details of the Fourier and Chebyshev (real) projections are given in [4] and discussions of the other projections are given for one variable in [5], [6], [7], and [8].

(i) Fourier Projection: $F_{\underline{n}}$ (Order $\underline{n}$ in $\underline{\theta}$)

$$(F_{\underline{n}} f)(\underline{\theta}) = \sum_{\underline{k}=-\underline{n}}^{\underline{n}} a_{\underline{k}}\, e^{i\underline{k}.\underline{\theta}} \tag{3}$$

$$\text{where} \quad a_{\underline{k}} = (2\pi)^{-N} \int_{X^0} f(\underline{u})\, e^{-i\underline{k}.\underline{u}}\, \underline{du} \tag{4}$$

(ii) Taylor Projection: $S_{\underline{n}}$ (Degree $\underline{n}$ in $\underline{z}$)

$$(S_{\underline{n}} f)(\underline{z}) = \sum_{\underline{k}=\underline{0}}^{\underline{n}} c_{\underline{k}} z_1^{k_1} \cdots z_N^{k_N} \tag{5}$$

$$\text{where } c_{\underline{k}} = (2\pi i)^{-N} \int_{C_r} \frac{f(\underline{t})}{(t_1)^{k_1+1} \cdots (t_N)^{k_N+1}} \, \underline{dt} \quad (0 \leqslant r_j \leqslant \rho_j) \tag{6}$$

(iii) Laurent Projection: $B_{\underline{m},\underline{n}}$ (Degrees $\underline{m}$ in $\underline{z}^{-1}$, $\underline{n}$ in $\underline{z}$)

$$(B_{\underline{m},\underline{n}} f)(\underline{z}) = \sum_{\underline{k}=-\underline{m}}^{\underline{n}} c_{\underline{k}} z_1^{k_1} \cdots z_N^{k_N} \tag{7}$$

$$\text{where } c_{\underline{k}} = (2\pi i)^{-N} \int_{C_r} \frac{f(\underline{t})}{(t_1)^{k_1+1} \cdots (t_N)^{k_N+1}} \, \underline{dt} \quad (\rho_j^{-1} \leqslant r_j \leqslant \rho_j) \tag{8}$$

(iv) Real Chebyshev 1st Kind Projection: $G_{\underline{n}}$ (Degree $\underline{n}$ in $\underline{x}$)

$$(G_{\underline{n}} g)(\underline{x}) = (F_{\underline{n}} f)(\underline{\theta}) = \sum_{\underline{k}=\underline{0}}^{\underline{n}} \hat{a}_{\underline{k}} T_{k_1}(x_1) \cdots T_{k_N}(x_N), \tag{9}$$

where $\hat{a}_k$ are cosine coefficients in $F_{\underline{n}} f$, $T_{k_j}(x_j) = \cos k_j \theta_j$,

and $f(\underline{\theta}) = g(\underline{x})$ for $x_j = \cos \theta_j$ $(j = 1, \ldots, N)$. (10)

Hence $\|G_{\underline{n}}\|_\infty \leqslant \|F_{\underline{n}}\|_\infty$ (11)

(v) Complex Chebyshev 1st Kind Projection: $G_{\underline{n}}^+$ (Degree $\underline{n}$ in $\underline{w}$)

$$(G_{\underline{n}}^+ g)(\underline{w}) = 2^{-N} (B_{\underline{n},\underline{n}} f)(\underline{z}) = {\sum_{\underline{k}=\underline{0}}^{\underline{n}}}' c_{\underline{k}} T_{k_1}(w_1) \cdots T_{k_N}(w_n) \tag{12}$$

where $c_{\underline{k}}$ are given by (8) (for $1 < r_j \leqslant \rho_j$),

$f(\underline{z}) = 2^N g(\underline{w})$ for $w_j = \frac{1}{2}(z_j + z_j^{-1})$ (13)

and $T_{k_j}(w_j) = \frac{1}{2}\left[(z_j)^{k_j} + (z_j)^{-k_j}\right]$.

As before $\|G_n^+\|_\infty \leqslant \|B_{n,n}\|_\infty$ (14)

(vi) Real Chebyshev 2nd Kind Projection: $H_{\underline{n}-\underline{1}}$ (Degree $\underline{n} - \underline{1}$ in $\underline{x}$)

$$(H_{\underline{n}-\underline{1}} h)(\underline{x}) = (F_{\underline{n}} f)(\underline{\theta}) / (\sin \theta_1 \sin \theta_2 \cdots \sin \theta_N)$$

$$= \sum_{\underline{k}=\underline{1}}^{\underline{n}} \tilde{a}_{\underline{k}} U_{k_1-1}(x_1) \cdots U_{k_N-1}(x_N) . \tag{15}$$

$\tilde{a}_{\underline{k}}$ are sine coefficients in $F_{\underline{n}} f$, $U_{k_j-1}(x_j) = \sin(k_j\theta_j)/\sin\theta_j$, and $f(\underline{\theta}) = \sin\theta_1 \sin\theta_2 \ldots \sin\theta_N\, h(\underline{x})$ for $x_j = \cos\theta_j$. (16)

Since $\underline{dx} = (-1)^N \sin\theta_1 \sin\theta_2 \ldots \sin\theta_N\, \underline{d\theta}$,

$$\|h\|_1 = \int_X |h|\, \underline{dx} = \int_{X^0} |f|\, \underline{d\theta} = \|f\|_1 \text{ from (16)},$$

and $\|H_{\underline{n}-\underline{1}}h\|_1 = \|F_{\underline{n}}f\|_1$ similarly from (15).

Hence $\|H_{\underline{n}-\underline{1}}\|_1 \leqslant \|F_{\underline{n}}\|_1$ (17)

(vii) <u>Complex Chebyshev 2nd Kind Projection:</u> $H^+_{\underline{n}-\underline{1}}$ (Degree $\underline{n} - \underline{1}$ in $\underline{w}$)

$$(H^+_{\underline{n}-\underline{1}}h)(\underline{w}) = (B_{\underline{n},\underline{n}}f)(\underline{z}) \Big/ \prod_{j=1}^{N} (z_j - z_j^{-1})$$

$$= \sum_{\underline{k}=\underline{1}}^{\underline{n}} c_{\underline{k}}\, U_{k_1-1}(w_1) \ldots U_{k_N-1}(w_N) \qquad (18)$$

where $c_{\underline{k}}$ are given by (8) (for $1 < r_j \leqslant \rho_j$),

$$f(\underline{z}) = \prod_{j=1}^{N} (z_j - z_j^{-1}) \,.\, h(\underline{w}) \text{ for } w_j = \tfrac{1}{2}(z_j + z_j^{-1}), \qquad (19)$$

and $U_{k_j-1}(w_j) = \left[(z_j)^{k_j} - (z_j)^{-k_j}\right] / (z_j - z_j^{-1})$.

Since $\underline{dw} = \prod [z_j^{-1}.\tfrac{1}{2}(z_j - z_j^{-1})].\underline{dz}$, (19) gives

$$\|h\|_1^{\xi_r} = \int_{\xi_r} |h|\, |\underline{dw}| = \prod_j (2r_j)^{-1}. \int_{C_r} |f|.|\underline{dz}| =$$

$$= \prod_j (2r_j)^{-1}.\|f\|_1^{C_r};$$

and $\|H^+_{\underline{n}-\underline{1}}h\|_1^{\xi_r} = \prod_j (2r_j)^{-1}.\|B_{n,n}f\|_1^{C_r}$ similarly from (18).

Hence $\|H^+_{\underline{n}-\underline{1}}\|_1^{\xi_r} \leqslant \|B_{\underline{n},\underline{n}}\|_1^{C_r}$ (20)

4. Minimal L_∞ Projections by Series

<u>Theorem 4.1</u> $F_{\underline{n}}$ is a minimal L_∞ projection from $C^0(X^0)$ to $\Pi^0_{\underline{n}}$

<u>Theorem 4.2</u> $S_{\underline{n}}$ is a minimal L_∞ projection from $A(D)$ to $\Pi^+_{\underline{n}}$

<u>Theorem 4.3</u> $B_{\underline{m},\underline{n}}$ is a minimal L_∞ projection from $A(R)$ to $\Pi^+_{\underline{m},\underline{n}}$

These are generalisations of results of LOZINSKI [9] (for a proof see [10]), GEDDES and MASON [5], and MASON [6], all of which are based on kernel formulae of a type introduced by BERMAN [11]. For example:

Lemma Let P be any projection of A(R) to $\Pi^{+}_{\underline{m},\underline{n}}$, and define the shift $E_{\underline{t}}$ and inverse shift $E_{\underline{t}*}$, for $\underline{t} = (t_1, t_2, \ldots, t_N)$, by

$$(E_{\underline{t}}f)(\underline{z}) = f(t_1z_1, \ldots, t_Nz_N),\quad (E_{\underline{t}*}f)(\underline{z}) = f(t_1^{-1}z_1, \ldots, t_N^{-1}z_N)$$

Then $B_{\underline{m},\underline{n}} = \psi$, where

$$(\psi f)(\underline{z}) = \left(\frac{1}{2\pi i}\right)^N \int_{C_1} (E_{\underline{t}*}\ P\ E_{\underline{t}}f)(\underline{z})\ \frac{dt_1}{t_1} \cdots \frac{dt_N}{t_N} \tag{21}$$

Proof of 4.3 From the Lemma, since $\|E_{\underline{t}}\| = \|E_{\underline{t}*}\| = 1$,

$$\|\psi f\| \leqslant (2\pi)^{-N} \int_{C_1} \|E_{\underline{t}*}\| \ .\ \|P\|\ .\ \|E_{\underline{t}}\|\ \|f\|\ \underline{dt} = \|P\|\ .\ \|f\|$$

Thus $\|B_{\underline{m},\underline{n}}\| \leqslant \|P\|$, and the result follows.

Proof of Lemma It is sufficient to consider, for all k_j, $f = \Pi(z_j)^{k_j}$

(i) For $-m_j \leqslant k_j \leqslant n_j$, $f \in \Pi^{+}_{\underline{m},\underline{n}}$, so that $B_{\underline{m},\underline{n}}f = f$. Also $E_{\underline{t}*}\ P\ E_{\underline{t}}f = E_{\underline{t}*}(E_{\underline{t}}f) = f$, since $E_{\underline{t}}f \in \Pi^{+}_{\underline{m},\underline{n}}$.

From (21), $(\psi f)(\underline{z}) = (2\pi i)^{-N} f(\underline{z}). \prod \int_{C_1} \frac{dt_j}{t_j} = f(\underline{z})$

Thus $B_{\underline{m},\underline{n}}f = \psi f$

(ii) For $k_N > n_N$ or $k_N < -m_N$ (i.e. at least one index outside $-m_j \leqslant k_j \leqslant n_j$).

Here $(B_{\underline{m},\underline{n}}f)(\underline{z}) = 0$.

Suppose $(Pf)(\underline{w}) = \sum_{\underline{j}=-\underline{m}}^{\underline{n}} b_{\underline{j}}\ (w_1)^{j_1} \ldots (w_N)^{j_N}$ (for some $b_{\underline{j}}$)

Then (c.f. [6]) the innermost integral in the multiple integral (21) defining $(\psi f)(\underline{z})$ is

$$\int_{|t_N|=1} (t_N)^{k_N} \prod_{j_N=-m_N}^{n_N} b_{\underline{j}}(z_N/t_N)^{j_N}\ .\ t_N^{-1}\ dt_N$$

The integrand is either a polynomial in t_N $(k_N > n_N)$ or the product of t_N^{-2} and a polynomial in t_N^{-1} $(k_N < -m_N)$. Thus the complete integral, and hence $(\psi f)(\underline{z})$, is zero. Hence $B_{\underline{m},\underline{n}} f = \psi f$ again. Q.E.D.

5. Near-best L_∞ Approximations by Series

The approximations $F_{\underline{n}}f$, $S_{\underline{n}}f$, $B_{\underline{n},\underline{n}}f$, $G_{\underline{n}}g$ defined in §3 are all near-best L_∞ approximations within relative distances asymptotically proportional to $\prod \log n_j$.

<u>Theorem 5.1</u> $\|F_{\underline{n}}\|_\infty = \|G_{\underline{n}}\|_\infty = \Lambda_{\underline{n}} = \prod_{j=1}^{N} \lambda_{n_j}$

where $\lambda_p = \frac{1}{\pi}\int_0^\pi \frac{|\sin(p+\frac{1}{2})x|}{\sin\frac{1}{2}x}\,dx \sim \frac{4}{\pi^2}\log p.$

<u>Theorem 5.2</u> $\|S_{\underline{n}}\|_\infty = \prod_{j=1}^{N} \gamma_{n_j}$

where $\gamma_p = 1 + \left(\frac{1}{2}\right)^2 + \left(\frac{1.3}{2.4}\right)^2 + \ldots + \left[\frac{1.3\ \ldots\ (2p-1)}{2.4\ \ldots\ 2p}\right]^2 \sim \frac{1}{\pi}\log p.$

<u>Theorem 5.3</u> $\|B_{\underline{n},\underline{n}}\|_\infty \leqslant \Lambda_{\underline{n}}$, $\|G^+_{\underline{n}}\|_\infty \leqslant \Lambda_{\underline{n}}$

These are multivariate generalisations of results due, respectively, to FEJER [12], LANDAU (see [13]), and MASON [6] and GEDDES [14].
The inequalities (11) and (14) enable the results for $G_{\underline{n}}$ and $G^+_{\underline{n}}$ to be deduced.
For example:

<u>Proof of 5.1</u> From (3) and (4)

$$(F_{\underline{n}}f)(\underline{\theta}) = (2\pi)^{-N}\int_{X^0} f(\underline{\theta}+\underline{t}) \prod_{j=1}^{N} \frac{\sin(n_j+\frac{1}{2})t_j}{\sin\frac{1}{2}t_j}\,\underline{dt}$$

Thus $$\|F_{\underline{n}}f\|_\infty \leqslant \|f\|_\infty \cdot \prod_{j=1}^{N}\left[\pi^{-1}\int_0^\pi \frac{|\sin(n_j+\frac{1}{2})t_j|}{\sin\frac{1}{2}t_j}\,dt_j\right]$$
$$= \|f\|_\infty \cdot \Lambda_{\underline{n}}$$

This bound is attained (at $\underline{\theta} = \underline{0}$) for $f(\underline{\theta})$ arbitrarily close to the function

$$f^*(\underline{\theta}) = \operatorname{sgn}\prod_{j=1}^{N}\frac{\sin(n_j+\frac{1}{2})\theta_j}{\sin\frac{1}{2}\theta_j}\,. \tag{22}$$

Hence $\|F_{\underline{n}}\|_\infty = \Lambda_{\underline{n}}$.

Also f^* is a function of $\cos\theta_j$ and hence

$$f^*(\underline{\theta}) = g^*(\underline{x}) \quad \text{for } x_j = \cos\theta_j.$$

Now, from (9), (10), (11) above

$$\|G_{\underline{n}}g\|_\infty \leqslant \|g\|_\infty \cdot \Lambda_{\underline{n}}.$$

Choosing g arbitrarily close to g^*, the bound is attained.

Hence $\|G_{\underline{n}}\|_\infty = \Lambda_{\underline{n}}$. Q.E.D.

6. Minimal L_1 Projections by Series

Theorem 6.1 $F_{\underline{n}}$ is a minimal L_1 projection from $L_1^0(X^0)$ to $\Pi_{\underline{n}}^0$

Theorem 6.2 $S_{\underline{n}}$ is a minimal L_1 projection from A(D) to $\Pi_{\underline{n}}^+$

Theorem 6.3 $B_{\underline{m},\underline{n}}$ is a minimal L_1 projection from A(R) to $\Pi_{\underline{m},\underline{n}}^+$

These are multivariate generalisations of univariate results due to LAMBERT [15] for $F_{\underline{n}}$ and MASON [16] for $S_{\underline{n}}$ and $B_{\underline{m},\underline{n}}$. The proof of 6.1 is as follows (and the proofs of 6.2 and 6.3 follow similar lines).

Proof of 6.1 For any projection P of $L_1^0(X^0)$ to $\Pi_{\underline{n}}^0$, Berman's kernel formula gives

$$(F_{\underline{n}}f)(\underline{\theta}) = (2\pi)^{-N}\int_{X^0} (E_{\underline{t}^*}\, P\, E_{\underline{t}}f)(\underline{\theta})\, \underline{dt}$$

where $E_{\underline{t}}f(\underline{\theta}) = f(\underline{\theta}+\underline{t})$ and $E_{\underline{t}^*}f(\underline{\theta}) = f(\underline{\theta}-\underline{t})$.

$$\text{Hence } \|F_{\underline{n}}f\|_1 = (2\pi)^{-N}\int_{X^0}\underline{d\theta}\left|\int_{X^0}(E_{\underline{t}^*}\, P\, E_{\underline{t}}f)(\underline{\theta})\,\underline{dt}\right|$$

$$\leqslant (2\pi)^{-N}\int_{X^0}\underline{dt}\int_{X^0}|(E_{\underline{t}^*}\, P\, E_{\underline{t}}f)(\underline{\theta})|\,\underline{d\theta}$$

$$= \|E_{\underline{t}^*}\, P\, E_{\underline{t}}f\|_1 \leqslant \|E_{\underline{t}^*}\|_1 \cdot \|P\|_1 \cdot \|E_{\underline{t}}\|_1 \cdot \|f\|_1 = \|P\|_1\,\|f\|_1$$

Thus $\|F_{\underline{n}}\|_1 \leqslant \|P\|_1$ and the result follows.

7. Near-best L_1 Approximations by Series

The following results establish that $F_{\underline{n}}f$, $S_{\underline{n}}f$, $B_{\underline{m},\underline{n}}f$ and $H_{\underline{n}-\underline{1}}h$ are near-best in L_1 within relative distances asymptotic to $\Pi \log n_j$.

Theorem 7.1 $\|F_{\underline{n}}\|_1 \leqslant \Lambda_{\underline{n}}$ and $\|H_{\underline{n}-\underline{1}}\|_1 \leqslant \Lambda_{\underline{n}}$

Theorem 7.2 $\|S_{\underline{n}}\|_1^{C_r} \leqslant \Lambda_{\underline{n}}$, and $\|S_{\underline{n}}\|_1^D \leqslant \Lambda_{\underline{n}}$

Theorem 7.3 If $m_j + n_j$ is even for all j, then

$$\|B_{\underline{m},\underline{n}}\|_1^{C_r} \leqslant \Lambda_{\underline{\ell}}, \quad \|B_{\underline{m},\underline{n}}\|_1^R \leqslant \Lambda_{\underline{\ell}}$$

$$\|H^+_{\underline{n}-\underline{1}}\|_1^{\xi_r} \leqslant \Lambda_{\underline{n}} \quad \|H^+_{\underline{n}-\underline{1}}\|_1^E \leqslant \Lambda_{\underline{n}}$$

where $\underline{\ell} = \frac{1}{2}(\underline{m} + \underline{n})$, $W(\underline{z}) = 1$ in R, $W(\underline{w}) = \prod |w_j^2 - 1|^{-\frac{1}{2}}$ in E.

These generalise results of FREILICH and MASON [7] and MASON [6], [8].

The proof of 7.3, which follows [6] (for one variable and $\underline{m} = \underline{n}$), is given below for $B_{\underline{m},\underline{n}}$ only. Results for H and H^+ follow immediately from inequalities (17) and (20) (see also [6]).

Proof of 7.3 From (7) and (8),

$$(B_{\underline{m},\underline{n}} f)(\underline{z}) = (2\pi i)^{-N} \int_{C_1} \prod_{j=1}^{N} \left[\frac{s_j^{m_j+n_j+1} - 1}{s_j^{n_j+1} \quad (s_j - 1)} \right] . f(z_1 s_1, \ldots, z_N s_N) \, \underline{ds}$$

Hence $$\|B_{\underline{m},\underline{n}}\|_1^{C_r} = \int_{C_r} (2\pi)^{-N} \left| \int_{C_1} \prod [\ldots] . f(z_1 s_1, \ldots) \, \underline{ds} \right| \underline{dz}$$

$$\leqslant \prod_j (2\pi)^{-1} \int_{C_1} \left| \frac{s_j^{m_j+n_j+1} - 1}{s_j - 1} \right| \underline{ds} . \int_{C_r} |f(z_1 s_1, \ldots)| \, \underline{dz}$$

$$= \Lambda_{\underline{\ell}} . \|f\|_1^{C_r} \quad \text{(compare [8])}$$

This establishes the contour result.

Integrating over r_j from ρ_j^{-1} to ρ_j, and noting that $dS_j = dr_j . |dz_j|$, (where dS_j denotes an element of area in variable z_j),

$$\|B_{\underline{m},\underline{n}} f\|_1^R \leqslant \Lambda_{\ell} . \|f\|_1^R .$$

The domain result follows.

8. Interpolation Projections

Interpolation projections F*, S*, B*, G*, H*, G+*, may be defined which are closely related to the projections F, S, B, G, H, G^+. (Note, however, that spaces such as $L_1(X)$ are inappropriate in an interpolation context.) A tensor product of interpolation points is assumed throughout; and $\rho_j = \rho$.

F* : projection from $C^0(X^0)$ to $\Pi^0_{\underline{n}}$ by interpolation at zeros of $\sin(n_j + \frac{1}{2})\theta_j$ (i.e. $\theta_j = \frac{2k\pi}{2n_j+1}$ for $k = 0, 1, \ldots, 2n_j$).

S* : projection from A(D) to $\Pi^+_{\underline{n}-\underline{1}}$ by interpolation at zeros of $(z_j)^{n_j} = (\rho)^{n_j}$

B* ; projection from A(R) to $\Pi^+_{\underline{n},\underline{n}-\underline{1}}$ by interpolation at zeros of $(z_j)^{n_j} = (\rho)^{n_j}$ and $(z_j)^{n_j} = -(\rho)^{-n_j}$.

G* : projection from C(X) to $\Pi_{\underline{n}-\underline{1}}$ by interpolation at zeros of $T_{n_j}(x_j)$.

H* : projection from C(X) to $\Pi_{\underline{n}-\underline{1}}$ by interpolation at zeros of $U_{n_j}(x_j)$.

G+*: projection from A(E) to $\Pi_{\underline{n}-\underline{1}}$ by interpolation at images of zeros of $(z_j)^{n_j} = i(\rho)^{n_j}$ under $w_j = \frac{1}{2}(z_j + z_j^{-1})$.

In real variables, if Pf is a polynomial of degree n_j-1 in x_j interpolating at points $x_j^{(i)}$ ($i = 1, \ldots, n_j$), then from the Lagrange interpolation formula (see [2] and [4]):

$$\|P\|_\infty = \sup_{\underline{x}} \prod_{j=1}^{N} \sum_{i=1}^{n_j} |\ell_i^{(j)}(x_j)| \tag{23}$$

where
$$\ell_i^{(j)}(x_j) = \prod_{\substack{k=1\\k\neq i}}^{n_j} \left(\frac{x_j - x_j^{(k)}}{x_j^{(i)} - x_j^{(k)}}\right) \tag{24}$$

Theoretical multivariate results on near-best approximations are here given only for L_∞ and real variables.

Theorem 8.1 $\|F^*\|_\infty = \prod_{j=1}^{N} \tau_{n_j}$

where
$$\tau_p = \frac{1}{2p+1}\left[1 + 2\sum_{k=1}^{p} \operatorname{cosec}\frac{(2k-1)\pi}{(2p+1)2}\right] \sim \frac{2}{\pi}\log p .$$

Theorem 8.2 $\|G^*\|_\infty = \prod_{j=1}^{N} \sigma_{n_j}$

where $\sigma_p = \frac{1}{p} \sum_{k=1}^{p} \cot \frac{(2k-1)\pi}{4p} \sim \frac{2}{\pi} \log p.$

Theorem 8.3 $\|H^*\|_\infty = \prod_{j=1}^{N} \sigma^*_{n_j}$

where $\sigma^*_p = \alpha_p$ (p even)

and $\alpha_p - (p-1)^{-2} < \sigma^*_p < \alpha_p$ (p odd),

with $\alpha_p = \frac{1}{p-1} \sum_{k=1}^{p-1} \frac{\cot(2k-1)\pi}{4(p-1)} \sim \frac{2}{\pi} \log p.$

These results are generalisations of results of EHLICH and ZELLER [17] (8.1, 8.2, 8.3) and POWELL [18] (8.2).

Theorem 8.2 is proved in [4] and Theorems 8.1, 8.3 may be proved similarly.

Two univariate results are also relevant. Firstly ($N=1$, $n_1 = n$) F^* is minimal in L_∞ from $C^0(X^0)$ to Π^0_{n-1} (DE BOOR and PINKUS [19]) and secondly S^* is near-best in L_∞ within a relative distance asymptotic to $2\pi^{-1} \log n$ (GEDDES and MASON [5]). Although results for B^* and G^{+*} are not given above, MASON [20] and GEDDES [21] have covered certain limiting cases as well as computing numerical values of projection norms, and this will be discussed further in §11.

9. Convergence in Norm

By imposing, if necessary, slight restrictions on the functions f, it is possible to establish convergence in norm (as m_j and $n_j \to \infty$, all j) in all cases where near-best theorems have been given above.

Theorem 9.1 $S_{\underline{n}}f$, $B_{\underline{m},\underline{n}}f$, and $G^+_{\underline{n}}f$ converge uniformly (i.e. in L_∞) on D, R, and E for f in $A(\bar{D})$, $A(\bar{R})$, $A(E)$, respectively.

Theorem 9.2 S^*f, $G^{+*}f$ converge in L_∞ on the closed interior of D, E if f is analytic on and in D, E, respectively.

Theorem 9.3 $F_{\underline{n}}f$, $G_{\underline{n}}f$, G^*f converge in L_∞ on X^0, X, X, respectively, provided f is Dini-Lipschitz continuous, in the sense that

$$\sum_j \omega_j(\delta_j) \prod_j \log \delta_j \to 0 \text{ as } \{\delta_j\} \to 0$$

where ω_j is a partial modulus of continuity on the variable x_j.

Theorem 9.4 $H_n f$, $H_n^+ f$ converge in L_1 on X, E for f in $L_2(X)$, $A(E)$, respectively ($W = \prod|w_j^2 - 1|^{-\frac{1}{2}}$ for L_1 on E).

Theorem 9.1 is a classical result, G_n^+ being a partial sum of a Faber series (KOVARI and POMMERENKE [22] give a more general result).

Theorem 9.2 follows from a general result of FEJER [23]. Theorem 9.3, given by MASON [4] is an immediate consequence of a Jackson-type theorem for multivariate approximation (see [24])).

Proof of 9.4 (i) $F_n f$ converges in L_1 on X^0, since it converges in L_2. The L_1 convergence of $H_n f$ follows from inequality (17).

(ii) $B_{n,n} f$ converges in L_1 on C_r, (compare [4]). The L_1 convergence of $H_n^+ f$ is valid on ξ_r from (20), and hence on E (with weight W) by integration over r_j. Q.E.D.

10. Other Domains and Methods

It is to be noted that some of the complex results above may be generalised to other domains via Faber series and mappings. Some general results are given in [22] and a discussion of series/interpolation covering this topic is given by MASON [25].

Other more specialised approaches for near-best approximation have been followed by, for example, MASON [26], LEWANOWICZ [27] and BRUTMAN [28]. Also TREFETHEN and GUTKNECHT [29] have studied the "Carathéodory-Fejer method" in the context of near-best rational approximation, and ELLACOTT and GUTKNECHT [30] have extended this approach in the context of Faber series.

11. Two Bivariate Problems

We conclude with two open bivariate problems concerning functions of one state variable x, z, or w.

Firstly, it has already been shown that results in L_1 for a circular or elliptical domain may be obtained from results for a contour, by integrating over a spatial variable r. This is discussed in detail in [6]. Is this method extendable to more general domains, perhaps by incorporating a mapping?

Secondly, the L_∞ complex interpolation projection norm for a ring or ellipse is dependent on a spatial variable in addition to z (or w). GEDDES [21] computed numerical values of G^{+*}, as defined in §8. He deduced that $\|G^{+*}\|_\infty \leqslant 4$ for $n \leqslant 10$ and $1 \leqslant \rho < \infty$, and we conjecture that

$$\|G^{+*}\|_\infty \sim C_1(\rho) \log n \text{ for some } C_1(\rho).$$

Also MASON [20] computed values of B*, as defined in §8, deduced that $\|B^*\|_\infty \leqslant 5$ for $n \leqslant 10$ and $1 \leqslant \rho < \infty$, and conjectured that

$$\|B^*\|_\infty \sim C_2(\rho) \log n \quad \text{for some } C_2(\rho).$$

Are these conjectures correct, and, if so, what are $C_1(\rho)$ and $C_2(\rho)$?

12. References

1 Mason, J.C. Orthogonal polynomial approximation methods in numerical analysis. In: "Approximation Theory", A. Talbot (Ed.), Academic Press. London, 1970, pp 7-33.

2 Cheney, E.W. and Price, K.H. "Minimal projections". Ibid pp 261-290.

3 Gunning, R.C. and Rossi, H. "Analytic functions of several complex variables", Prentice-Hall, New Jersey, 1965.

4 Mason, J.C. Near-best multivariate approximation by Fourier series, Chebyshev series, and Chebyshev interpolation. J. of Approx. Th. 28 (1980), 349-358.

5 Geddes, K.O. and Mason, J.C. Polynomial approximation by projections on the unit circle. SIAM J. Numer. Anal. 12 (1975), 111-120.

6 Mason, J.C. Near-best L_∞ and L_1 approximations to analytic functions on two-dimensional regions. In: "Multivariate Approximation", D.C. Handscomb (Ed.), Academic Press, London, 1978, pp 115-135.

7 Freilich, J.H. and Mason, J.C. Best and near-best L_1 approximations by Fourier series and Chebyshev series. J. of Approx. Th. 4 (1971), 183-193.

8 Mason, J.C. Near-best L_1 approximations on circular and elliptical contours. J. of Approx. Th. 24 (1978), 330-343.

9 Lozinski, S.M. On a class of linear operators. Dokl. Akad. Nauk, SSR 61 (1948), 193-196 (Russian).

10 Cheney, E.W. "Introduction to Approximation Theory", McGraw Hill, New York, 1966.

11 Berman, D.L. On the impossibility of constructing a linear operator furnishing an approximation within the order of the best approximation. Dokl. Akad. Nauk. SSSR 120 (1958), 1175-1177 (Russian).

12 Féjer, L. Lebesguesche Konstanten und divergente Fourierreihen. Crelle J. reine angew. Math. 138 (1910), 22-53.

13 Dienes, P. "The Taylor Series - an Introduction to the Theory of Functions of a Complex Variable", Oxford, 1931.

14 Geddes, K.O. Near-minimax polynomial approximation in an elliptical region. SIAM J. Numer. Anal. 15 (1978), 1225-1233.

15 Lambert, P.V. On the minimum norm property of the Fourier series in L^1-spaces. Bull. de la Soc. Math. de Belgique 21 (1969), 370-391.

16 Mason, J.C. Minimal L_1 projections by Fourier, Taylor, and Laurent series. RMCS Dept. of Maths. Report 82/1 (1982) (submitted for publication).
17 Ehlich, H. and Zeller, K. Auswertung der Normen von Interpolationsoperatoren. Math. Annalen 164 (1966), 105-112.
18 Powell, M.J.D. On the maximum errors of polynomial approximations defined by interpolation and least squares criteria. Computer J. 9 (1967), 404-407.
19 de Boor, C. and Pinkus, A. Proof of the conjectures of Bernstein and Erdös concerning the optimal nodes for polynomial interpolation. J. of Approx. Th. 24 (1978), 289-303.
20 Mason, J.C. Near-minimax interpolation by a polynomial in z and z^{-1} on a circular annulus. IMA J. Numer. Anal. 1 (1981), 359-367.
21 Geddes, K.O. Chebyshev nodes for interpolation on a class of ellipses. In: "Theory of Approximation with Applications", A. Law and B. Sahney (Eds.), Academic Press, London, 1976, pp 155-170.
22 Kovari, T. and Pommerenke, Ch. On Faber polynomials and Faber expansions. Math. Z. 99 (1967), 193-206.
23 Féjer, L. Interpolation und konforme Abbildung. Göttinger Nachrichten (1918), 319-331.
24 Handscomb, D.C. "Methods of Numerical Approximation", Pergamon, 1966.
25 Mason, J.C. Recent advances in near-best approximation. In: "Approximation Theory III", E.W. Cheney (Ed.), Academic Press, 1980, pp 629-636.
26 Mason, J.C. Some methods of near-minimax approximation using Laguerre polynomials. SIAM J. Numer. Anal. 10 (1973), 470-477.
27 Lewanowicz, S. Properties of some polynomial projections. Bull de l'Acad. Polonaise des Sciences 27 (1978), 727-732.
28 Brutman, L. On the Lebesgue function for polynomial interpolation. SIAM J. Numer. Anal. 15 (1978), 694-704.
29 Trefethen, L. and Gutknecht, M. The Carathéodory-Féjer method for real rational approximation. Numer. Anal. Project Manuscript NA-81-15 (1981), Stanford University.
30 Ellacott, S.W. and Gutknecht, M.H. The polynomial Carathéodory-Féjer method for Jordan regions. Res. Report 82-02 (1982), Seminar für Angew. Math., E.T.H., Zürich.

Dr John C. Mason, Department of Mathematics and Ballistics, Royal Military College of Science, Shrivenham, Swindon, Wilts, SN6 8LA, England.

SHARP "A PRIORI" ERROR BOUNDS FOR POLYNOMIAL APPROXIMATION IN SOBOLEV SPACES

Jean Meinguet

Institut de Mathématique Pure et Appliquée, Université Catholique de Louvain, Louvain-la-Neuve, Belgium.

1. Introduction

In the matter of *quantitative* "a priori" error estimation for *pointwise* and *mean-square* polynomial approximation problems in Sobolev spaces, we are naturally interested in *key estimates* of the general form

$$|v-Pv|_{I} \leqslant e_{o}\ |v|_{II} \text{ for all } v \in H^{m}(\Omega);$$

here P is a linear projector onto polynomials, while $|.|_{I}$ and $|.|_{II}$ denote seminorms involving appropriately related subsets of (generalized) partial derivatives of v. As for e_o, the so-called *optimal error coefficient*, it is a numerical constant (with respect to v) we want to estimate *quantitatively*. It should be realized once and for all that the otherwise classical *Bramble-Hilbert lemma*, which strictly speaking is a tool for *qualitative* error analysis, cannot be of any help in that connection.

Unlike the *characterization* of e_o, which actually requires to solve unduly complicated eigenvalue or boundary value problems, the actual determination of reasonably sharp bounds for e_o is quite feasible (this is even rather trivial for *lower bounds*). This will be shown concretely in Section 2, which indeed is devoted to a rather complete treatment of a typical example of application.

As already analyzed elsewhere (see e.g. [7, 8]), it turns out that *realistic upper bounds* can be obtained (at reasonable cost !) for certain *generic constants* which tend to pervade the modern literature on error estimation (typically in spline analysis, in connection with the rate of convergence of the finite element method, see e.g. [2], Theorems 5, 6). It is the main purpose of this paper to present quite significant improvements of our former results [7, 8], the underlying key estimates being obtained by manipulating (most carefully !) the remainder term in an *averaged Taylor series* (serving as *standard representation formula* in $H^m(\Omega)$, at least whenever $\Omega \subset \mathbb{R}^n$ is bounded and convex). Pointwise and mean-square approximation are considered in Section 3 and in Section 4, respectively. It must be emphasized that our "main result" (see Theorem 2 in Section 3) is so sharp that it can be regarded as optimal.

2. A Typical Example of Application

2.1 The Problem

Find explicitly a (realistic) estimate of the smallest constants $c_o(x)$ and d_o for which, respectively,

(1) $$|(v-Tv)(x)| \leqslant c_o(x)|v|_2 \text{ for } x \in \overline{\Omega} \text{ and all } v \in H^2(\Omega),$$

(2) $$|v-Tv|_1 \leqslant d_o|v|_2 \text{ for all } v \in H^2(\Omega),$$

it being understood that :

- $\Omega \subset \mathbb{R}^2$ is the interior of a nondegenerate 2-simplex (i.e., triangle) of prescribed vertices x^i, $1 \leqslant i \leqslant 3$.

- $|.|_1$ and $|.|_2$ are (rotation invariant) seminorms on the classical Sobolev space $H^2(\Omega)$, to be interpreted according to the general definition

$$(3) \qquad |v|_j := \{ \sum_{i_1,\ldots,i_j=1}^{n} \int_\Omega |\partial^j v(x)/\partial x_{i_1} \ldots \partial x_{i_j}|^2 dx\}^{1/2}$$

we shall need later on (for $\Omega \subset \mathbb{R}^n$, $v \in H^m(\Omega)$, $0 \leqslant j \leqslant m$); for $m \geqslant 1$, the kernel of $|.|_m$, restricted to any connected component of Ω, is clearly the vector space P_{m-1} of all polynomials in n variables of (total) degree $\leqslant m-1$.

- Tv, the so-called Lagrange P_1-interpolant of v at the vertices of the triangle, is simply the affine function of x_1, x_2 :

$$(4) \qquad (Tv)(x) := \sum_{i=1}^{3} v(x^i)p_i(x) \text{ for any } x = (x_1,x_2) \in \mathbb{R}^2,$$

where $p_i(x)$ denotes the i-th barycentric coordinate of x with respect to the three points x^i.

Remark. Problems of this type arise most naturally in connection with the *nodal finite element method*, where indeed piecewise polynomial interpolants over triangulated domains play a leading role. As a matter of fact, they can arise more generally whenever T is a *polynomial preserving operator*, i.e., not necessarily of interpolation type. Of course, for the sake of simplicity, we consider here only the finite element called *Courant's triangle*.

2.2 Theoretical Results

2.2.1 The Optimal Error Coefficients $c_o(x)$.

By using the *Sobolev imbedding theorem* and the *Rellich compactness lemma*, it is easy to prove that the vector space

$$(5) \qquad V := (I-T)(H^2(\Omega)) \equiv \{v \in H^2(\Omega): v(x^i) = 0 \text{ for } i=1,2,3\}$$

(where I denotes the identity mapping), provided with the semi-inner product $(.,.)_2$ underlying the quadratic seminorm $|.|_2$, is a *functional Hilbert subspace of* $C^0(\bar{\Omega})$ (equipped with the topology of uniform convergence). In other words, V is a Hilbert space of bona fide (uniformly continuous) functions on Ω which is such that, for each $x \in \Omega$, the *evaluation functional* $\delta_x : v \mapsto v(x)$ on V is linear and bounded. Hence the basic *representation formula in* $H^2(\Omega)$:

(6) $$v(x) = (Tv)(x) + (v,k_x)_2 \text{ for } x \in \Omega \text{ and all } v \in H^2(\Omega),$$

where $k_x \in V$ denotes the (necessarily unique) *Fréchet-Riesz representer of the* 2-*dimensional Dirac measure* δ_x on V. It immediately follows, by Schwarz's inequality, that

(7) $$c_0(x) = |k_x|_2 \equiv [k(x,x)]^{1/2},$$

where the function k on $\Omega \times \Omega$, defined by $k(x,y) := k_x(y) \equiv (k_x, k_y)_2$, is the so-called *reproducing kernel* or *kernel function* of V.

We proceed now to show that k_x, for each $x \in \Omega$, can be interpreted as the *unique* solution of a *boundary value problem.* Let $<.,.>$ denote the *duality pairing* between $\mathcal{D}'(\Omega)$ (i.e., the space of Schwartz distributions in Ω) and $\mathcal{D}(\Omega)$ (i.e., the space of infinitely differentiable functions with compact support in Ω). Since, by definition of the differentiation for distributions, we have

$$(v,k_x)_2 = <\Delta^2 \bar{k}_x, v> \text{ for } x \in \Omega \text{ and all } v \in \mathcal{D}(\Omega),$$

it immediately follows from (6) (restricted to $\mathcal{D}(\Omega) \subset H^2(\Omega)$) that, for each $x \in \Omega$, k_x must be a solution of the (distributional) *partial differential equation*

(8a) $$\Delta^2 k_x = \delta_x \text{ in } \mathcal{D}'(\Omega),$$

such that

(8b) $\qquad k_x \in V,$

which involves the *essential boundary conditions* : $k_x(x^i) = 0$ for i=1,2,3. Now $\mathcal{D}(\Omega)$ is *not dense* in $H^2(\Omega)$, so that we have still to exploit the fact that (6) holds for all $v \in H^2(\Omega)$. By taking the "scalar product" $<.,.>$ of (8a) with $v \in V$ and comparing with (6) (trivially restricted to V), we get indeed an additional relation, viz.,

(9) $\qquad (v,k_x)_2 = <\Delta^2 \bar{k}_x, v> \text{ for } x \in \Omega \text{ and all } v \in V,$

which can be interpreted as playing the role of the *natural boundary conditions* :

(8c) $\qquad \partial_\nu(\Delta k_x) + \partial_{\nu\tau\tau} k_x = 0 \text{ on } \partial\Omega - \{x^1, x^2, x^3\},$

(8d) $\qquad \partial_{\nu\nu} k_x = 0 \text{ on } \partial\Omega - \{x^1, x^2, x^3\};$

here τ (resp. ν) denotes the unit tangential vector (resp. unit outer normal vector) existing everywhere (except at the vertices) along $\partial\Omega$ oriented in the usual way (the positive sense keeps Ω on the left).

The (at least formal) equivalence of (8c,d) and (9) readily follows from the comparison of (9) with the relatively sophisticated identity (13) (where u is to be replaced by k_x, the double integral on the right by $\bar{v}(x)$, and of course $v \in V$); it must be emphasized that (8a,c,d) *may be interpreted in the classical sense*, the data (i.e., k_x, v and $\partial\Omega$) being smooth enough to warrant the underlying integrations by parts. As a matter of fact, (8a) shows that, *for each* $x \in \Omega$, k_x in $\mathbb{R}^2 - \{x\}$ *can only differ by an entire function* (viz., any biharmonic function) *from the well known fundamental solution of* Δ^2 :

$$e_x(y) := (8\pi)^{-1} |y-x|^2 \ln |y-x| \text{ for } y \neq x.$$

Another interesting consequence of this structural result is that $k(x,y)$ is uniformly continuous on $\Omega \times \Omega$ and can therefore be extended by continuity to $\overline{\Omega} \times \overline{\Omega}$; it finally follows that the basic result (7) holds for each $x \in \overline{\Omega}$.

2.2.2 The Optimal Error Coefficient d_o. From (2), it is clear that

$$(10) \qquad d_o := \sup_{v \in V, |v|_2 \leq 1} |v|_1 .$$

Since V, as defined by (5), is a Hilbert space, it follows from the Rellich compactness lemma that the set of admissible v's in (10) is a *compact* subset of $H^1(\Omega)$ (see [11], p.420), so that there must exist in V a function u, necessarily of unit norm $|\cdot|_2$, for which the supremum is attained. It is well known that such a *stationary point formulation* (of Rayleigh-Ritz type) is equivalent to the *weak formulation* (of Galerkin type) : d_o is the greatest possible scalar for which there exists a (nonzero) function $u \in V$ such that

$$(11) \qquad (u,v)_2 - d_o^{-2}(u,v)_1 = 0 \text{ for all } v \in V.$$

By an argument very similar to the one used in Section 2.2.1, it will be shown presently that this variational problem strictly amounts to finding the *smallest possible eigenvalue* d_o^{-2} and the associated eigenfunction(s) u for the (self-adjoint, positive definite) *eigenvalue problem* :

$$(12a) \qquad \Delta^2 v + \lambda \Delta v = 0 \text{ in } \mathcal{D}'(\Omega),$$

$$(12b) \qquad v \in V,$$

$$(12c) \qquad \partial_\nu(\Delta v) + \partial_{\nu\tau\tau} v + \lambda \partial_\nu v = 0 \text{ on } \partial\Omega - \{x^1,x^2,x^3\},$$

$$(12d) \qquad \partial_{\nu\nu} v = 0 \text{ on } \partial\Omega - \{x^1,x^2,x^3\},$$

where (12b) and (12c,d) play the role of *essential* and *natural* boundary conditions, respectively; physically speaking, (12a,b, c,d) can be interpreted as the *buckling problem* for a triangular plate fixed only at the vertices.

Now, for each $\lambda > 0$, the operator $\Delta^2 + \lambda\Delta$ is *analytic-hypoelliptic*, so that every solution in $\mathbb{R}^2$ of (12a) must be an *entire* function; since $\partial\Omega$ is smooth (except at the three points x^i), it is clear that (12a,c,d) *can be interpreted in the classical sense.* As for the asserted equivalence of (11) and (12), it essentially follows from two basic identities (where s denotes the curvilinear abscissa along $\partial\Omega$), to wit :

- the (first) *Green formula* :

$$(u,v)_1 = -\iint_\Omega (\Delta u)\overline{v}\, dx + \int_{\partial\Omega} (\partial_\nu u)\overline{v}\, ds,$$

which holds for all $u \in H^2(\Omega)$, $v \in H^1(\Omega)$.

- a (hardly known !) identity (of *Rayleigh-Green* type) specifically adapted to the triangle $\overline{\Omega}$:

$$(13) \qquad (u,v)_2 = \iint_\Omega (\Delta^2 u)\overline{v}\, dx - \int_{\partial\Omega} [\partial_\nu(\Delta u) + \partial_{\nu\tau\tau}u]\overline{v}\, ds$$
$$+ \int_{\partial\Omega} (\partial_{\nu\nu}u)(\partial_\nu\overline{v})ds - \sum_{i=1}^{3} [\partial_{\nu\tau}u(x^i+) - \partial_{\nu\tau}u(x^i-)]\overline{v}(x^i),$$

which holds for all $u \in H^4(\Omega)$, $v \in H^2(\Omega)$; the detailed proof is too long for being reproduced here (it essentially amounts to a succession of integrations by parts).

2.2.3 Conclusion. In spite of the extreme simplicity of Courant's triangle among the potential applications, the above complete characterizations of the optimal error coefficients $c_o(x)$ and d_o, from the respective solutions of "classical" boundary value and eigenvalue problems, prove to be so complicated that there is no hope whatever to derive for them closed form expressions. Hence a strongly motivated need for methods of truly practical value for finding realistic (upper and lower) bounds for such theoretical quantities.

2.3 Practical Results

2.3.1 Finding Lower Bounds is Easy ! Indeed, from the definition (10) of d_o, it is clear that the *Rayleigh quotient* $|v-Tv|_1/|v|_2$, for any particular $v \in H^2(\Omega)$ such that $|v|_2 \neq 0$, must yield a lower bound for d_o. Needless to say, a similar conclusion holds for $c_o(x)$, in view of the alternative definition

$$c_o(x) = \sup|(v-Tv)(x)|/|v|_2 \text{ for each } x \in \overline{\Omega},$$

which is implied by (6), (7).

Consider now, by way of example, the error coefficient d_o corresponding to the *standard rectangular triangle* (of vertices (0,0), (1,0) and (0,1)). It is not very difficult to prove, by explicit computation, that

$$\sup_{v \in P_2,\, |v|_2 \neq 0} |v-Tv|_1^2/|v|_2^2 = (3+5^{1/2})/24 = 0.21817...,$$

the supremum being attained for

$$v(x) := x_1x_2 - ((5^{1/2}-1)/4)(x_1^2+x_2^2);$$

hence it follows that

$$(14) \qquad d_o^2 > 0.218, \quad d_o > 0.467.$$

It seems that a factor 1/2 was forgotten in [10] (see p.113), where indeed it is claimed (without detailed proof, however) that $0.4353... := (0.6598)^2$ is the lower bound for d_o^2 that can be obtained by working (essentially) as described above.

2.3.2 Finding Upper Bounds is Quite Feasible. As analyzed in detail in [6, 7, 8, 9], the first stage of the practical method we have developed, for bounding from above the remainders corresponding to polynomial preserving linear approximation operators, consists of selecting as *representation formula* an appropriate substitute for the familiar Taylor formula (which indeed cannot be used as such in Sobolev spaces). A most natural choice here is the variant of *averaged Taylor series* which will be considered hereafter in Theorem 1 (see formulas (20a,b,c) for $m = n = 2$ and $|\alpha| = 0$, Ω being the prescribed triangle).

The second stage consists of rewriting the remainder to be estimated as a finite sum of *standard* linear mappings, each of them depending *continuously* on one and the same total derivative (such standard mappings are typically certain partial derivatives, possibly composed with pointwise multiplication by given functions or integration with respect to given measures). For the concrete application considered here, this requirement is satisfied, for example, by the trivial decomposition

$$(15)\qquad v - Tv = (V_2D^2v) - \sum_{i=1}^{3} (V_2D^2v)(x^i)p_i \quad \text{for all } v \in H^2(\Omega),$$

from which it immediately follows that

$$(16a)\qquad |(v-Tv)(x)| \leqslant |(V_2D^2v)(x)| + \sum_{i=1}^{3} |(V_2D^2v)(x^i)|\,|p_i(x)| \quad \text{for } x \in \bar{\Omega},$$

$$(16b)\qquad |v-Tv|_1 \leqslant |V_2D^2v|_1 + \sum_{i=1}^{3} |(V_2D^2v)(x^i)|\,|p_i|_1,$$

for all $v \in H^2(\Omega)$. It should be noted that there do often exist several possibilities in the matter of "interesting" decompositions of specific remainders.

The final stage amounts to determining realistic upper bounds of the appropriate type for each term in the finite sums just obtained, either directly or by making use of suitable *key estimates*. As regards specifically (16a,b), the following appraisals

$$(17a)\qquad |(V_2D^2v)(x)| \leqslant 12^{-1/2}\, S^{-1/2} \max_{i=1,2,3} \|x-x^i\|^2 |v|_2 \quad \text{for } x \in \bar{\Omega},$$

(17b) $\qquad |V_2 D^2 v|_1 \leqslant h|v|_2,$

are known to hold for all $v \in H^2(\Omega)$ (see Theorem 2, formulas (24a,b), and Theorem 3, formulas (29b,c), respectively). Now the three barycentric coordinates $p_i(x)$ of any $x \in \bar{\Omega}$ are nonnegative numbers of sum identically equal to 1, so that we immediately get from (16a, 17a) the final result (for all $v \in H^2(\Omega)$) :

(18) $$|(v-Tv)(x)| \leqslant 3^{-1/2}\, S^{-1/2}\, \{1 + \frac{\max_{i=1,2,3} \|x-x^i\|^2}{h^2}\} \frac{h^2|v|_2}{2!},\quad x \in \bar{\Omega},$$

whose comparison with (1) yields an upper bound of practical value for $c_o(x)$. On the other hand, it can be proved (by elementary geometry !) that

$$\sum_{i=1}^{3} |p_i|_1 = 2S^{1/2}/r,$$

where r denotes the Euclidean diameter of the inscribed sphere of $\bar{\Omega}$, so that (16b) and (17a,b) finally yield

(19a) $$|v-Tv|_1 \leqslant (1+3^{-1/2}\, h/r)h|v|_2 < (3^{-1/2}+\sin\theta)(h^2|v|_2/r),$$

for all $v \in H^2(\Omega)$; here θ denotes the smallest angle of the triangle $\bar{\Omega}$, so that we have a fortiori (since $\sin\theta \leqslant 3^{1/2}/2$) the result

(19b) $$|v-Tv|_1 < (5/2)3^{-1/2}(h^2|v|_2/r) \text{ for all } v \in H^2(\Omega),$$

whose comparison with (2) yields an upper bound of practical value for d_o; it may be interesting to note that the inequality $r/h < \sin\theta$, which underlies the last inequality in (19a), is actually sharper than the one mentioned in [2] (see p.185).

<u>Remark</u>. Compared with former results, (19a,b) repre-

sents a most *significant improvement.* Indeed, $(5/2)3^{-1/2}$ = = 1.4433..., whereas the similar error coefficients in [1] and in [7, 8] were 3 and $2^{3/2}$ = 2.8284..., respectively; on the other hand, although $1.444(h^2/r)$ is much greater than the (remarkably sharp) value 0.81 obtained in [10] *by applying a highly specific method to the standard rectangular triangle*, our new error coefficient 1.444 is again significantly better than the value 2.23 which would follow from 0.81 for an *arbitrary triangle* by a "*change of scale*" analysis (in this respect, see Theorem 1.2 in [10]).

Let us finally mention the very recent result (which answers a question asked by C.A. Micchelli at the present Oberwolfach Conference) : *as regards the theoretical error coefficient associated with appraisals of the form* (19b), *the greatest possible lower bound that can be obtained by using only polynomials of exact degree* 2 *and for arbitrary isosceles triangles is rigorously* 1/6 (the proof can be made similar to the one outlined in Section 2.3.1).

3. Sharp Key Estimates in $H^m(\Omega)$ for Pointwise Approximation

3.1 A Suitable Representation Formula in Sobolev Spaces

Unlike other mathematicians also interested in *quantitative* error analysis, we have adopted once and for all (in [7, 8] already) a specially simple variant of *averaged Taylor series* (see Theorem 1 hereafter) as standard representation formula (with integral expression of the remainder) in Sobolev spaces. Except in complicated geometrical situations, this apparently quite rigid choice proves amply justified for reasons of practical convenience and compares favourably (specially in regard to possibilities of closed form integrations) with such popular alternatives as the *Kowalewski-Ciarlet-Wagschal formula* (used in [1, 5]) and, above all, the more general *Sobolev representation formula* (advocated, for example, in [3, 4]). On the other hand, theoretically speaking, there does not seem to exist any convincing reason for any particular representation formula to perform systematically better.

Theorem 1. *Let* Ω *be a bounded open convex set in* $\mathbb{R}^n$ *with Lebesgue measure* S. *If* $v \in H^m(\Omega)$ *(with* m *integer* ≥ 1) *and* $x \in \overline{\Omega}$, *then every distributional derivative of order* $< m$ *of* v, *say* $\partial^\alpha v$ *for* $\alpha := (\alpha_1,\ldots,\alpha_n) \in \mathbb{N}^n$ *and* $|\alpha| := \alpha_1+\ldots+\alpha_n < m$, *can be expressed in the form*

$$(20a) \qquad \partial^\alpha v(x) \overset{\text{a.e.}}{=} P_{m-|\alpha|-1}(\partial^\alpha v)(x) + V_{m-|\alpha|}D^{m-|\alpha|}(\partial^\alpha v)(x),$$

where

$$(20b) \qquad P_{m-|\alpha|-1}(\partial^\alpha v)(x) := S^{-1}\sum_{|\beta|=0}^{m-|\alpha|-1}\int_\Omega \frac{\partial^{\alpha+\beta}v(a)(x-a)^\beta}{\beta!}da$$

is a polynomial (in n *variables) of (total) degree* $\leq m-|\alpha|-1$ *and*

$$(20c) \qquad V_{m-|\alpha|}D^{m-|\alpha|}(\partial^\alpha v)(x)$$

$$\overset{\text{a.e.}}{:=} \frac{S^{-1}}{(m-|\alpha|-1)!}\int_\Omega\int_0^1 (1-t)^{m-|\alpha|-1}D^{m-|\alpha|}(\partial^\alpha v)(a+t(x-a)).(x-a)^{m-|\alpha|}dt\,da.$$

Moreover, if

$$(21) \qquad |\alpha| < m - n/2,$$

then $\partial^\alpha v$ *is uniformly continuous on* $\overline{\Omega}$ *and* (20a,c) *holds accordingly everywhere in* $\overline{\Omega}$.

This theorem (where D denotes the *total derivative* operator, to be interpreted componentwise in the distributional sense) readily follows from the classical *Taylor formula of the* $(m-|\alpha|)$*-th order about the point* $a \in \overline{\Omega}$ (with integral expression of the remainder), which holds at every $x \in \overline{\Omega}$ for every $\partial^\alpha v$ with $v \in C^m(\overline{\Omega})$, by integration over Ω with respect to the Lebesgue measure da (other measures could be used here, though at the expense of simplicity). For definiteness, the reader is reminded of the following *matrix interpretation* : every expression of the type $D^j f(y).(x-a)^j$ (where $y \in \mathbb{R}^n$ and $j \in \mathbb{N}$) can be interpreted simply as the product of the row matrix, formed with the coordinates $\partial^j f(y)/\partial y_{i_1} \ldots \partial y_{i_j}$ (where the subscripts run independently from 1 to n) of the (completely symmetric) covariant tensor $D^j f(y)$, and of the column matrix, formed with the coordinates of the j-fold Kronecker product $(x-a)^j$, these coordinates being taken with respect to the canonical basis of $\mathbb{R}^{(n^j)}$ and linearly ordered in some consistent way (the lexicographic one, for example). Since $C^m(\overline{\Omega})$ is dense in $H^m(\Omega)$, equipped with the natural norm $\|.\|_m := (|.|_0^2 + \ldots + |.|_m^2)^{1/2}$ for $|.|_j$ defined by (3), this Taylor representation formula can be extended by continuity, which finally yields (20a,b,c); indeed, according to the Sobolev imbedding theorem, the completed space is contained (with a continuous injection) in $C^0(\overline{\Omega})$ if (21) is satisfied, in the Lebesgue space $L^1(\Omega)$ otherwise.

As easily verified, it turns out that :

- $P_{m-|\alpha|-1}$ is a *continuous* linear projector of $H^{m-|\alpha|}(\Omega)$ onto the space of polynomials of degree $\leqslant m-1-|\alpha|$.
- $V_{m-|\alpha|}$ is a linear *right inverse* of $D^{m-|\alpha|}$ (regarded as a surjection, from $H^{m-|\alpha|}(\Omega)$ onto its range).
- the following *commutativity relations* hold :

$$\text{(22a)} \qquad \partial^\alpha(P_{m-1}v) = P_{m-|\alpha|-1}(\partial^\alpha v)$$

and accordingly

$$\text{(22b)} \qquad \partial^\alpha(V_m D^m v) \overset{a.e.}{=} \partial^\alpha v - P_{m-|\alpha|-1}(\partial^\alpha v) \equiv V_{m-|\alpha|} D^{m-|\alpha|}(\partial^\alpha v).$$

Remark. The foregoing actually shows that (20a,b,c) holds a.e. over the set of all points x with respect to which $\overline{\Omega}$ is *star-shaped* (whether Ω is convex or not !).

3.2 The Main Result

It consists of the following sharp *key estimates*, whose

proof (given hereafter) shows that they are even *optimal* (in a *relative* sense, to be explained later).

Theorem 2. *Let Ω be a bounded open set in $\mathbb{R}^n$, with Lebesgue measure* S *and Euclidean diameter* h, *such that $\overline{\Omega}$ is star-shaped with respect to every point in a subset* X. *If* $v \in H^m(\Omega)$ *(with* m *integer* ≥ 1) *and* $x \in X$, *then we have, for every multi-index* $\alpha := (\alpha_1,\dots,\alpha_n)$ *such that* $|\alpha| := \alpha_1 + \dots + \alpha_n < m-n/2$, *the sharp appraisal :*

$$(23) \qquad |\partial^\alpha (V_m D^m v)(x)|$$

$$\leq \frac{S^{-1}}{(m-|\alpha|)!}\left\{\frac{m-|\alpha|}{4(m-|\alpha|)^2-n^2}\int_{\partial\Omega}\|x-z\|^{2(m-|\alpha|)+n} d\omega_x(z)\right\}^{1/2}|\partial^\alpha v|_{m-|\alpha|},$$

where $d\omega_x(z)$ *means the elementary solid angle with vertex* x *and in the direction specified by the variable point* $z \in \partial\Omega$. *A reasonable simplification leads to the result of more practical value :*

$$(24a) \qquad |\partial^\alpha (V_m D^m v)(x)| \leq c(x)\frac{h^{m-j}|\partial^\alpha v|_{m-j}}{(m-j)!},$$

where $j := |\alpha| < m-n/2$ *and*

$$(24b) \qquad c(x) := \left\{\frac{(m-j)nS^{-1}}{4(m-j)^2-n^2}\right\}^{1/2} \frac{\sup_{z\in\partial\Omega}\|z-x\|^{m-j}}{h^{m-j}},$$

from which the global estimate

$$(24c) \qquad \|D^j (V_m D^m v)(x)\| \leq c(x)\frac{h^{m-j}|v|_m}{(m-j)!}$$

directly follows.

In view of the commutativity relations (22b), an explicit proof for the particular case $|\alpha| = 0$ is all that is needed here.

Consider then the expression (20c) for $(V_m D^m v)(x)$. By the change of variables $a \mapsto y := x+(1-t)(a-x) \equiv a+t(x-a)$, this "double" integral over the *cylinder* $\Omega \times (0,1)$ in the *Euclidean space* $\mathbb{R}^n \times \mathbb{R}$ is transformed into an integral over the $(n+1)$-dimensional *cone* with base $\Omega \times \{0\}$ and vertex $\{x\} \times \{1\}$, viz.,

$$(25a) \qquad (V_m D^m v)(x) = \frac{S^{-1}}{(m-1)!} \int_0^1 (1-t)^{-n-1} \{\int_{\Omega(x,1-t)} D^m v(y).$$
$$(x-y)^m dy\} dt;$$

here the region of integration $\Omega(x,1-t) := \{y=x+(1-t)(a-x) : a \in \Omega\}$ can be interpreted as the cross section of that cone cut by the hyperplane $\mathbb{R}^n \times \{t\}$ for $0 \leqslant t \leqslant 1$ or, equivalently, as the image of $\Omega \times \{0\}$ in $\mathbb{R}^n \times \{0\}$ under the contraction mapping of center $\{x\} \times \{0\}$ and ratio $1-t$. Next, by interchanging the order of integration in (25a), we obtain

$$(25b) \qquad (V_m D^m v)(x) = \frac{S^{-1}}{(m-1)!} \int_\Omega D^m v(y).K(x,y)dy,$$

the (tensor-valued) kernel $K(x,y)$ being defined as

$$(25c) \qquad K(x,y) := \{\int_0^{\tau_x(y)} (1-t)^{-n-1} dt\} (x-y)^m,$$

where $\tau_x(y)$ means the altitude of the upper boundary surface of the cone expressed as a function of the point $\{y\} \times \{0\}$ ranging over the base; by elementary geometry, it is clear that

$$(25d) \qquad \tau_x(y) := \|z-y\| / \|z-x\|,$$

where z denotes the intersection with $\partial\Omega$ of the line emanating from x and passing through y. Finally, by applying Schwarz's inequality in $L^2(\Omega)$ to each term of the inner product in (25b), we get the appraisal

$$(26) \qquad |(V_m D^m v)(x)| \leqslant \frac{S^{-1}}{(m-1)!} \{\int_\Omega \|K(x,y)\|^2 dy\}^{1/2} |v|_m,$$

which proves to be exactly (23) (for $|\alpha| = 0$)! Indeed, since $\|(x-y)^m\|^2 \equiv \|x-y\|^{2m}$, the integrand in (26) is a radial function;

hence, by changing to spherical polar coordinates, the volume element dy becomes simply $\rho^{n-1} d\rho \, d\omega_x(z)$, where $\rho := \|y-x\|$ is the radial coordinate while $d\omega_x(z)$ stands for the elementary solid angle or element of hypersurface area on the unit (n-1)-dimensional sphere centered at x, so that we finally obtain

$$\int_\Omega \|K(x,y)\|^2 dy = \int_{\partial\Omega} \int_0^{\|z-x\|} \left\{\frac{(\|z-x\|/\rho)^n - 1}{n}\right\}^2 \rho^{2m+n-1} d\rho \, d\omega_x(z),$$

which actually completes the proof of (23) (by computing explicitly the first integral). As for the results (24a,b,c), they directly follow from (23) by application of the most remarkable (though hardly known !) identity :

(27a) $$\int_\Omega \|y-x\|^s dy = (n+s)^{-1} \int_{\partial\Omega} \|z-x\|^{n+s} d\omega_x(z),$$

which holds for Re s > -n (as easily verified by changing again to polar coordinates), whose specialization (for s=0)

(27b) $$S = n^{-1} \int_{\partial\Omega} \|z-x\|^n d\omega_x(z)$$

is particularly worth mentioning here.

The quite surprising fact that the appraisal (23) strictly follows, by only one application of Schwarz's inequality in $L^2(\Omega)$, from the integral representation (25b,c,d) of $(V_m D^m v)(x)$ implied by Theorem 1, explains in which precise sense it may be regarded as an *optimal* result. As analyzed in Section 2.2.1 in connection with a concrete example, results that would be optimal in the absolute sense are obviously out of practical reach, since obtaining them would amount to solving extremely complicated boundary value problems. Though the kernel K(x,y), as defined above, is not the m-th total derivative (of a so-called *Rodrigues function*), which indeed would be the form theoretically required for being directly involved in any *Peano kernel theorem*, formulas (25b,c,d) can often be used to derive interesting *Peano-like integral representations for remainders* together with realistic upper bounds.

Remark. Compared with our former results, *the improvement* achieved here *is quite significant* : indeed, the ratio of $c(x)$, as defined by (24b), to the similar error coefficient in [7, 8] (see e.g. [8], p.181, formula (3.8)) is bounded from above by

$$\frac{1}{2}\left\{\frac{2n}{2(m-j)+n}\right\}^{1/2} \frac{\sup_{z\in\partial\Omega}\|z-x\|^{m-j}}{h^{m-j}},$$

which not only is always $< 1/2$ (since j in Theorem 2 must be $< m-n/2$) but might be as small as $(1/2)^{m-j+1}$ (whenever x is *center* of Ω).

4. Key Estimates in $H^m(\Omega)$ for Mean-Square Approximation

Theorem 3. *Let Ω be a bounded open convex set in $\mathbb{R}^n$ with Lebesgue measure* S *and Euclidean diameter* h. *If* $v \in H^m(\Omega)$ *(with* m *integer* ≥ 1), *then we have, for every multi-index* $\alpha := (\alpha_1,\ldots,\alpha_n)$ *such that* $|\alpha| := \alpha_1 + \ldots + \alpha_n < m$, *the following appraisal* :

$$(28)\qquad |\partial^\alpha(V_m D^m v)|_0 \leq \frac{S^{-1/2}}{(m-|\alpha|)!}\{(m-|\alpha|)\sup_{x\in\Omega}\int_\Omega \|x-a\|^{2(m-|\alpha|)}da$$
$$\int_0^1 (1-t)^{m-|\alpha|-1}\min(t^{-n},(1-t)^{-n})dt\}^{1/2}|\partial^\alpha v|_{m-|\alpha|}.$$

A reasonable simplification leads to the result of more practical value :

$$(29a)\qquad |\partial^\alpha(V_m D^m v)|_0 \leq d\,\frac{h^{m-j}|\partial^\alpha v|_{m-j}}{(m-j)!},$$

where $j := |\alpha| < m$ *and*

$$(29b)\qquad d := \left\{\frac{(m-j)n}{2(m-j)+n}\int_0^1 (1-t)^{m-j-1}\min(t^{-n},(1-t)^{-n})dt\right\}^{1/2},$$

from which the global estimate

(29c) $$|V_m D^m v|_j \leq d \frac{h^{m-j}|v|_m}{(m-j)!}$$

directly follows.

However, for $j := |\alpha| < m-n/2$, *somewhat sharper appraisals are known, for example,*

(30) $$|V_m D^m v|_j \leq \left\{\frac{(m-j)n}{4(m-j)^2-n^2}\right\}^{1/2} \frac{h^{m-j}|v|_m}{(m-j)!}$$

as it trivially results from Theorem 2.

Here again, owing to the commutativity relations (22b), we are justified in proving (28) and (29) only for $|\alpha| = 0$. However, as regards the expected sharpness of these appraisals, the situation is apparently less satisfactory than it was throughout Section 3.2, in so far as more than just one application of basic integral inequalities seems to be here indispensable.

Among the various approaches we have experienced, it seems that the most interesting is the following one. Consider the expression (20c) for $(V_m D^m v)(x)$, which holds only a.e. if $m \leq n/2$. By applying first the Cauchy-Schwarz inequality under the integral sign and regarding then the resulting "double" integral over the cylinder $\Omega \times (0,1)$ in the Euclidean space $\mathbb{R}^n \times \mathbb{R}$ as an inner product with respect to the measure $(1-t)^{m-1}$ dt da, we readily get the intermediate result

$$|(V_m D^m v)(x)|^2 \overset{a.e.}{\leq} m(S^{-1}/m!)^2 \int_\Omega \|x-a\|^{2m} da$$
$$\int_\Omega \int_0^1 \|D^m v(a+t(x-a))\|^2 (1-t)^{m-1} dt\, da$$

by making use of *Schwarz's inequality*. Next, simply by integrating with respect to x over Ω and applying then *Hölder's integral inequality*, we find

$$|V_m D^m v|_0^2 \leq m(S^{-1}/m!)^2 \sup_{x\in\Omega} \int_\Omega \|x-a\|^{2m} da$$

$$\int_0^1 (1-t)^{m-1}\{\int_\Omega\int_\Omega \|D^m v(a+t(x-a))\|^2 \, da\, dx\}dt.$$

The proof of (28) is completed by noticing that the "double" integral in the last result can be rewritten, and accordingly bounded from above, *either* in the form

$$\int_\Omega\{\int_{\Omega(a,t)} \|D^m v(y)\|^2 t^{-n}\, dy\}da \leqslant t^{-n} S|v|_m^2$$

owing to the change of variables $x \mapsto y := a+t(x-a)$ for $t > 0$, *or* in the form

$$\int_\Omega\{\int_{\Omega(x,1-t)} \|D^m v(y)\|^2 (1-t)^{-n}\, dy\}dx \leqslant (1-t)^{-n} S|v|_m^2$$

owing to the change of variables $a \mapsto y := x+(1-t)(a-x)$ for $t < 1$. On the other hand, in view of (27a,b), the integral $\int_\Omega \|x-a\|^{2m}\, da$ is bounded from above by $(nS/(2m+n)) \sup_{z\in\partial\Omega} \|z-x\|^{2m}$, so that its required supremum with respect to x ranging over Ω is bounded from above by $h^{2m}\, nS/(2m+n)$; this essentially completes the proof of (29a,b,c).

Remark. The ratio of d, as defined by (29b), to the similar error coefficient in [7, 8] (see e.g. [8], p.182, formula (3.9b) where 2p is to be replaced by m-j-1) is equal to $\{n/[2(m-j)+n]\}^{1/2}$. As for the ratio of d to the error coefficient in (11), it is bounded from above by

$$2\,\{\frac{m-j-n/2}{m-j}\int_{1/2}^1 t^{-n}dt\}^{1/2},$$

as it follows from *Chebyshev's inequality*.

References

[1] Arcangéli, R. et Gout, J.L. (1976) Sur l'évaluation de l'erreur d'interpolation de Lagrange dans un ouvert de $\mathbb{R}^n$. R.A.I.R.O. Analyse numérique 10, 5-27.

[2] Ciarlet, P.G. and Raviart, P.A. (1972) General Lagrange and

Hermite interpolation in $\mathbb{R}^n$ with applications to finite element methods. Arch. Rational Mech. Anal. 46, 177-199.
[3] Dupont, T. and Scott, R. (1978) Constructive polynomial approximation in Sobolev spaces. In : de Boor, C. and Golub, G.H. (eds), Recent advances in numerical analysis (Academic Press, New York), 31-44.
[4] Dupont, T. and Scott, R. (1979) Polynomial approximation of functions in Sobolev spaces. Report No. 79-13, ICASE, NASA Langley Research Center, Hampton.
[5] Gout, J.L. (1977) Estimation de l'erreur d'interpolation d'Hermite dans $\mathbb{R}^n$. Numer. Math. 28, 407-429.
[6] Meinguet, J. (1975) Realistic estimates for generic constants in multivariate pointwise approximation. In : Miller, J.J.H. (ed.), Topics in numerical analysis II (Academic Press, London), 89-107.
[7] Meinguet, J. (1977) Structure et estimations de coefficients d'erreurs. R.A.I.R.O. Analyse numérique 11, 355-368.
[8] Meinguet, J. (1978) A practical method for estimating approximation errors in Sobolev spaces. In : Handscomb, D.C. (ed.), Multivariate approximation (Academic Press, London), 169-187.
[9] Meinguet, J. and Descloux, J. (1977) An operator-theoretical approach to error estimation. Numer. Math. 27, 307-326.
[10] Natterer, F. (1975) Berechenbare Fehlerschranken für die Methode der Finiten Elemente. International Series of Numerical Mathematics 28 (Birkhäuser Verlag, Basel), 109-121.
[11] Schwartz, L. (1970) Analyse : Topologie générale et analyse fonctionnelle (Hermann, Paris).

Prof. Dr. Jean Meinguet, Institut de Mathématique Pure et Appliquée, Université Catholique de Louvain, Chemin du Cyclotron 2, B-1348 Louvain-la-Neuve, Belgium.

AN IMMEDIATE CONSTRUCTION OF NUMERICAL INTEGRATION AND DIFFERENTIATION FORMULAE

H. Michael Möller

FernUniversität Hagen, W. Germany

In this paper we prove a general version of the Max Noether theorem. Then this theorem from algebraic geometry is used to obtain numerical integration and differentiation formulae with a moderate number of evaluation points. Some examples illustrate the immediate construction.

1. Introduction

The condition, that a numerical integration formula

$$I(f) \approx \sum_{k=1}^{N} A_k f(y_k) \quad , \text{ I integral ,}$$

is exact for all polynomials of (total) degree $\leq d$, leads to a nonlinear system of equations in $y_1,\dots,y_N$, $A_1,\dots,A_N$. Since the number if equations generally differs from the number of unknowns, a solution can easily be found only for small d or if the system can be reduced to a small one by symmetry arguments, see ENGELS (1980) of MYSOVSKIH (1981). In the following, we present for arbitrary d and arbitrary I a method for obtaining special formulae with moderate numbers N, where the nonlinear system is splitted in a nonlinear system with n eq. in n unknowns yielding the y_k's and then a linear one for the A_k's. This method is ready for automatic computation as we show by some examples.

Our method is essentially based on the Max Noether theorem. In this theorem the common zeros of n polynomials in n variables are considered. We prove it for zeros of arbitrary multiplicity and arbitrary n, because in literature we found only weaker versions and some of them (especially for $n > 2$) even without proof. In the recent book of I.P. MYSOVSKIH (1981) references are given, the n-dimensional version for simple nodes is proved, and a version for multiple nodes and arbitrary n is formulated.

Using the Max Noether theorem, some authors have already constructed numerical integration formulae (cf. the references in MYSOVSKIH's book), but they used only special versions of the theorem. We demonstrate the principle of construction for more general instances and show by some examples the immediate and automatic computation of numerical integration and differentiation formulae.

2. Max Noether's theorem and the construction of formulae

For nonnegative integers k let P_k denote the linear space of real polynomials of total degree $\leq k$ considered as functions on D, $D \subseteq \mathbb{R}^n$ containing sufficiently many points. Let $P := \bigcup_{k \geq 0} P_k$. G_k denotes the subspace of P_k, which consists of the even (odd) polynomials of P_k for k even (or odd resp.).

The problem of constructing a formula of degree d for a given $L \in P'$ - it is an integration formula if L is an integral, or a differentiation formula if $L = L^* \circ D$, where D is a differential operator and L^* a point evaluation functional - consists in finding appropriate functionals $L_1, \dots, L_N \in P_d'$, e.g. point evaluation functionals or even of type $L^* \circ D$ as above, and weights $A_1, \dots, A_N \in \mathbb{R}$, such that

$$(1) \qquad L(f) = \sum_{i=1}^{N} A_i L_i(f) \quad \text{for all} \quad f \in P_d .$$

Using an elementary argument of linear algebra, this problem can be splitted into two subproblems

Problem 1: Find appropriate $L_1,\dots,L_N \in P_d'$:

(2) $f \in P_r$, $L_i(f) = 0$, $i = 1,\dots,N \Rightarrow L(f) = 0$.

Problem 2: For given $L_1,\dots,L_N \in P_d'$ compute $A_1,\dots,A_N$ in (1).

Problem 2 is only a linear one and will not be treated further. For (2) we need more informations on polynomials,which satisfy the left hand statement of (2). These are available by means of the Max Noether theorem for some sets $\{L_1,\dots,L_N\}$ as follows.

Definition: Using the usual multiindex notation we define for given functionals $D_1,\dots,D_s$,

(3) $$D_i(f) := \sum_{|\alpha|\le k_i} c_\alpha^{(i)} \frac{1}{\alpha!} \frac{\partial^{|\alpha|}}{\partial x^\alpha} f(y_i) \quad \text{for all } f \in P ,$$

the *derived functionals* $D_i^{(\beta)}$, β multiindex, by

(4) $$D_i^{(\beta)}(f) := \sum_{|\alpha|\le k_i} c_\alpha^{(i)} \frac{1}{(\alpha-\beta)!} \frac{\partial^{|\alpha-1|}}{\partial x^{\alpha-\beta}} f(y_i) \text{ for all } f\in P,$$

where the summation is extended only over multiindices α satisfying $\alpha_\nu \ge \beta_\nu, \nu = 1,\dots,n$. (Hence $D_i^{(\beta)} = 0$ holds for $|\beta|>k_i$.) We define also in analogy to MÖLLER (1976)

$$H(D_1,\dots,D_s) := \text{span}\{D_i^{(\beta)} \mid i=1,\dots,s; \beta \text{ arbitrary multiindex}\}.$$

Now we are able to present the announced general version of the Max Noether theorem.

Theorem 1. *If polynomials* $\varphi_1,\dots,\varphi_n$ *with degrees* $\mu_1,\dots,\mu_n$ *resp. satisfy*

(5) $$\lim_{t\to 0} t^{\mu_i}\varphi_i\left(\frac{x_1}{t},\dots,\frac{x_n}{t}\right) = 0, i=1,\dots,n \Rightarrow (x_1,\dots,x_n) = 0,$$

then functionals $D_1,\dots,D_s$ (3) *exist with* $\dim H(D_1,\dots,D_s) = \mu_1\cdots\mu_n$ *such that the following statements are equivalent for all* $f \in P$

(i) $\exists g_1,\ldots,g_n \in P\colon\ f = \sum_{i=1}^{n} g_i\varphi_i\ ,\ \max_{i=1}^{n} \deg g_i\varphi_i = \deg f$.

(ii) $D \in H\ (D_1,\ldots,D_s) \Rightarrow D(f) = 0$.

Proof. Apart of the statements for the degrees, (i) means that f belongs to the ideal $(\varphi_1,\ldots,\varphi_n)$. The degree statement follows from (5) as shown by the author (1979,theorem 2).In analogy to MÖLLER (1977), the remaining assertions of the theorem can be derived from GRÖBNER (1970, chapt. IV, § 2).

Remark 1. Theorem 2 is still nonconstructive because of the D_i's. The points $y_1,\ldots,y_s$ corresponding to $D_1,\ldots,D_s$ by (3) are(the) common zeros of $\varphi_1,\ldots,\varphi_n$, because the evaluation functionals corresponding to $y_1,\ldots,y_s$ belong to $H(D_1,\ldots,D_s)$, and the D_i's ($D_i^{(\beta)}$'s resp.) are evaluations of differential operators of highest(lower resp.) order, which vanish for $\varphi_1,\ldots,\varphi_n$, cf. MÖLLER (1976). Thus the detailed study of the common zeros of $\varphi_1,\ldots,\varphi_n$ leads to constructive variants of the Max Noether theorem. For example if $\det(\frac{\partial\varphi_j}{\partial x_k}(y_i))_{j,k} \neq 0$, then $D_i(f) = f(y_i)$, because then no differential operator of order 1 evaluated at y_i can satisfy (ii) for $\varphi_1,\ldots,\varphi_n$. A simple necessary and sufficient condition for $D_i(f) = f(y_i)$, $i=1,\ldots,s$, is $y_i \neq y_j$ for $i \neq j$ and $s = \mu_1\cdots\mu_n$, because then $H(D_1,\ldots,D_s) = \bigoplus_{i=1}^{s} H(D_i)$ yielding dim $H(D_i) = 1$ and conversely, cf. MÖLLER (1977).

Remark 2. If each of the φ_i's is even or odd and if $f \in G_r$ for some r, then in (i) it many be assumed, that $g_i\varphi_i \in G_r$ holds for $i = 1,\ldots,n$. Because the splitting of g_i in even and odd polynomials, $g_i = g_i^+ + g_i^-$, leads to $\Sigma g_i^+\varphi_i = f$ and $\Sigma g_i^-\varphi_i = 0$, we have w.l.o.g. $g_i^- = 0$, $i = 1,\ldots,n$, and hence $g_i^+\varphi_i \in G_r$.

Theorem 1 together with appropriate orthogonality conditions solves problem 1 and hence allows the construction of formulae:

Theorem 2. *Let* $V = P_d$ *or* $= G_d$ *and* $L\colon V \to \mathbb{R}$ *linear. Let* $\varphi_1,\ldots,\varphi_n \in P$ *satisfy* (5) *and* $H(D_1,\ldots,D_s)$ *as in theorem 1. If* $y_i^* \in D$, $c_i^* \in \mathbb{R}$, $i = 1,\ldots,n$, *exist with*

(E1) $g_i\varphi_i \in V \Rightarrow L(g_i\varphi_i) = c_i^* g(y_i^*)$

(E2) $c_j^*\varphi_i(y_j^*) = 0$ if $i \neq j$, $c_i^* \neq 0 \Rightarrow \varphi_i(y_i^*) = 1$,

then a formula exact for V *exists,*

$$(6) \qquad L(f) = \sum_{i=1}^{n} c_i^* \; f(y_i^*) + \sum_{i=1}^{m} A_i L_i(f) \quad \text{for all } f \in V.$$

Here $A_1,\ldots,A_m \in \mathbb{R}$ *and* $L_1,\ldots,L_m$ *constitute a basis of* $H(D_1,\ldots,D_s)$.

Proof. We only show, that the implication (2) holds for the functional $L^*(f) := L(f) - \Sigma c_j^* \; f(y_j^*)$. Let $f \in V$, $L_i(f) = 0$, $i = 1,\ldots,m$. Then, by theorem 1 (ev. modified by remark 2) $f = \Sigma g_i\varphi_i$, $g_i\varphi_i \in V$. Using (E1) and (E2),

$$\begin{aligned} L^*(f) &= \sum_1^n L^*(g_i\varphi_i) \\ &= \sum_{i=1}^{n} c_i^* \; g(y_i^*) - \sum_{i=1}^{n}\sum_{j=1}^{n} c_j^* g_i(y_j^*)\varphi_i(y_j^*) \\ &= \sum_{i=1}^{n} c_i^* \; g(y_i^*) - \sum_{i=1}^{n} c_i^* g_i(y_i^*)\varphi_i(y_i^*) = 0. \qquad \& \end{aligned}$$

The conditions (E1) and (E2) are discussed extensively by MÖLLER (1979) for positive L. Herefrom we mention only, if L is strictly positive and $V = P_d$ or G_d, then for $c_i^* = 0, i=1,\ldots,n$, (E2) is satisfied and (E1) means, that $\varphi_1,\ldots,\varphi_n$ are d-orthogonal, and for $d = 2m$ and $c_i^* \neq 0, i = 1,\ldots,n$, the φ_i's are representers of point evaluation functionals, apart of normalizations, in the inner product space P_m or G_m,

$$g(y_i^*) = L(g \; c_i^{*-1}\varphi_i) = (g, c_i^{*-1}\varphi_i) \text{ for all } g \in P_m \text{ or } G_m.$$

If Φ is the reproducing kernel of P_m or G_m, then it follows

$$(7) \qquad \varphi_i(x) = c_i^* \; \Phi\,(x, y_i^*) \;, \; i = 1,\ldots,n \;.$$

If L is of type (3) and $c_i^* = 0$, $i = 1,\ldots,n$, then by (E1) we get $L \in H(D_1,\ldots,D_s)$. Hence in this case, a formula for L leads to a linear dependence over V for elements of $H(D_1,\ldots,D_s)$. Its construction by means of theorem 2 would be only a detour. But example 1 will show, that th. 2 is useful for the construction of differentiation formulae if $c_i^* \neq 0$, $i = 1,\ldots,n$.

3. The computation of formulae

Example 1. Let $V = G_4$ and

$$L(f) := \left(\frac{\partial^4 f}{\partial x^4} + 2\frac{\partial^4 f}{\partial x^2 \partial y^2} + \frac{\partial^4 f}{\partial y^4}\right)\Bigg|_{(x,y) = (0,0)}$$

We choose $\varphi_1, \varphi_2 \in G_3$. The simplest formulae are obtained then for

$$\varphi_1(x,y) = A(x^3 - j^2h^2x), \quad \varphi_2(x,y) = A(y^3 - j^2h^2y),$$
$$y_1^* = (ih,0), \quad y_2^* = (0,ih), \quad i \neq j, \; i,j \in \mathbb{N}.$$

(E1) and (E2) hold iff $1 = Aih^2(i^2 - j^2)$, $24A = c_1^* ih = c_2^* ih$. φ_1, φ_2 have exactly 9 different simple zeros

$$(0,0), \; (0, \pm jh), \; (\pm jh, 0), \; (\pm jh, \pm jh).$$

Hence by theorem 1 with remark 1 and theorem 2 a formula for L exists exact for all $f \in G_4$ using evaluations at y_1^*, y_2^* and at the 9 common zeros of φ_1, φ_2. By a symmetry consideration and a simple computation, we get even a formula exact for all $f \in P_5$:

$$\begin{aligned} h^4 Lf = {} & \frac{12}{i^2(i^2 - j^2)}\{f(ih, 0) + f(-ih,0) + f(0,ih) + f(0,-ih)\} \\ & + \frac{2}{j^4}\{f(jh,jh) + f(-jh,jh) + f(jh,-jh) + f(-jh,-jh)\} \\ & + \frac{4i^2 + 8j^2}{j^4(j^2 - i^2)}\{f(jh,0) + f(-jh,0) + f(0,jh) + f(0,-jh)\} \\ & + \frac{8i^2 + 48j^2}{j^4 i^2} f(0,0). \end{aligned}$$

The book of JAIN (1979) contains many differentiation formulae but it mentions of our presented class only the formula for $i = 2$, $j = 1$, thus we believe the others are new.

In the following, we want to demonstrate, how th. 2 can be used to construct automatically integration formulae. We show it exemplary for some $D \subseteq \mathbb{R}^3$ and $V = G_4$.

Let $f_1, \ldots, f_r$ denote a basis for G_2 and $L: G_4 \to \mathbb{R}$ strictly positive. Then we proceed as follows.

STEP 1: Compute the inverse of the Gram matrix $G = (L(f_i f_j))$

STEP 2: Implement the reproducing kernel function

$\phi\ (a,x) := (f_1(a),\ldots,f_r(a))G^{-1}(f_1(x),\ldots,f_r(x))^t$.

STEP 3: Fix successively $y_i^* \in D, i = 1,\ldots,n$, satisfying

$\phi\ (y_j^*,y_i^*) = 0\ ,\ j = 1,\ldots,i-1$.

(Then by symmetry of ϕ, $\phi(y_i^*,y_j^*) = 0$ holds for all $i \neq j$).

STEP 4: Define $\varphi_i := \phi(y_i^*,.)/\phi(y_i^*,y_i^*)$, verify (5), and compute the common zeros of $\varphi_1,\ldots,\varphi_n$ or compute $H(D_1,\ldots,D_s)$, if not all zeros are simple (cf. Remark 1).

STEP 5: Compute the formula (6) with $c_i^* = 1/\phi(y_i^*,y_i^*) > 0$.

Since in the following examples, we consider only functionals L satisfying f odd $\Rightarrow L(f) = 0$, we obtain in addition a second formula by the mapping $x \to -x$. The arithmetic mean of these formulae is then exact for odd polynomials and has hence degree 5. All the obtained 5 degree formulae have a minimal number of nodes, see MYSOVSKIH (1981, § 9).

Example 2. Let

$$L(f) = \int_D w(x) f(x) dx \quad \text{for all } f \in P.$$

Computing STEP 1 and 2 by hand, starting STEP 3 with $y_1^* = (0,0,0)$, we obtained automatically (rounded to 10 digits) the following formulae of degree 5

$$L(f) = w_1 f(0,0,0) + \sum_2^7 w_i \{f(y_i) + f(-y_i)\}\ ,$$

$$w_i := c_i^*/2,\ y_i := y_i^*,\ i = 2,3;\ w_i := A_i/2,\ i = 4,\ldots,7.$$

Example 2.1: $D=[-1,1]^3$, $w(x) = 1$.

$y_1^{(i)}$	$y_2^{(i)}$	$y_3^{(i)}$	w_i
0	0	0	+2.105263158'-1
+7.333333333'-1	+6.666666667'-1	+5.333333333'-1	+7.934492827'-2
-8.289146920'-1	-3.439917721'-1	+6.791441385'-1	+6.956793890'-2
-4.277506485'-1	+7.187129602'-1	+7.530921127'-1	+7.498264731'-2
-1.685464367'-1	-9.928188125'-1	+5.025630019'-1	+5.283052368'-2
+4.503417734'-1	-3.156399463'-1	+9.819523299'-1	+5.500320092'-2
+8.892222036'-1	-6.532031568'-1	+2.219823760'-1	+6.300760302'-2

Example 2.2: $D=[-1,1]^3$, $w(x)=\sqrt{1-x_1^2}\cdot\sqrt{1-x_2^2}\cdot\sqrt{1-x_3^2}$

$y_1^{(i)}$	$y_2^{(i)}$	$y_3^{(i)}$	w_i
0	0	0	+2.500000000'-1
-9.000000000'-1	-1.000000000'-1	+4.242640687'-1	+5.258287060'-2
-5.438613967'-1	+8.131957904'-1	+2.071892557'-1	+6.087863911'-2
-2.962925624'-1	-7.379236292'-1	+6.063657600'-1	+6.653116007'-2
-2.301984534'-1	+3.578796302'-1	+9.049479777'-1	+5.252719973'-2
+6.538728179'-1	-3.658761584'-1	+6.622574837'-1	+6.998289055'-2
+5.971944464'-1	+6.632517195'-1	+4.510609158'-1	+7.249723994'-2

Example 2.3: $D=[-1,1]^3$, $w(x)=1/(\sqrt{1-x_1^2}\ \sqrt{1-x_2^2}\ \sqrt{1-x_3^2})$

0	0	0	+1.428571429'-1
-9.000000000'-1	0	+9.695359715'-1	+5.978527795'-2
-9.515424475'-1	+8.908523545'-1	+2.257189693'-1	+6.378848714'-2
-6.600291130'-1	-9.721452010'-1	+6.076966991'-1	+7.762937476'-2
-1.008232973'-1	+9.282228774'-1	+9.371429734'-1	+6.091970806'-2
+6.525623272'-1	-6.945255090'-1	+9.174948100'-1	+8.531234942'-2
+9.488673754'-1	+6.316469165'-1	+6.713217386'-1	+8.113578124'-2

Example 2.4: $D=\mathbb{R}^3$, $w(x)=\exp(-\|x\|_2)$

0	0	0	+6.000000000'-1
+7.333333333'-1	+6.666666667'-1	+5.386815179'+0	+3.333333333'-2
-3.537983456'+0	-2.578664658'+0	+3.291376861'+0	+3.333333333'-2
-3.659664867'+0	+3.137547387'+0	+2.600509462'+0	+3.333333333'-2
+1.926295764'+0	-4.319227132'+0	+2.762908180'+0	+3.333333333'-2
+1.729411106'+0	+4.929798244'+0	+1.645061246'+0	+3.333333333'-2
+5.181724636'+0	+3.212581448'-1	+1.745429174'+0	+3.333333333'-2

Example 2.5: $D=\mathbb{R}^3$, $w(x)=\exp(-\|x\|_2^2)$

0	0	0	+4.000000000'-1
+7.333333333'-1	+6.666666667'-1	+1.231981241'+0	+5.000000000'-2
-1.248760266'+0	-9.121825776'-1	+3.294248682'-1	+5.000000000'-2
-1.955608306'-1	+1.558579464'+0	+1.805159697'-1	+5.000000000'-2
-9.163985808'-1	+4.630951646'-1	+1.202396153'+0	+5.000000000'-2
+8.242028544'-2	-8.603482533'-1	+1.324012001'+0	+5.000000000'-2
+1.420562044'+0	-5.827969681'-1	+3.772945453'-1	+5.000000000'-2

Example 2.6: $D=\{x\in\mathbb{R}^3\ ;\ \|x\|_2=1\}$, $w(x)=1$

$\varphi_1(x):=x_1^2+x_2^2+x_3^2-1$, formally $w_1:=0$

-	-	-	0
0	+1.000000000'+0	0	+8.333333333'-2
+8.944271910'-1	+4.472135955'-1	0	+8.333333333'-2
-7.236067977'-1	+4.472135955'-1	+5.257311121'-1	+8.333333333'-2
-2.763932023'-1	-4.472135955'-1	+8.506508084'-1	+8.333333333'-2
+2.763932023'-1	+4.472135955'-1	+8.506508084'-1	+8.333333333'-2
+7.236067977'-1	-4.472135955'-1	+5.257311121'-1	+8.333333333'-2

4. Conclusion

For the computation of the nodes of a formula of fixed degree a nonlinear system arises. In contrast to the general case we presented here only systems, where the number of eq. and unknowns are equal. For the solution of these systems a scale of methods is available, see ORTEGA and RHEINBOLDT (1970). At least for small d and n the computation is easily performed. For instances the critical STEP 4 for the formulae in ex. 2 required at most 0.7 sec. This has to be balanced against the fact, that in some instances formulae of the same degree with less nodes may exist. But as ex. 2 shows, for small d even the minimal number of nodes is reached by our method.

ACKNOWLEDGEMENT. The programming work to obtain the formulae of ex. 2 was done by J. Brinker; we are very much indebted to him.

REFERENCES

Engels, H. (1980) Numerical quadrature and cubature (Academic Press, London).

Jain, M.K. (1979) Numerical solution of differential equations (Wiley Eastern Ltd, New Delhi).

Möller, H.M. (1976) Mehrdimensionale Hermite-Interpolation und numerische Integration, Math. Z. 148, 107-118.

Möller, H.M. (1979) The construction of cubature formulae and ideals of principal classes. In: Multivariate approximation theory, ed. W. Schempp and K. Zeller, ISNM 51 (Birkhäuser Verlag Basel).

Mysovskih, I.P. (1981) Interpoljacionnye kubaturnye formuly (Nauka , Moskva).

Ortega, J.M. and Rheinboldt, W.C. (1970) Iterative solution of nonlinear equations in several variables (Academic Press, New York).

H.M. Möller
FB Mathematik
FernUniversität
Postfach 940
D 5800 Hagen 1
W. Germany

ON THE APPROXIMATION BY MULTIPLE ORTHOGONAL SERIES

F. Móricz

Bolyai Institute, University of Szeged, Hungary

1. Introduction

Let N^d be the set of d-tuples $k = (k_1,\ldots,k_d)$ with positive integers for coordinates, where d is a fixed positive integer. As usual, we write $2^k = (2^{k_1},\ldots,2^{k_d})$, $1 = (1,\ldots,1)$, and $k \leq n$ iff $k_j \leq n_j$ for each $j = 1,\ldots,d$.

Let (X,F,μ) be an arbitrary positive measure space and $\{\phi_k(x) : k \in N^d\}$ an orthonormal system defined on X. We shall consider the d-multiple orthogonal series

$$(1.1)\qquad \sum_{k\in N^d} a_k\phi_k(x) = \sum_{k_1=1}^{\infty} \cdots \sum_{k_d=1}^{\infty} a_{k_1,\ldots,k_d}\phi_{k_1,\ldots,k_d}(x),$$

where $\{a_k : k \in N^d\}$ is a d-multiple sequence of real numbers (coefficients), for which

$$(1.2)\qquad \sum_{k\in N^d} a_k^2 < \infty.$$

By the well-known Riesz-Fischer theorem there exists a function $f(x) \in L^2(X,F,\mu)$ such that series (1.1) is the generalized Fourier series of $f(x)$ with respect to $\{\phi_k(x)\}$ and the rectangular partial sums of (1.1) defined by

$$s_n(x) = \sum_{1\leq k\leq n} a_k\phi_k(x) = \sum_{k_1=1}^{n_1} \dots \sum_{k_d=1}^{n_d} a_k\phi_k(x) \qquad (n \in N^d)$$

converge to f(x) in the metric of $L^2(X,F,\mu)$:

$$\int_X [s_n(x)-f(x)]^2 d\mu(x) \to 0 \quad \text{as} \quad \min_{1\leq j\leq d} n_j \to \infty.$$

It is a fundamental fact that condition (1.2) does not ensure the pointwise convergence of $s_n(x)$ to f(x) almost everywhere on X (in abbreviation: a.e.).

By the extension of the famous Rademacher-Menšov theorem, proved by a number of authors (see, e.g. [1] for d = 2, [9] for d ≧ 2 etc.), if

$$(1.3) \qquad \sum_{k\in N^d} a_k^2 \prod_{j=1}^{d} [\log(k_j+1)]^2 < \infty,$$

then $s_n(x)$ regularly converges a.e. (see [7] for d = 2 and [10] for d ≧ 2), a fortiori, converges to f(x) a.e. in the sense of Pringsheim, too.

It is a simple consequence that if $0 = n_j(1) \leq n_j(2) \leq \dots$ is a sequence of integers and $n_j(p) \to \infty$ as $p \to \infty$ for each j = 1,...,d, and if

$$\sum_{p\in N^d} \Big(\sum_{k_1=n_1(p_1)+1}^{n_1(p_1+1)} \dots \sum_{k_d=n_d(p_d)+1}^{n_d(p_d+1)} a_k^2\Big) \prod_{j=1}^{d} [\log(p_j+1)]^2 < \infty,$$

then the d-multiple subsequence $\{s_{n_1(p_1),\dots,n_d(p_d)}(x) : p = (p_1,\dots,p_d) \in N^d\}$ of the rectangular partial sums of (1.1) regularly converges a.e. (The empty sums $\sum_{k=n+1}^{n}$ if any are taken to equal 0.)

The case $n_j(p) = 0$ for p = 1 and 2^{p-2} for p = 2,3,... and for j = 1,...,d is of special interest: If

(1.4) $$\sum_{k \in N^d} a_k^2 \prod_{j=1}^{d} [\log\log (k_j+3)]^2 < \infty,$$

then $s_{2^{p_1},\ldots,2^{p_d}}(x)$ regularly converges a.e.

Denote by $\sigma_n(x)$ the first arithmetic means of $s_k(x)$:

$$\sigma_n(x) = \left(\prod_{j=1}^{d} n_j^{-1}\right) \sum_{k_1=1}^{n_1} \ldots \sum_{k_d=1}^{n_d} s_k(x) =$$

$$= \sum_{k_1=1}^{n_1} \ldots \sum_{k_d=1}^{n_d} \left[\prod_{j=1}^{d} \left(1-\frac{k_j-1}{n_j}\right)\right] a_k \phi_k(x) \qquad (n \in N^d).$$

For $d \geq 2$ the a.e. equiconvergence of $s_{2^p}(x)$ and $\sigma_{2^p}(x)$ is no longer true, which is the case for $d = 1$ (see, e.g. [2, p. 118]). In spite of this fact, under condition (1.4) the means $\sigma_n(x)$ do converge to $f(x)$ a.e. (see [5] for $d = 2$).

2. Approximation by $s_n(x)$ and $\sigma_n(x)$

Let $\{\kappa(n) : n \in N^d\}$ and $\{\lambda(n) : n \in N^d\}$ be two d--multiple sequences of real numbers, $\lambda(n) \neq 0$ if $\min_{1 \leq j \leq d} n_j$ is large enough. We set

$$\kappa(n) = o\{\lambda(n)\}$$

if there exists a constant C such that

$$|\kappa(n)| \leq C|\lambda(n)| \qquad (n \in N^d)$$

and

$$\frac{\kappa(n)}{\lambda(n)} \to 0 \qquad \text{as} \qquad \min_{1 \leq j \leq d} n_j \to \infty.$$

Here and in the sequel C denotes a positive constant, not necessarily the same at each occurrence. Furthermore, $\{\lambda(n)\}$ is said to be nondecreasing if

$\lambda(n_1,\ldots,n_{j-1},n_j,n_{j+1},\ldots,n_d) \leqq \lambda(n_1,\ldots,n_{j-1},n_j+1,$ $n_{j+1},\ldots,n_d)$ for each $j = 1,\ldots,d$ and $n \in N^d$.

In the introduction it was already mentioned that (1.3) and (1.4) are sufficient conditions for the a.e. convergence of $s_n(x)$ and $\sigma_n(x)$ to $f(x)$, respectively. Now replacing (1.3) by a stronger condition, one can even state an approximation rate for the deviation $s_n(x)-f(x)$. The following theorem is a generalization of [14, Satz 1].

Theorem 1. If $\{\lambda(n) : n \in N^d\}$ is a nondecreasing sequence of positive numbers, for which

$$(2.1)\qquad \lambda(n) \to \infty \qquad \text{as} \qquad \max_{1\leqq j\leqq d} n_j \to \infty$$

and

$$(2.2)\qquad \sum_{k\in N^d} a_k^2\lambda^2(k) \prod_{j=1}^{d} [\log (k_j+1)]^2 < \infty,$$

then

$$s_n(x)-f(x) = o_x\{\max_{1\leqq j\leqq d} \lambda^{-1}(1,\ldots,1,n_j,1,\ldots,1)\} \qquad \text{a.e.}$$

The proof of Theorem 1 is based on the extended Rademacher-Menšov theorem and a d-multiple Abel transformation (concerning the latter, see [6] and also [11]).

Problem 1. It seems to be very likely, but proved not yet, that under certain conditions ensuring the "regular increase" of $\{\lambda(n)\}$, the condition

$$(2.3)\qquad \sum_{k\in N^d} a_k^2\lambda^2(k) \prod_{j=1}^{d} [\log\log (k_j+3)]^2 < \infty$$

implies the statement

$$(2.4)\qquad \sigma_n(x)-f(x) = o_x\{\max_{1\leqq j\leqq d} \lambda^{-1}(1,\ldots,1,n_j,1,\ldots,1)\} \qquad \text{a.e.}$$

Now the main point is that if $\lambda(k)$ is of the particular form $\prod_{j=1}^{d} k_j^{2\gamma_j}$ with $0 < \gamma_j < 1$ for $j = 1,\ldots,d$, then one can delete the factor $\prod_{j=1}^{d} [\log\log(k_j+3)]^2$ from condition (2.3) without spoiling conclusion (2.4). More precisely, the following theorem is valid.

Theorem 2. If

$$\sum_{k\in N^d} a_k^2 \prod_{j=1}^{d} k_j^{2\gamma_j} < \infty \tag{2.5}$$

with some $0 < \gamma_j < 1$ for each $j = 1,\ldots,d$, then

$$\sigma_n(x) - f(x) = o_x\{\max_{1\leq j\leq d} n_j^{-\gamma_j}\} \quad \text{a.e.} \tag{2.6}$$

For the special case $d = 1$, Theorem 2 was proved in [8].

The proof of Theorem 2 runs in great lines as follows. Without loss of generality one may assume that $a_k = 0$ if $k_j = 1$ for at least one $j = 1,\ldots,d$. First, by (2.5),

$$\sum_{n_1=0}^{\infty} \cdots \sum_{n_d=0}^{\infty} \left(\prod_{j=1}^{d} 2^{2n_j\gamma_j}\right) \int_X \Big[\sum_{k_1=2^{n_1}+1}^{2^{n_1+1}} \cdots \sum_{k_d=2^{n_d}+1}^{2^{n_d+1}} a_k\phi_k(x)\Big]^2 d\mu(x) < \infty,$$

whence B. Levi's theorem implies

$$\left(\prod_{j=1}^{d} 2^{n_j\gamma_j}\right) \sum_{k_1=2^{n_1}+1}^{2^{n_1+1}} \cdots \sum_{k_d=2^{n_d}+1}^{2^{n_d+1}} a_k\phi_k(x) \to 0 \quad \text{a.e.}$$

as $\max_{1\leq j\leq d} n_j \to \infty$. From here it follows that

$$s_{2^{p_1},\ldots,2^{p_d}}(x) - f(x) = o_x\{\max_{1\leq j\leq d} 2^{-p_j\gamma_j}\} \quad \text{a.e.} \tag{2.7}$$

Then one proves that

$$\sigma_{2^{p_1},\ldots,2^{p_d}}(x) - s_{2^{p_1},\ldots,2^{p_d}}(x) = o_x\{\max_{1\leq j\leq d} 2^{-p_j\gamma_j}\} \quad \text{a.e.,} \tag{2.8}$$

while using the following representation of the difference in the left-hand side:

$$(2.9)\qquad \sum_{k_1=1}^{2^{p_1}} \cdots \sum_{k_d=1}^{2^{p_d}} \{- \sum_{j=1}^{d} \frac{k_j-1}{2^{p_j}} + \sum_{1\leq j_1<j_2\leq d} \frac{(k_{j_1}-1)(k_{j_2}-1)}{2^{p_{j_1}} 2^{p_{j_2}}} - \cdots$$

$$\cdots + (-1)^d \prod_{j=1}^{d} \frac{k_j-1}{2^{p_j}}\} a_k \phi_k(x).$$

Finally, one shows that

$$(2.10)\qquad \max_{2^{p_1}<n_1\leq 2^{p_1+1}} \cdots \max_{2^{p_d}<n_d\leq 2^{p_d+1}} |\sigma_{n_1,\ldots,n_d}(x) -$$

$$-\sigma_{2^{p_1},\ldots,2^{p_d}}(x)| = o_x\{\max_{1\leq j\leq d} 2^{-p_j\gamma_j}\} \qquad \text{a.e.}$$

The trick of how to estimate this d-multiple maximum is illuminated in the special case d = 2 as follows:

$$\max_{2^{p_1}<n_1\leq 2^{p_1+1}} \max_{2^{p_2}<n_2\leq 2^{p_2+1}} |\sigma_{n_1,n_2}(x) - \sigma_{2^{p_1},2^{p_2}}(x)| \leq$$

$$\leq \max_{2^{p_1}<n_1\leq 2^{p_1+1}} \max_{2^{p_2}<n_2\leq 2^{p_2+1}} |\sigma_{n_1,n_2}(x) - \sigma_{n_1,2^{p_2}}(x) -$$

$$-\sigma_{2^{p_1},n_2}(x) + \sigma_{2^{p_1},2^{p_2}}(x)| + \max_{2^{p_1}<n_1\leq 2^{p_1+1}} |\sigma_{n_1,2^{p_2}}(x) -$$

$$-\sigma_{2^{p_1},2^{p_2}}(x)| + \max_{2^{p_2}<n_2\leq 2^{p_2+1}} |\sigma_{2^{p_1},n_2}(x) - \sigma_{2^{p_1},2^{p_2}}(x)| =$$

$$= M_1(x) + M_2(x) + M_3(x).$$

Now the Cauchy inequality yields

$$M_1^2(x) \leq \sum_{k_1=2^{p_1}+1}^{2^{p_1+1}} \sum_{k_2=2^{p_2}+1}^{2^{p_2+1}} k_1^{2\gamma_1+1} k_2^{2\gamma_2+1} [\sigma_{k_1,k_2}(x)-$$

$$-\sigma_{k_1,k_2-1}(x)-\sigma_{k_1-1,k_2}(x)+\sigma_{k_1-1,k_2-1}(x)]^2$$

$$\sum_{k_1=2^{p_1}+1}^{2^{p_1+1}} \sum_{k_2=2^{p_2}+1}^{2^{p_2+1}} k_1^{-2\gamma_1-1} k_2^{-2\gamma_2-1} .$$

Since the second factor here is $O\{2^{-2p_1\gamma_1} 2^{-2p_2\gamma_2}\}$, it is enough to demonstrate that the first factor on the right-hand side tends to 0 a.e. as $\max(p_1,p_2) \to \infty$. As a result we get that

$$M_1(x) = o_x\{2^{-p_1\gamma_1} 2^{-p_2\gamma_2}\} \qquad \text{a.e.}$$

Analogously,

$$M_2(x) = o_x\{2^{-p_1\gamma_1}\} \quad \text{and} \quad M_3(x) = o_x\{2^{-p_2\gamma_2}\} \quad \text{a.e.}$$

Relations (2.7), (2.8) and (2.10) clearly provides (2.6).

Assuming that $n = (n_1,\dots,n_d)$ tends restrictedly to ∞, one can obtain essentially better rate of approximation. In other words, the assumption is that the n_j tend to ∞ in such a way that all the ratios n_j/n_k $(j,k = 1,\dots,d)$ remain bounded.

<u>Theorem 3.</u> If

(2.11) $$\sum_{k\in N^d} a_k^2 [\max_{1\leq j\leq d} k_j]^{2\gamma} < \infty$$

with some $0 < \gamma < 1$, then for every $\theta > 1$

(2.12) $$\max_{n_2:\theta^{-1}\leq n_2/n_1\leq\theta} \cdots \max_{n_d:\theta^{-1}\leq n_d/n_1\leq\theta} |\sigma_n(x)-f(x)| =$$

$$= o_x\{n_1^{-\gamma}\} \qquad \text{a.e.}$$

In comparison with Theorem 2, condition (2.11) is not so restrictive as condition (2.5) in the case $\gamma_1 = \ldots = \gamma_d = \gamma$. On the other hand, the rate of approximation in statement (2.12) is not worse than in (2.6).

The proof of Theorem 3 can be carried out in a similar manner as that of Theorem 2 sketched above. The detailed proofs of both theorems will appear in [12].

3. Strong approximation by $s_k(x)$

It is a trivial consequence of statement (2.12) that

$$\Big(\prod_{j=1}^{d} n_j^{-1}\Big) \sum_{k_1=1}^{n_1} \ldots \sum_{k_d=1}^{n_d} [s_{k_1,\ldots,k_d}(x) - f(x)] = o_x\{n_1^{-\gamma}\} \quad \text{a.e.}$$

provided

$$\theta^{-1} \leq n_j/n_k \leq \theta \qquad (j,k = 1,\ldots,d),$$

where $\theta > 1$ is fixed.

The following theorem indicates that the mean value of $s_k(x)-f(x)$ is of $o_x\{n_1^{-\gamma}\}$, not because of the cancellation of positive and negative terms, but because the indices $k = (k_1,\ldots,k_d)$ for which $|s_k(x)-f(x)|$ is not small are sparse.

Theorem 4. If (2.11) is satisfied with some $0 < \gamma < 1/2$, then for every $\theta > 1$

$$(3.1) \qquad n_1^{-d} \sum_{k_1=1}^{n_1} \sum_{k_2:\theta^{-1}\leq k_2/k_1\leq\theta} \ldots \sum_{k_d:\theta^{-1}\leq k_d/k_1\leq\theta} [s_k(x)-f(x)]^2 = o_x\{n_1^{-2\gamma}\} \qquad \text{a.e.}$$

Following [3], this type of approximation is called strong approximation. In particular, from (3.1) it immediately follows that

$$n_1^{-d} \sum_{k_1=1}^{n_1} \sum_{k_2:\theta^{-1}\leq k_2/k_1\leq\theta} \cdots \sum_{k_d:\theta^{-1}\leq k_d/k_1\leq\theta} |s_k(x)-f(x)| = \\ = o_x\{n_1^{-\gamma}\} \quad \text{a.e.}$$

For the special case d = 1, Theorem 4 was proved in [13].

The proof of Theorem 4 is done in two steps. First, by Theorem 3,

$$(3.2) \qquad n_1^{-d} \sum_{k_1=1}^{n_1} \sum_{k_2:\theta^{-1}\leq k_2/k_1\leq\theta} \cdots \sum_{k_d:\theta^{-1}\leq k_d/k_1\leq\theta} [\sigma_k(x)-f(x)]^2 = \\ = o_x\{n_1^{-2\gamma}\} \quad \text{a.e.}$$

Second, making use of a representation corresponding to (2.9), one can deduce that

$$(3.3) \qquad n_1^{-d} \sum_{k_1=1}^{n_1} \sum_{k_2:\theta^{-1}\leq k_2/k_1\leq\theta} \cdots \sum_{k_d:\theta^{-1}\leq k_d/k_1\leq\theta} [s_k(x)-\sigma_k(x)]^2 = \\ = o_x\{n_1^{-2\gamma}\} \quad \text{a.e.}$$

Inequalities (3.2) and (3.3) obviously lead to (3.1). As to the details cf. [12].

In addition, the proof of (3.3) remains valid for $\gamma = 0$, too. This special case deserves some interest in itself.

<u>Theorem 5.</u> If condition (1.2) is satisfied, $\sigma_n(x)$ converges to f(x) as $\min_{1\leq j\leq d} n_j \to \infty$ and bounded a.e., then for every $\theta > 1$ the left-hand side of (3.1) is $o_x\{1\}$ a.e.

For the special case d = 1, Theorem 5 was proved in [4].

<u>Problem 2.</u> It is very unlikely, although a counter-example has not yet been given, that under conditions of Theorem 4

$$n_1^{-1} \sum_{k_1=1}^{n_1} \Big(\max_{k_2:\theta^{-1}\leq k_2/k_1\leq\theta} \cdots \max_{k_d:\theta^{-1}\leq k_d/k_1\leq\theta} [s_k(x)-f(x)]^2 \Big) \neq$$

$$\neq o_x\{n_1^{-2\gamma}\} \qquad \text{a.e.}$$

Problem 3. Suppose condition (2.5) is satisfied with $\gamma_1 = \ldots = \gamma_d = \gamma$, $0 < \gamma < 1/2$. It is an open question whether the relation

$$\Big(\prod_{j=1}^{d} n_j^{-1}\Big) \sum_{k_1=1}^{n_1} \ldots \sum_{k_d=1}^{n_d} [s_k(x)-f(x)]^2 = o_x\{\min_{1\le j\le d} n_j^{-2\gamma}\} \qquad \text{a.e.}$$

holds true or not.

4. Approximation by square partial sums

In the case when $n_1 = \ldots = n_d = n$ $s_{n,\ldots,n}(x)$ is called the square partial sum of the d-multiple series (1.1). Denote by $\tau_r(x)$ the first arithmetic mean of the square partial sums:

$$\tau_r(x) = r^{-1} \sum_{j=1}^{r} s_{j,\ldots,j}(x) =$$

$$= \sum_{j=1}^{r} \Big(1-\frac{j-1}{r}\Big) \sum_{k:\max(k_1,\ldots,k_d)=j} a_k \phi_k(x).$$

Here and in the sequel $r \in N$.

The results of [14], [8], [13] and [4] pertaining to the case $d = 1$ are extended as follows.

Theorem 6a. If $\{\lambda(r): r = 1,2,\ldots\}$ is a nondecreasing sequence of positive numbers tending to ∞ and if

$$\sum_{r=1}^{\infty} \Big(\sum_{k:\max(k_1,\ldots,k_d)=r} a_k^2\Big) \lambda^2(r)[\log (r+1)]^2 < \infty,$$

then

$$s_{r,\ldots,r}(x) = \sum_{k_1=1}^{r} \ldots \sum_{k_d=1}^{r} a_k \phi_k(x) = o_x\{\lambda^{-1}(r)\} \qquad \text{a.e.}$$

Theorem 6b. If $\{\lambda(r)\}$ is a nondecreasing sequence of positive numbers tending to ∞ such that

$$\lambda(r^2) \leqq C\lambda(r) \qquad (r = 1,2,\ldots),$$

and if

$$\sum_{r=1}^{\infty} \Big(\sum_{k:\max(k_1,\ldots,k_d)=r} a_k^2 \Big) \lambda^2(r) [\log\log(r+3)]^2 < \infty,$$

then

$$\tau_r(x) - f(x) = o_x\{\lambda^{-1}(r)\} \qquad \text{a.e.}$$

Theorem 7. If

$$\sum_{r=1}^{\infty} \Big(\sum_{k:\max(k_1,\ldots,k_d)=r} a_k^2 \Big) r^{2\gamma} < \infty \tag{4.1}$$

with some $0 < \gamma < 1$, then

$$\tau_r(x) - f(x) = o_x\{r^{-\gamma}\} \qquad \text{a.e.}$$

Theorem 8. If condition (4.1) is satisfied with some $0 < \gamma < 1/2$, then

$$r^{-1} \sum_{j=1}^{r} [s_{j,\ldots,j}(x) - f(x)]^2 = o_x\{r^{-2\gamma}\} \qquad \text{a.e.} \tag{4.2}$$

Theorem 9. If condition (1.2) is satisfied and $\sigma_r(x)$ converges to $f(x)$ a.e., then the left-hand side of (4.2) is $o_x\{1\}$ a.e.

References

[1] Agnew, P.R. (1932) On double orthogonal series. Proc. London Math. Soc., Ser. 2, 33, 420-434.

[2] Alexits, G. (1961) Convergence problems of orthogonal series (Pergamon, Oxford).

[3] Alexits, G. (1964) Über die Approximation im starken Sinne. Approximationstheorie (Proc. Conf. Oberwolfach 1963), 89-95 (Birkhäuser, Basel).

[4] Borgen, S. (1928) Über (C,1)-Summierbarkeit von Reihen orthogonaler Funktionen. Math. Annalen 98, 125-150.

[5] Csernyák, L. (1968) Bemerkung zur Arbeit von V.S. Fedulov "Über die Summierbarkeit der doppelten Orthogonalreihen". Publ. Math. Debrecen 15, 95-98.

[6] Hardy, G.H. (1903-1904) On the convergence of certain multiple series. Proc. London Math. Soc., Ser. 2, 1, 124-128.

[7] Hardy, G.H. (1916-1919) On the convergence of certain multiple series. Proc. Cambridge Philosoph. Soc. 19, 86-95.

[8] Leindler, L. (1964) Über die punktweise Konvergenz von Summationsverfahren allgemeiner Orthogonalreihen. Approximationstheorie (Proc. Conf. Oberwolfach 1963), 239-244 (Birkhäuser, Basel).

[9] Móricz, F. (1978) Multiparameter strong laws of large numbers. I (Second order moment restrictions). Acta Sci. Math. (Szeged) 40, 143-156.

[10] Móricz, F. (1979) On the convergence in a restricted sense of multiple series. Analysis Math. 5, 135-147.

[11] Móricz, F. (1981) The Kronecker lemmas for multiple series and some applications. Acta Math. Acad. Sci. Hungar. 37, 39-50.

[12] Móricz, F. (1983) Approximation by rectangular partial sums and Cesàro means of multiple orthogonal series. Tôhoku Math. J., to appear.

[13] Sunouchi, G. (1967) Strong approximation by Fourier series and orthogonal series. Indian J. Math. 9, 237-246.

[14] Tandori, K. (1959) Über die orthogonalen Funktionen. VII (Approximationssätze). Acta Sci. Math. (Szeged) 20, 19-24.

Prof. Ferenc Móricz, Bolyai Institute, University of Szeged, 6720 Szeged, Aradi vértanúk tere 1, Hungary.

QUANTITATIVE THEOREMS ON APPROXIMATION PROCESSES OF POSITIVE LINEAR OPERATORS

Toshihiko Nishishiraho
Department of Mathematics
Ryukyu University
Okinawa, Japan

We establish a theorem of Korovkin type for approximation processes of positive linear operators and give a quantitative version of this result. The most typical examples of these processes in question are summation processes of positive linear operators which can be obtained by very general summability methods including convergence, almost convergence and so on. Also, an example of applications is given by the Bernstein-Lototsky-Schnabl functions on compact convex subsets of a real locally convex Hausdorff vector space.

1. Introduction

Let X be a compact Hausdorff space and let B(X) denote the Banach lattice of all real-valued bounded functions on X with the supremum norm $||\cdot||$. C(X) denotes the closed sublattice of B(X) consisting of all real-valued continuous functions on X. Here we are concerned with approximation processes defined as follows (cf. [14]):

DEFINITION 1. *Let* A *be a linear subspace of* B(X) *and let* $\{T_{\alpha,\lambda};\ \alpha\in D,\ \lambda\in\Lambda\}$ *be a family of bounded linear operators of* A *into* B(X), *where* D *is a directed set and* Λ *is an arbitrary index set. The family* $\{T_{\alpha,\lambda}\}$ *is said to be an approximation process on* A *if for every* $f\in A$,

$$\lim_{\alpha} ||T_{\alpha,\lambda}(f) - f|| = 0 \quad \textit{uniformly in } \lambda\in\Lambda.$$

In this paper we establish a theorem of Korovkin type with respect to this convergence behaviour for positive linear operators and give a quantitative version of this result under suitable conditions.

Such problems are now classical for the usual convergence in C[a, b], with [a, b] a finite closed interval of the real line; an excellent source for references and a systematic treatment of quantitative Korovkin theorems for positive linear operators in C[a, b] can be found in the book of DeVore [3]. Also, for the multi-dimensional case see Censor [2], and for an infinite dimensional case see the author [13].

Concerning the almost convergence introduced by Lorentz [9], in C[a, b] they were studied by King and Swetits [8] and by Mohapatra [12], whose results were recently generalized by Swetits [17] to a general summability method considered by Bell [1] (cf. [10]) which includes F_A-summability of Lorentz [9], A_B-summability of Mazhar and Siddiqi [11] and order summability of Jurkat and Peyerimhoff [6, 7] and so on.

The quantitative theory for linear approximation processes of convolution operators in an arbitrary Banach space setting is treated by the author [14].

The second problem is studied in the setting of a real locally convex Hausdorff vector space (cf. [13]). Our results sharpen the generalized Korovkin-type convergence theorem with a quantitative estimation of the rate of convergence on the test systems. They also give an estimation of the rate of convergence of various summation processes of positive linear operators, which can be induced by the method of $\mathcal{A}$-summability introduced by the author [15], which recovers that by Bell [1] (cf. [10]). Consequently, we obtain a generalization of the results of Swetits [17] to an infinite dimensional case. Also, an example of applications is given by the Bernstein-Lototsky-Schnabl functions on compact convex subsets of a real locally convex Hausdorff vector space (cf. [5], [16]).

2. A Convergence Theorem: A Korovkin-Type Theorem

Let G be a subset of C(X) which separates the points of X. Throughout this paper, let A be a linear subspace of B(X) which contains 1_X, G and G^2, where 1_X denotes the unit function on X defined by $1_X(x) = 1$ for all $x \in X$ and $G^2 = \{g^2; g \in G\}$, and let $\{T_{\alpha,\lambda}; \alpha \in D, \lambda \in \Lambda\}$ be a family of positive linear operators of A into B(X).

Given a positive linear operator T of A into B(X), a function $g \in G$ and a point $x \in X$, we define

$$\mu(x; T, g) = T((g - g(x)1_X)^2)(x).$$

Then we have

THEOREM 1. *If for every* $g \in G$,

$$\lim_\alpha \|\mu(\cdot; T_{\alpha,\lambda}, g)\| = 0 \quad \textit{uniformly in } \lambda \in \Lambda$$

and

$$\lim_\alpha \|T_{\alpha,\lambda}(1_X) - 1_X\| = 0 \quad \textit{uniformly in } \lambda \in \Lambda,$$

then for every $f \in A \cap C(X)$,

$$\lim_\alpha \|T_{\alpha,\lambda}(f) - f\| = 0 \quad \textit{uniformly in } \lambda \in \Lambda. \tag{1}$$

In particular, if for every $g \in G$ *and for* $i = 0, 1, 2$,

$$\lim_\alpha \|T_{\alpha,\lambda}(g^i) - g^i\| = 0 \quad \textit{uniformly in } \lambda \in \Lambda, \tag{2}$$

then (1) *holds*.

PROOF. This is carried out by adapting the method of proof of Theorem 1 of [13], and so we omit the details.

Theorem 1 asserts that if A is contained in C(X), then $\{T_{\alpha,\lambda}\}$ is an approximation process on A if and only if (2) holds for every $g \in G$ and for $i = 0, 1, 2$.

REMARK 1. Theorem 1 can be reformulated with respect to pointwise convergence and the following localization principle holds: Let $x \in X$. Suppose that for every $g \in G$,

$$\lim_\alpha \mu(x; T_{\alpha,\lambda}, g) = 0 \quad \text{uniformly in } \lambda \in \Lambda.$$

If f_1 and f_2 are two functions in A such that $f_1 = f_2$ on a neighborhood of x, then

$$\lim_\alpha T_{\alpha,\lambda}(f_1)(x) = \lim_\alpha T_{\alpha,\lambda}(f_2)(x) \quad \text{uniformly in } \lambda \in \Lambda.$$

3. A Quantitative Theorem

Let E be a real locally convex Hausdorff vector space and E' its dual space. In this section, it will be assumed that X is a compact convex subset of E and $G = \{h|_X; h \in E'\}$, where $h|_X$ denotes the restriction of h to X and that for each $\alpha \in D$,

$$\sup\{\|T_{\alpha,\lambda}(1_X)\|; \lambda \in \Lambda\} < \infty.$$

Here we give a quantitative version of Theorem 1. For this we need some definitions and lemmas.

DEFINITION 2. *Let* $f \in B(X)$, $\{g_1, g_2, \cdots, g_m\}$ *a finite subset of* G *and* δ *a non-negative real number. We define*

$$\omega(f; g_1, g_2, \cdots, g_m, \delta) = \sup\{|f(x) - f(y)|; x, y \in X, \sum_{i=1}^m (g_i(x) - g_i(y))^2 \leqq \delta^2\}$$

and

$$\omega(f, \delta) = \inf\{\omega(f; g_1, g_2, \cdots, g_m, \delta); g_1, g_2, \cdots, g_m \in G, \|\sum_{i=1}^m g_i^2\| = 1, m = 1, 2, \cdots\},$$

which are called the modulus of continuity of f *with respect to* $g_1, g_2, \cdots, g_m$ *and the total modulus of continuity of* f, *respectively*.

REMARK 2. If X is a compact convex subset of the m-dimensional real Euclidean space and g_i is the i-th coordinate function on X, i.e., $g_i(x_1, x_2, \cdots, x_m) = x_i$, then $\omega(f; g_1, g_2, \cdots, g_m, \delta)$ reduces to the usual modulus of continuity of f (cf. [2], [13; Remark 2]).

DEFINITION 3. *For each* $\alpha \in D$ *and* $f \in A$, *we define*

$$\omega_\alpha(f) = \inf\{\omega(f; g_1, g_2, \cdots, g_m, \mu_\alpha(g_1, g_2, \cdots, g_m)^{1/2});$$
$$g_1, g_2, \cdots, g_m \in G, \mu_\alpha(g_1, g_2, \cdots, g_m) > 0, m = 1, 2, 3, \cdots\},$$

where

$$\mu_\alpha(g_1, g_2, \cdots, g_m) = \sup\{\| \sum_{i=1}^m \mu(\cdot; T_{\alpha,\lambda}, g_i)\|; \lambda \in \Lambda\}.$$

The moduli of continuity have the following fundamental properties:

LEMMA 1. *Let* $f \in B(X)$ *and* $\{g_1, g_2, \cdots, g_m\}$ *a finite subset of* G.

(i) $\omega(f; g_1, g_2, \cdots, g_m, \cdot)$ *is a non-decreasing function on* $[0, \infty)$.

(ii) $\omega(f; g_1, g_2, \cdots, g_m, \xi\delta) \leqq (1 + \xi)\omega(f; g_1, g_2, \cdots, g_m, \delta)$ *for each* $\xi, \delta \geqq 0$.

(iii) *If* $f \in C(X)$, *then* $\lim_{\delta \to 0+}\omega(f, \delta) = 0$.

LEMMA 2. *If* $\lim_\alpha \mu_\alpha(g) = 0$ *for every* $g \in G$, *then* $\lim_\alpha \omega_\alpha(f) = 0$ *for every* $f \in A \cap C(X)$.

LEMMA 3. *Let* T *be a positive linear operator of* A *into* $B(X)$. *Let* $x \in X$, $f \in A$ *and* $\{g_1, g_2, \cdots, g_m\}$ *a finite subset of* G. *Then, for each* $\delta > 0$, *we have*

$$|T(f)(x) - f(x)T(1_X)(x)| \leqq \{T(1_X)(x) + \delta^{-2}\sum_{i=1}^m \mu(x; T, g_i)\}$$
$$\times\ \omega(f; g_1, g_2, \cdots, g_m, \delta).$$

PROOF. Let y be an arbitrary point of X. If $\Sigma_{i=1}^m(g_i(x) - g_i(y))^2 > \delta^2$, then it follows from (ii) of Lemma 1 that

$$|f(y) - f(x)| \leqq \omega(f; g_1, g_2, \cdots, g_m, \delta)$$
$$\times\ \{1 + \delta^{-1}(\sum_{i=1}^m (g_i(y) - g_i(x))^2)^{1/2}\}$$
$$\leqq \omega(f; g_1, g_2, \cdots, g_m, \delta)$$
$$\times\ \{1 + \delta^{-2}\sum_{i=1}^m (g_i(y) - g_i(x))^2\}.$$

If $\Sigma_{i=1}^{m}(g_i(y) - g_i(x))^2 \leqq \delta^2$, then

$$|f(y) - f(x)| \leqq \omega(f; g_1, g_2, \cdots, g_m, \delta).$$

Consequently, we have

$$|f - f(x)1_X| \leqq \omega(f; g_1, g_2, \cdots, g_m, \delta) \times \{1_X + \delta^{-2}\sum_{i=1}^{m}(g_i - g_i(x)1_X)^2\}.$$

Thus, since T is positive and linear, we have

$$|T(f) - f(x)T(1_X)| \leqq \omega(f; g_1, g_2, \cdots, g_m, \delta) \times \{T(1_X) + \delta^{-2}\sum_{i=1}^{m}T((g_i - g_i(x)1_X)^2)\},$$

which implies the desired result.

For each $f \in A$ and $\alpha \in D$, let

$$|||T_\alpha(f) - f||| = \sup\{||T_{\alpha,\lambda}(f) - f||; \lambda \in \Lambda\}.$$

Note that $\{T_{\alpha,\lambda}\}$ is an approximation process on A if and only if $\lim_\alpha |||T_\alpha(f) - f||| = 0$ for all $f \in A$.

We are now in a position to recast Theorem 1 in a quantitative form as follows.

THEOREM 2. *For all* $f \in A$ *and all* $\alpha \in D$, *we have*

$$|||T_\alpha(f) - f||| \leqq ||f||\,|||T_\alpha(1_X) - 1_X||| + C_\alpha\omega_\alpha(f), \tag{3}$$

where

$$C_\alpha = \sup\{||T_{\alpha,\lambda}(1_X) + 1_X||; \lambda \in \Lambda\}.$$

In particular, if $T_{\alpha,\lambda}(1_X) = 1_X$ *for all* $\alpha \in D$ *and all* $\lambda \in \Lambda$, *then* (3) *reduces to*

$$|||T_\alpha(f) - f||| \leqq 2\omega_\alpha(f).$$

PROOF. Obviously, we have

$$|||T_\alpha(f) - f||| \leqq ||f||\,|||T_\alpha(1_X) - 1_X||| + K_\alpha(f), \tag{4}$$

where

$$K_\alpha(f) = \sup\{\|T_{\alpha,\lambda}(f) - fT_{\alpha,\lambda}(1_X)\|;\ \lambda \in \Lambda\}.$$

Let $\{g_1, g_2, \cdots, g_m\}$ be a finite subset of G with $\mu_\alpha(g_1, g_2, \cdots, g_m) > 0$. Taking $T = T_{\alpha,\lambda}$ in Lemma 3, we conclude

$$(5)\qquad |T_{\alpha,\lambda}(f) - fT_{\alpha,\lambda}(1_X)| \leqq \omega(f; g_1, g_2, \cdots, g_m, \delta) \times \{T_{\alpha,\lambda}(1_X) + \delta^{-2}\mu_\alpha(g_1, g_2, \cdots, g_m)1_X\}.$$

Setting $\delta = \mu_\alpha(g_1, g_2, \cdots, g_m)^{1/2}$ in (5) and taking the norm, we establish

$$\|T_{\alpha,\lambda}(f) - fT_{\alpha,\lambda}(1_X)\| \leqq \|T_{\alpha,\lambda}(1_X) + 1_X\| \times \omega(f; g_1, g_2, \cdots, g_m, \mu_\alpha(g_1, g_2, \cdots, g_m)^{1/2}).$$

Thus we have

$$K_\alpha(f) \leqq C_\alpha \omega(f; g_1, g_2, \cdots, g_m, \mu_\alpha(g_1, g_2, \cdots, g_m)^{1/2}),$$

and so $K_\alpha(f) \leqq C_\alpha \omega_\alpha(f)$, which yields (3) by (4).

COROLLARY 1. *Let* $\{g_1, g_2, \cdots, g_m\}$ *be a finite subset of* G *and* $(\xi_\alpha)_{\alpha \in D}$ *a net of positive real numbers converging to zero. If*

$$|||T_\alpha(1_X) - 1_X||| = O(\xi_\alpha^2) \quad \textit{and} \quad |||T_\alpha(g_j^i) - g_j^i||| = O(\xi_\alpha^2)$$

for all $\alpha \in D$, $i = 1, 2$ *and* $j = 1, 2, \cdots, m$, *then there exists a constant* $C > 0$ *such that for all* $f \in A$ *and all* $\alpha \in D$,

$$(6)\qquad |||T_\alpha(f) - f||| \leqq C(1 + \|f\|) \times \{\xi_\alpha^2 + \omega(f; g_1, g_2, \cdots, g_m, \xi_\alpha)\}.$$

In particular, if for all $\alpha \in D$, $\lambda \in \Lambda$ *and all* $g \in G$, $T_{\alpha,\lambda}(1_X) = 1_X$ *and* $T_{\alpha,\lambda}(g) = g$ *and if there exists a constant* $K > 0$ *such that* $\|\Sigma_{i=1}^n (T_{\alpha,\lambda}(h_i^2) - h_i^2)\| \leqq K\xi_{\alpha,\lambda}\|\Sigma_{i=1}^n h_i^2\|$ *for all* $\alpha \in D$, $\lambda \in \Lambda$ *and for any finite subset* $\{h_1, h_2, \cdots, h_n\}$ *of* G, *where* $\{\xi_{\alpha,\lambda};\ \alpha \in D, \lambda \in \Lambda\}$ *is a family of positive real numbers with*

$\lim_\alpha \xi_{\alpha,\lambda} = 0$ *uniformly in* $\lambda \in \Lambda$, *then* (6) *reduces to*

$$|||T_\alpha(f) - f||| \leqq 2\omega(f, K^{1/2}\xi_\alpha)$$

with

$$\xi_\alpha = (\sup\{\xi_{\alpha,\lambda};\ \lambda \in \Lambda\})^{1/2}.$$

REMARK 3. It should be noted that results analogous to Theorem 3 and Corollary 3 in [13] are obtained for the process $\{T_{\alpha,\lambda}\}$. We also note that all the results obtained in this section can be reformulated with respect to pointwise convergence, which is the situation to which Remark 1 refers.

4. $\mathcal{A}$-Summation Processes of Positive Linear Operators

Let N denote the set of all non-negative integers. In view of the concept of $\mathcal{A}$-summability introduced by the author [15], we make the following definition.

DEFINITION 4. *Let* $\mathcal{A} = \{A^{(\lambda)};\ \lambda \in \Lambda\}$ *be a family of infinite matrices* $A^{(\lambda)} = (a_{nm}^{(\lambda)})_{n,m\in N}$ *of real numbers. A sequence* $\{L_m\}_{m\in N}$ *of bounded linear operators of* A *into* B(X) *is said to be an* $\mathcal{A}$-*summation process on* A *if* $\{L_m(f)\}$ *is* $\mathcal{A}$-*summable to* f *for every* $f \in A$, i.e.,

$$(7) \qquad \lim_{n\to\infty} \|\sum_{m=0}^{\infty} a_{nm}^{(\lambda)} L_m(f) - f\| = 0 \quad \textit{uniformly in } \lambda \in \Lambda,$$

where it is assumed that the series in (7) *converge for each* n, λ *and* f.

We shall now mention some examples.

(1°) Given a matrix B, if $A^{(\lambda)} = B$ for all $\lambda \in \Lambda$, then $\mathcal{A}$-summability is just matrix summability by B.

(2°) Let $Q = \{q^{(\lambda)};\ \lambda \in \Lambda\}$ be a family of sequences $q^{(\lambda)} = \{q_m^{(\lambda)}\}_{m\in N}$ of real numbers such that for each n and λ

$$Q_n^{(\lambda)} = q_0^{(\lambda)} + q_1^{(\lambda)} + \cdots + q_n^{(\lambda)} \neq 0.$$

Let

$$a_{nm}^{(\lambda)} = q_{n-m}^{(\lambda)}/Q_n^{(\lambda)} \quad \text{for } 0 \leqq m \leqq n$$
$$= 0 \quad \text{for } m > n.$$

Then $\mathcal{A}$-summability is called a (N, Q)-aummability. Clearly if, for a sequence $\{q_m\}$ of non-negative real numbers with $q_0 > 0$, one takes $q_m^{(\lambda)} = q_m$ for each m and λ, then (N, Q)-summability reduces to the Nörlund summability. Another example is the following: Let Λ be a subset of $(0, \infty)$ and $q_m^{(\lambda)} = A_m^{(\lambda - 1)}$ for each $m \in N$, $\lambda \in \Lambda$, where

$$A_m^{(\kappa)} = \binom{m+\kappa}{m} = \{(\kappa+1)(\kappa+2)\cdots(\kappa+m)\}/m!, \quad \kappa > -1.$$

(3°) Let Λ be a subset of $[0, \infty)$. Fix $\kappa > -1$ and let

$$a_{nm}^{(\lambda)} = A_{n-m}^{(\kappa+\lambda)}/A_m^{(\kappa+\lambda)} \quad \text{for } 0 \leqq m \leqq n$$
$$= 0 \quad \text{for } m > n.$$

In particular, if $\Lambda = \{0\}$, then $\mathcal{A}$-summability reduces to the Cesàro summability (C, κ).

(4°) Let Λ be a subset of $[0, 1]$ and let

$$a_{nm}^{(\lambda)} = \binom{n}{m}\lambda^m(1-\lambda)^{n-m} \quad \text{for } 0 \leqq m \leqq n$$
$$= 0 \quad \text{for } m > n.$$

(5°) Let Λ be a subset of $[0, \infty)$ and

$$a_{nm}^{(\lambda)} = \exp(-n\lambda)(n\lambda)^m/m!.$$

(6°) If one takes $\Lambda = N$, then $\mathcal{A}$-summability reduces to that by Bell [1] (cf. [10]). This method includes F-summability (almost convergence method) and F_A-summability of Lorentz [9], A_B-summability of Mazhar and Siddiqi [11] and order summability of Jurkat and Peyerimhoff [6, 7], respectively.

Concerning detailed statements for $\mathcal{A}$-summability methods in the setting of arbitrary Banach spaces one may consult [15; Sec. 4].

From now on let $\mathcal{A} = \{(a_{nm}^{(\lambda)})_{n,m\in N}; \lambda \in \Lambda\}$ be a family of infinite matrices of non-negative real numbers and $\{L_m\}_{m\in N}$ a sequence of positive linear operators of A into B(X) such that for each $n \in N$ and $\lambda \in \Lambda$,

$$(8) \qquad \sum_{m=0}^{\infty} a_{nm}^{(\lambda)} \| L_m(1_X) \| < \infty.$$

For each $f \in A$, $n \in N$ and $\lambda \in \Lambda$ let

$$T_{n,\lambda}(f) = \sum_{m=0}^{\infty} a_{nm}^{(\lambda)} L_m(f),$$

which is well-defined by (8) and belongs to B(X).

Consequently, under the above setting all the results obtained in the preceding sections are applicable to the family $\{T_{n,\lambda}\}$, with D = N. In particular, the results corresponding to Theorems 1 and 2 extend Theorem 4 of King and Swetits [8] and Theorem 1 of Swetits [17] (cf. [12; Theorem 1]), respectively to the setting of arbitrary compact Hausdorff spaces and more general $\mathcal{A}$-summability methods. Also, as an immediate consequence of Corollary 1 we have the following which is more convenient for latter applications.

COROLLARY 2. _Let_ X _and_ G _be as in Section_ 3. _Suppose that_ $\Sigma_{m=0}^{\infty} a_{nm}^{(\lambda)} = 1$ _for all_ $n \in N$, $\lambda \in \Lambda$ _and that_ $L_m(1_X) = 1_X$ _and_ $L_m(g) = g$ _for all_ $m \in N$ _and all_ $g \in G$. _If there exists a constant_ $K > 0$ _such that_ $\| \Sigma_{i=1}^{k} (L_m(g_i^2) - g_i^2) \| \leq K\theta_m \| \Sigma_{i=1}^{k} g_i^2 \|$ _for all_ $m \in N$ _and for all_ $\{g_1, g_2, \cdots, g_k\} \subset G$, _where_ $\{\theta_m\}$ _is a sequence of positive real numbers which is_ $\mathcal{A}$-_summable to zero, then we have_

$$|||T_n(f) - f||| \leq 2\omega(f, K^{1/2}\xi_n)$$

for all $n \in N$ _and all_ $f \in A$, _where_

$$\xi_n = (\sup\{\sum_{m=0}^{\infty} a_{nm}^{(\lambda)} \theta_m; \lambda \in \Lambda\})^{1/2}.$$

5. Bernstein-Lototsky-Schnabl operators

Let H be a linear subspace of C(X) containing 1_X and T a Markov operator on C(X), i.e., a positive linear operator of C(X) into itself with $T(1_X) = 1_X$. Given a point $x \in X$, a Radon probability measure ν_x on X is called a T(H)-representing measure for x if

$$T(h)(x) = \int_X h d\nu_x$$

for all $h \in H$ (cf. [4]).

From now on let X be as in Section 3, and let A(X) denote the space of all real-valued continuous affine functions on X. Let $V = \{V_n\}_{n\geq 1}$ be a sequence of Markov operators on C(X), $\mathcal{U}^V = \{\nu_{x,n};\ n \geq 1, x \in X\}$ a family of Radon probability measures on X such that $\nu_{x,n}$ is a $V_n(A(X))$-representing measure for x, $P = (p_{nj})_{n,j\geq 1}$ an infinite lower triangular stochastic matrix, $\mathcal{Y} = \{y_x;\ x \in X\}$ a family of points of X and $\rho = \{\rho_n\}_{n\geq 1}$ a sequence of functions mapping X into [0, 1]. Then we define

$$\nu^{(V)}_{x,n,\rho} = \rho_n(x)\nu_{x,n} + (1 - \rho_n(x))\varepsilon_{y_x} \circ V_n,$$

where ε_t denotes the point mass of t, and

$$\pi_{n,P}: X^n \to X \quad \text{by} \quad (x_1, x_2, \cdots, x_n) \to \sum_{j\geq 1} p_{nj}x_j.$$

DEFINITION 5. <u>Given a function</u> $f \in C(X)$, <u>the</u> n-<u>th Bernstein-Lototsky-Schnabl function of</u> f <u>on</u> X <u>with respect to</u> $\mathcal{U}^V$, P, $\mathcal{Y}$, <u>and</u> ρ <u>is defined by</u>

$$B_n(f)(x) = B^{(\mathcal{U}^V, \mathcal{Y})}_{n,P,\rho}(f)(x) = \int_{X^n} f \circ \pi_{n,P}\, d \bigotimes_{1\leq j\leq n} \nu^{(V)}_{x,j,\rho}.$$

REMARK 4. If one takes V = {I}, I the identity operator, then the original definition of Grossman [5] (cf. [16]) is obtained.

LEMMA 4. *If* g *belongs to* A(X), *then*

$$B_n(g)(x) = \Sigma_{j\geqq 1} p_{nj}\rho_j(x)V_j(g)(x) + \Sigma_{j\geqq 1}p_{nj}(1 - \rho_j(x))V_j(g)(y_x)$$

and

$$B_n(g^2)(x) = \{B_n(g)(x)\}^2 + \Sigma_{j\geqq 1}p_{nj}^2\rho_j(x)\nu_{x,j}(g^2)$$

$$+ \Sigma_{j\geqq 1}p_{nj}^2\rho_j(x)\{(V_j(g)(y_x))^2 - 2V_j(g)(x)V_j(g)(y_x)\}$$

$$- \sum_{j\geqq 1} p_{nj}^2\rho_j^2(x)\{(V_j(g)(x))^2 + (V_j(g)(y_x))^2 - 2V_j(g)(x)V_j(g)(y_x)\}.$$

This follows from immediately by computations.

Let $\mathcal{A} = \{(a_{nm}^{(\lambda)})_{n,m\in N}; \lambda \in \Lambda\}$ be a family of infinite matrices of non-negative real numbers such that $\Sigma_{m=0}^{\infty} a_{nm}^{(\lambda)} < \infty$ for each n and λ. For each $n \in N$, $\lambda \in \Lambda$ and $f \in C(X)$ let

$$A_{n,\lambda}(f) = a_{n0}^{(\lambda)} f + \sum_{m=1}^{\infty} a_{nm}^{(\lambda)} B_m(f) \tag{9}$$

and

$$|||A_n(f) - f||| = \sup\{||A_{n,\lambda}(f) - f||;\ \lambda \in \Lambda\},$$

which are well-defined since each B_m is a positive linear operator of C(X) into B(X) with $B_m(1_X) = 1_X$.

THEOREM 3. *If*

$$\lim_{n\to\infty} |||A_n(g^i) - g^i||| = 0$$

for every $g \in A(X)$ *and for* i = 1, 2, *then*

$$\lim_{n\to\infty} |||A_n(f) - f||| = 0$$

for every $f \in C(X)$.

PROOF. This follows from Theorem 1 and Lemma 4.

REMARK 5. If

$$\lim_{n\to\infty} \| \sum_{j\geq 1} p_{nj}^2 \rho_j \| = 0 \quad \text{and} \quad \lim_{n\to\infty} \| B_n(g) - g \| = 0$$

for all $g \in A(X)$, then for all $f \in C(X)$ we have

$$\lim_{n\to\infty} \| B_n(f) - f \| = 0. \tag{10}$$

In particular, if $V_n(g) = g$ for all $n \geq 1$ and all $g \in A(X)$,

$$\lim_{n\to\infty} \| \sum_{j\geq 1} p_{nj}^2 \rho_j \| = 0 \quad \text{and} \quad \lim_{n\to\infty} \| \sum_{j\geq 1} p_{nj} \rho_j - 1_X \| = 0,$$

then (10) holds. This result extends Theorem of Grossman [5] (cf. [16; Satz 1]).

THEOREM 4. _Let_ $\mathcal{A}$ _and_ $A_{n,\lambda}$ _be as in Corollary_ 2 _and_ (9), _respectively. Suppose that_ $V_n(g) = g$ _for all_ $n \geq 1$ _and all_ $g \in A(X)$. _Then the following statements hold:_

(i) _If_ $y_x = x$ _for all_ $x \in X$ _and if_

$$\lim_{n\to\infty} \| \sum_{m=1}^{\infty} a_{nm}^{(\lambda)} (\sum_{j\geq 1} p_{mj}^2 \rho_j) \| = 0 \quad \text{uniformly in } \lambda \in \Lambda,$$

then for all $n \in N$ _and all_ $f \in C(X)$ _we have_

$$|||A_n(f) - f||| \leq 2\omega(f, \alpha_n),$$

where

$$\alpha_n = (\sup\{ \| \sum_{m=1}^{\infty} a_{nm}^{(\lambda)} (\sum_{j\geq 1} p_{mj}^2 \rho_j) \| ; \lambda \in \Lambda\})^{1/2}.$$

(ii) _If_ $\rho_n = 1_X$ _for all_ $n \geq 1$ _and if_

$$\lim_{n\to\infty} \sum_{m=1}^{\infty} a_{nm}^{(\lambda)} (\sum_{j\geq 1} p_{mj}^2) = 0 \quad \text{uniformly in } \lambda \in \Lambda,$$

then for all $n \in N$ _and all_ $f \in C(X)$ _we have_

$$|||A_n(f) - f||| \leq 2\omega(f, \beta_n),$$

where

$$\beta_n = (\sup\{ \sum_{m=1}^{\infty} a_{nm}^{(\lambda)} (\sum_{j\geq 1} p_{mj}^2); \lambda \in \Lambda\})^{1/2}.$$

PROOF. This follows from Corollary 2 and Lemma 4.

REMARK 6. Theorems 3 and 4 and Remark 5 can be reformulated with respect to pointwise convergence and the localization principle is obtained.

We close with the following remark concerning the continuity of Bernstein-Lototsky-Schnabl functions.

REMARK 7. Suppose that in Definition 5 the following conditions are satisfied:

(i) For every $n \geqq 1$, the map $x \to \nu_{x,n}$ is weak*-continuous;

(ii) For every $n \geqq 1$, ρ_n is continuous;

(iii) The map $x \to y_x$ is continuous on X.

Then for every $n \geqq 1$ and for every $f \in C(X)$, $B_n(f)(x)$ is continuous on X (cf. [5; Proposition]).

REFERENCES

[1] Bell, H., Order summability and almost convergence. Proc. Amer. Math. Soc. 38(1973), 548-552.

[2] Censor, E., Quantitative results for positive linear approximation operators. J. Approximation Theory 4(1971), 442-450.

[3] DeVore, R. A., The approximation of continuous functions by positive linear operators. Lecture Notes in Math. 293, Springer-Verlag, Berlin/Heidelberg/New York, 1971.

[4] Grossman, M. W., Korovkin theorems for adapted spaces with respect to a positive linear operator. Math. Ann. 220(1976), 253-262.

[5] Grossman, M. W., Lototsky-Schnabl functions on compact convex subsets. J. Math. Anal. Appl. 55(1976), 525-530.

[6] Jurkat, W. B. - Peyerimhoff, A., Fourier effectiveness and order summability. J. Approximation Theory 4(1971), 231-244.

[7] Jurkat, W. B. - Peyerimhoff, A., Inclusion theorems and order summability. J. Approximation Theory 4(1971), 245-262.

[8] King, J. P. - Swetits, J. J., Positive linear operators and summability. Austral. J. Math. 11(1970), 281-290.

[9] Lorentz, G. G., A contribution to the theory of divergent sequences. Acta Math. 80(1948), 167-190.

[10] Maddox, I. J., On strong almost convergence. Math. Proc. Camb. Phil. Soc. 85(1979), 345-350.

[11] Mazhar, S. M. - Siddiqi, A. H., On F_A-summability and A_B-summability of a trigonometric sequence. Indian J. Math. 9(1967), 461-466.

[12] Mohapatra, R. N., Quantitative results on almost convergence of a sequence of positive linear operators. J. Approximation Theory 20(1977), 239-250.

[13] Nishishiraho, T., The degree of convergence of positive linear operators. Tôhoku Math. J. 29(1977), 81-89.

[14] Nishishiraho, T., Quantitative theorems on linear approximation processes of convolution operators in Banach spaces. Tôhoku Math. J. 33(1981), 109-126.

[15] Nishishiraho, T., Saturation of multiplier operators in Banach spaces. Tôhoku Math. J. 34(1982), 23-42.

[16] Schempp, W., Zur Lototsky-transformation über kompakten Räumen von Wahrscheinlichkeitsmassen. Manuscripta Math. 5(1971), 199-211.

[17] Swetits, J. J., On summability and positive linear operators. J. Approximation Theory 25(1979), 186-188.

Department of Mathematics
Ryukyu University
Nishihara-Cho, Okinawa, 903-01 Japan

SOME RELATIONSHIPS BETWEEN SURFACE SPLINES AND KRIGING

K. Salkauskas

Department of Mathematics and Statistics, The University of Calgary, Calgary, Alberta, Canada.

1. Introduction

The classical interpolation problem is quite simply stated: Let $x_1,\ldots,x_N$ be distinct points in an open set $\Omega \subset \mathbb{R}^n$, and let there be given the values $f_i = f(x_i)$, $i = 1,\ldots,N$, of a function $f\colon \Omega \to \mathbb{R}$, $f \in C^0(\Omega)$. Find a function $g \in X \subset C^k(\Omega)$, $k \geq 0$ and fixed a priori, with X a linear space of functions having dimension N, such that $g(x_i) = f(x_i)$. Obviously the problem can be posed in terms of other than the evaluation functionals. A considerable development has taken place in the last 10 years or so, coping to a large extent with the extension of the solution from $\mathbb{R}$ to $\mathbb{R}^n$. In particular, the contributions of DUCHON [1], MEINGUET [6], MATHERON [4] must be acknowledged, as they are responsible for significant evolution of the theory of splines to $\mathbb{R}^n$.

Kriging is a process based on statistical considerations, for the estimation of the value $f(x_0)$, $x_0 \in \Omega$, as a linear combination of known values f_i, at $x_i \in \Omega$, $i = 1,\ldots,N$. In the hands of MATHERON [4] and others, it has experienced substantial development. It appears not to have been interpreted as a method of interpolation until recently, because automatic contouring processes often required function values at points of a regular rectangular grid, and these were obtained by kriging from irregularly positioned data. Thus, the actual functional form of the kriging interpolant was almost never examined, for an essentially finite element process took over once the values on the grid were established. Needless to say, the final product did not usually interpolate the data!

An interpolation process will be developed here based as much as possible on finite-dimensional techniques, that will combine aspects of the universal kriging method with splines, while making use of the deeper results of both theories.

2. The Classical Solution

Let the functions $\varphi_i \in C^k(\Omega)$, $k = 1,\ldots,N$, be linearly independent, and let $X = \text{span}\{\varphi_i\}_{i=1}^N$. Then the interpolant $g \in X$ has the form $g = \sum_{i=1}^N \alpha_i\varphi_i$, and the imposition of the interpolation conditions yields the system

$$V\underset{\sim}{\alpha}^T = \underset{\sim}{f}^T \;,$$

where $V = [\varphi_j(x_i)]_{i,j=1}^N$, $\underset{\sim}{\alpha} = (\alpha_1,\ldots,\alpha_N)$, $\underset{\sim}{f} = (f_1,\ldots,f_N)$. If V is non-singular, then

$$g = (\varphi_1,\ldots,\varphi_N)V^{-1}f^T \;, \tag{2.1}$$

and the set of functions $\{\psi_i\}_{i=1}^N$ defined by $(\psi_1,\ldots,\psi_N) = (\varphi_1,\ldots,\varphi_N)V^{-1}$ is a dual basis in X with respect to the evaluation functionals $\delta_{(x_i)} \in X^*$, $i = 1,\ldots,N$, satisfying the cardinality or bi-orthonormality conditions

$$\psi_i(x_j) = \delta_{ij}, \quad i,j = 1,\ldots,N \;. \tag{2.2}$$

Of course, g can then be written in Lagrangean form as

$$g = \sum_{i=1}^{N} \psi_i f_i \; . \tag{2.3}$$

Concerning the invertibility of V, we recall the

Definition 2.1. A (fixed) set of functions $\{\varphi_i\}_{i=1}^N$ is unisolvent on Ω if $\det V \neq 0$ for any set of N distinct points $x_i \in \Omega$.

We then have

Theorem 2.1 (Haar). *If $n \geq 2$, then no set of functions can be unisolvent on Ω.*

It is important to observe that the proof of the theorem does not apply if the φ_i depend on the x_i; this fact will be exploited later. When V is non-singular, it is clear that the mapping $P: C^0(\Omega) \to X$ defined by (2.1) or (2.3) is a projector. In view of the interpolating properties of $g = Pf$, we call P an *interpolating projector*.

A generalization of the interpolation problem can be obtained by permitting $\dim X > N$. Uniqueness of interpolants, if they exist, is then assured by the imposition of constraints in addition to the interpolation conditions. Such a situation is considered briefly in the next section.

3. Boolean Sums of Projectors

In the design of an interpolating projector it is often desirable that it act as the identity on some subspace Z of $C^0(\Omega)$; for example, we may choose $Z = P_m$, the space of polynomials of degree $\leq m$ (restricted to Ω), of dimension $M = \binom{n+m}{n}$, $M < N$. At the same time an interpolating projector employing the functions φ_i of the previous section may be available and/or desirable. Let now $Y = \text{span}\{\varphi_i\}_{i=1}^N$.

Lemma 3.1. *Let* $X = Y \oplus Z$, *with* $Y \cap Z = 0$, *and* $\dim Y = N$, $\dim Z \leq N$. *If there exist an interpolating projector* $P: C^0(\Omega) \to Y$ *and a projector* $Q: C^0(\Omega) \to Z$, *such that* $QP = Q$, *then the Boolean sum* $P \oplus Q = P - PQ + Q$ *is an interpolating projector* $\Pi: C^0(\Omega) \to Y \oplus Z$ *(with the property that* $\Pi z = z$, $\forall z \in Z$*).*

The proof of this lemma is elementary. It follows from the Lemma that $f = P(f-Qf) + Qf$, which can, on occasion, be viewed as a trend $Qf \in Z$ to which is added the interpolant $P(f-Qf)$ of a residual. This interpretation can be misleading, however.

We conclude this section with several examples.

Ex. 1. $n = 1$, $\Omega = (a,b)$, $a < x_1 < \ldots < x_N < b$, $Y = \text{span}\{\varphi_i\}_{i=1}^N$, $\varphi_i(x) = |x-x_i|^3$. Since the φ_i have discontinuities in the third derivative at the x_i, they are linearly independent as well as being unisolvent on Ω. Hence there exists an interpolating projector $P: C^0(\Omega) \to Y$, and Pf is an interpolating twice differentiable piecewise cubic on Ω; it is not a natural cubic spline and does not preserve first degree polynomials. If Q is any projector from $C^0(\Omega)$ onto P_1, satisfying $QP = Q$, then $(P \oplus Q)f$ is an interpolating cubic spline which is in P_1 if f is. For a particular choice of Q, to be obtained later, $(P \oplus Q)f$ is a natural cubic spline. It is then the one-dimensional version of the Duchon thin-plate spline [1,6].

Ex. 2. Let $X = \left\{v \in C^0(\mathbb{R}^n): \frac{d^k v}{dx^k} \in L_2(\mathbb{R}^n),\ |k| = k_1 + \ldots + k_n = m,\ m > \frac{n}{2}\right\}$. Equip X with the semi-inner-product

$$(u,v) = \sum_{|k|=m} \int_{\mathbb{R}^n} \frac{d^k u}{dx^k} \cdot \frac{d^k v}{dx^k}\, dx .$$

DUCHON [1] and MEINGUET [5] have shown that there exists a unique interpolant $g \in P_{m-1} \oplus Y \subset X$ of least semi-norm, with

$$Y = \text{span}\{\varphi_i\}_{i=1}^N, \quad \varphi_i(x) = \Phi(x-x_i), \quad \Phi(x) = \begin{cases} r^{2m-n}\ln r, & n \text{ even}, \\ r^{2m-n}, & n \text{ odd}, \end{cases}$$

and r the Euclidean norm $\|x\|$ of x. Again, g is obtained by the action of a Boolean sum of projectors. When $n = 1$ and $m = 2$ we obtain the natural cubic spline of example 1.

Ex. 3. We sketch the fundamental ideas of the statistically motivated *Kriging* process, elaborated by MATHERON [4] and his associates. The name honours Dr. D. G. KRIGE whose 1951 paper [3] is perhaps the earliest in the development of geostatistics.

Suppose that f is a random function on Ω, and that the values f_i are those of a realization g of f. Assume further that $f = r+p$, where p is a random polynomial $\in P_m$. For any $x \in \Omega$, the functional $E(f) = f(x) - \sum_{i=1}^N \lambda_i f(x_i)$ is a random variable (a generalized random process). The coefficients λ_i, which depend on x, are selected so that the variance, var $E(f)$, of $E(f)$ is minimized, while $E(p) = 0$ if $p \in P_m$. The last condition ensures that only a covariance model for r need be known for determining var $E(f)$. Then $g(x) = \sum_{i=1}^N \lambda_i f_i$ is taken as a satisfactory estimate of the value of the realization g at x. The modelling process involves choosing a positive-definite function Φ, and setting $\varphi_i(x) = \Phi(x-x_i) = \text{cov}(f(x), f(x_i))$. It then turns out that the (non-random) function g which represents the realization of f is an element of $P_m \oplus Y$, where $Y = \text{span}\{\varphi_i\}_{i=1}^N$, and involves a Boolean sum of projectors. The functions Φ defined in Ex. 2 are suitable. In that case there is no distinction between the kriging process and the Duchon splines.

4. The Kriging Interpolant

In this section we shall construct interpolating

projectors as Boolean sums by methods paralleling those of Kriging, but in a non-statistical setting. We begin by assessing the error in approximating the unknown value $f(x_0)$ by a linear combination of the known values $f_i = f(x_i)$, $i = 1,\ldots,N$. To this end let

$$E(f) = f(x_0) - \sum_{i=1}^{N} \lambda_i f_i \ . \tag{4.1}$$

Since it seems reasonable to require that $E(f) = 0$ when the data exhibits constant or planar behaviour, at least, we impose on $\{\lambda_i\}_1^N$ the condition

$$E(f) = 0,\ \forall\ f \in P_m,\ \dim P_m < N \ . \tag{4.2}$$

Put $M = \dim P_m = \binom{m+n}{n}$ and let $b_1,\ldots,b_M$ be a basis for P_m. Then (4.2) is equivalent to $(-1,\lambda_1,\ldots,\lambda_N) \in N_{\widetilde{B}}$, the null-space of $\widetilde{B}$, which is the matrix

$$\widetilde{B} = \left[\begin{array}{c|ccc} b_1(x_0) & b_1(x_1) & \ldots & b_1(x_N) \\ \vdots & \vdots & & \\ b_M(x_0) & b_M(x_1) & \ldots & b_M(x_M) \end{array}\right] = \left[\begin{array}{c|c} \underset{\sim}{b}^T(x_0) & B^T \end{array}\right] \ . \tag{4.3}$$

We assume that $\{x_i\}_{i=1}^N$ contains a subset ω of M points such that $\{b_i\}_{i=1}^M$ is unisolvent on ω. Then $\widetilde{B}$ has full rank. In applications it will be assumed that M is quite small, usually $M = 3$, so that the unisolvency condition is not very restrictive. Now, it is always possible to write $f = p+r$, $p \in P_m$. Then $E(f) = E(r)$ in view of (4.2), the vector $(p(x_0),\ldots,p(x_N))$ being in the row space of $\widetilde{B}$. The assessment of $|E(r)|$ will be carried out by working entirely in $N_{\widetilde{B}}$, which we equip with a norm derived from an inner-product

$$(\underset{\sim}{u},\underset{\sim}{v})_W = \underset{\sim}{u} W \underset{\sim}{v}^T,\ \underset{\sim}{u},\underset{\sim}{v} \in N_B \ , \tag{4.4}$$

where W is an $(N+1) \times (N+1)$ non-singular symmetric matrix which is positive-definite on $N_{\widetilde{B}}$. Then

$$|E(f)| = |E(r)| = |(r(x_0),\ r(x_1),\ldots,r(x_N))W^{-1}W(-1,\lambda_1,\ldots,\lambda_N)^T|$$

$$\leq \|(r(x_0),\ldots,r(x_N)\|_{W^{-1}}\ \|(-1,\underset{\sim}{\lambda})\|_W,\ \underset{\sim}{\lambda} = (\lambda_1,\ldots,\lambda_N)\ ,$$

provided that $(r(x_0),\ldots,r(x_N))W^{-1} \in N_{\tilde{B}}$. It is easy to see that there exists a decomposition $f = p+r$ so that this condition is fulfilled. Its precise nature is not known in practice because $f(x_0)$ is not available. We now make the

Definition 4.1. The approximation $\sum_{i=1}^{N} \lambda_i f_i$ is best with respect to W if $\underset{\sim}{\lambda} = (\lambda_1,\ldots,\lambda_N)$ minimizes $\|(-1,\underset{\sim}{\lambda})\|_W$, while $(-1,\underset{\sim}{\lambda}) \in N_{\tilde{B}}$.

In the statistical context which forms the basis of kriging (Ex. 3, Sec. 3), the variance of $E(f)$ is given by $\|(-1,\underset{\sim}{\lambda})\|_W$, W being the matrix of covariances $\mathrm{cov}(f(x_i),f(x_j))$, $i,j = 0,1,\ldots,N$.

It is possible to choose W in such a way that the best approximation coincides with the value at x_0 of the Duchon spline, which is optimal in another sense (Ex. 2, Sec. 3).

The optimization which flows from Definition 4.1 is easily performed with the aid of Lagrange's multipliers $\mu_1,\ldots,\mu_M$, and yields the system

$$(4.5)\qquad \left[\begin{array}{c|c} V & B \\ \hline B^T & 0 \end{array}\right] \left[\begin{array}{c} \underset{\sim}{\lambda}^T \\ \hline \underset{\sim}{\mu}^T \end{array}\right] = \left[\begin{array}{c} \underset{\sim}{v}^T \\ \hline \underset{\sim}{b}^T(x_0) \end{array}\right]$$

where B and $\underset{\sim}{b}^T(x_0)$ are as in (4.3), and W has been partitioned so that

$$(4.6)\qquad W = \left[\begin{array}{c|c} u & \underset{\sim}{v} \\ \hline \underset{\sim}{v}^T & V \end{array}\right] .$$

This is the so-called Kriging system. The solution can be written as

$$(4.7)\qquad \begin{aligned} \underset{\sim}{\lambda}^T &= P\underset{\sim}{v}^T + Q\underset{\sim}{b}^T(x_0) \ , \\ P &= V^{-1} - [V^{-1}B][B^TV^{-1}B]^{-1}[B^TV^{-1}] \ , \\ Q &= [V^{-1}B][B^TV^{-1}B]^{-1} \ . \end{aligned}$$

It should be noted that W may depend on x_0, so that in general $P\underset{\sim}{v}^T$ as well as $Q\underset{\sim}{b}^T$ can be functions of position $x_0 \in \Omega$. With this in mind, define the projectors $\mathcal{P}$: $C^0(\Omega) \rightarrow C^0(\Omega)$ and $\mathcal{Q}$: $C^0(\Omega) \rightarrow P_m$ by

$$(4.8)\qquad \begin{aligned} \mathcal{P}f &= (f(x_1),\ldots,f(x_N))P\underset{\sim}{v}^T \ , \\ \mathcal{Q}f &= (f(x_1),\ldots,f(x_N))[V^{-1}B][B^TV^{-1}B]\underset{\sim}{b}^T, \ \underset{\sim}{b} = (b_1,\ldots,b_M) \ . \end{aligned}$$

$\mathcal{Q}f$ has the appearance of a least squares approximation to the data f_i with a full weight matrix V^{-1}. Consequently $\mathcal{Q}$ satisfies $\mathcal{QP} = \mathcal{P}$ whenever $\mathcal{P}$ is an interpolating projector. However, we have not insisted on V being positive-definite on $\mathbb{R}^N$, so that $\mathcal{Q}f$ should not be interpreted as a trend surface. It is not hard to see from (4.7) that

$$\sum_{i=1}^{N} \lambda_i f_i = (\mathcal{P} \oplus \mathcal{Q})f \in \text{span}\{\lambda_i\}_{i=1}^{N} \ .$$

Referring to (2.1), we find that $\mathcal{P}$ is an interpolating projector when $V^{-1}\underset{\sim}{v}^T$ is a vector consisting of functions satisfying the cardinality conditions (2.2). We therefore choose $\underset{\sim}{v} = (\varphi_1,\ldots,\varphi_N)$, as in section 2; V is then determined. The value of u in W(cf. (4.6)) is arbitrary. Provided that $\text{span}\{\varphi_i\}_{i=1}^{N} \cap P_m = 0$, and V is non-singular, Lemma 3.1 applies, and $\mathcal{P} \oplus \mathcal{Q}$ is an interpolating projector. For future convenience we make

Definition 4.2. An $(N+1) \times (N+1)$ matrix is conditionally positive-definite of order $m+1$ if it is positive-definite on $N_{\widetilde{B}}$, the null-space of $\widetilde{B}$, given by (4.3).

In this definition, the order refers to the order $m+1$

of a constant-coefficient homogeneous differential operator whose kernel is $P_m = \text{span}\{b_i\}_{i=1}^{M}$. Symmetric matrices satisfying Definition 4.2 as well as the cardinality requirements (2.2) can be shown to exist. They can be constructed from conditionally positive-definite functions which are reviewed in the next section.

5. Positive-definite and conditionally positive-definite functions

In order to complete the development of the interpolant $P \oplus Q$ of the previous section, it is required to determine a suitable conditionally positive-definite matrix of order $m+1$ having the form (4.6). Semi-definite matrices can be found by using some results from the theory of Generalized Functions to which we now turn. The following definitions and theorems can be found in GEL'FAND and VILENKIN [2].

Definition 5.1. A continuous function f is positive-definite if for any points $x_0,\dots,x_N \in \mathbb{R}^n$ and any complex numbers $\xi_0,\dots,\xi_N$,

$$\sum_{i,j=0}^{N} f(x_i - x_j)\, \xi_i \overline{\xi}_j \geq 0 .$$

Theorem 5.1 (Bochner). *A continuous function f is positive-definite if and only if it is the Fourier transform of a finite positive measure μ, that is,*

$$f(x) = \int_{\mathbb{R}^n} e^{i(\lambda,x)} d\mu(\lambda), \quad (\lambda,x) = \lambda_1 x_1 + \dots + \lambda_n x_n$$

Ex. 4. The function $\Phi(x) = \exp[-\|x\|^2/2\rho^2]$ is the

Fourier transform of the finite positive measure $(\rho > 0)$ $d\mu(\lambda) = \rho^n(2\pi)^{-n/2} \exp[-\rho^2\|\lambda\|^2/2]d\lambda$ and is a positive-definite function. Let $\varphi_i(x) = \Phi(x-x_i)$, $i = 1,\ldots,N$, and put $u = \Phi(0)$. Then W is completely determined and is a positive-semi-definite matrix. The resulting interpolation scheme has been discussed by SCHAGEN [8] who uses $m = 0$ and gives statistical advice for the choice of ρ.

Matrices which are positive-semi-definite on $N_{\tilde{B}}$ can be constructed by making use of conditionally positive-definite generalized functions of order $m+1$ on the space K of infinitely differentiable functions with bounded supports in $\mathbb{R}^n$.

A generalized function Φ on K is positive-definite if $\langle\Phi,\varphi*\varphi^*\rangle \geq 0$ for all $\varphi \in K$, with $\varphi^*(x) = \overline{\varphi(-x)}$, and $\langle\cdot,\cdot\rangle$ being the duality bracket between dual topological vector spaces. By the Bochner-Schwartz theorem [2], which is a generalization of Theorem 5.1, Φ is positive-definite if and only if it is the Fourier transform of a positive tempered measure.

Φ is conditionally-positive-definite of order $m+1$ if it is positive-definite for all test functions of the form $\varphi = D\psi$, $\psi \in K$ and D a linear homogeneous constant-coefficient differential operator of order $m+1$ having the form $\sum_{|k|=m+1} a_k(d^k/dx^k)$. If we write

$$E(f) = \int_{\Omega} f(x)d\lambda(x) \quad ,$$

with $\lambda(x)$ a measure with finite support in Ω, then $d\lambda(x)$ can be regarded as a limit of measures defined by test functions in K. The requirement that $(-1,\underset{\sim}{\lambda}) \in N_{\tilde{B}}$ is satisfied if these test functions are of the form $\varphi = D\psi$, then their moments of order $k < 2m+2$ vanish.

By Theorem 3, Ch. II of Gel'fand and Vilenkin, the conditionally positive-definite Φ has the form

$$\langle\Phi,\varphi\rangle = \int_{\Omega_0} \tilde{\varphi}(\lambda)d\mu(\lambda) + \sum_{|k|=2m+2} a_k \frac{\tilde{\varphi}^{(k)}(0)}{k!} \quad ,$$

where Ω_0 is the complement of the origin, $a_k = \nu_k(0)$,

$d\nu_k(x) = x^k d\mu(x)$, the measure μ is positive, tempered and such that $\int_{0<|\lambda|<1} |\lambda|^{2m+2}\, d\lambda$ converges.

When such generalized functions exist as ordinary functions they yield, by the same construction as in Ex. 4, matrices W which are suitable if (strictly) positive-definite on $N_{\tilde{B}}$. An example is discussed by MATHERON [4], who suggests that the functions $\Phi(x) = (-1)^{p+1}\|x\|^{2p+1}$, $p = 0,1,\ldots,m$ are each suitable, as are certain linear combinations of them. The coefficient $(-1)^{p+1}$ is not important for the construction of an interpolant. When $2m-n+2 = 2p+1$, the resulting interpolants coincide with the Duchon spline.

As well, for certain values of the constants A and B, the functions $\Phi(x) = A\|x\|^{2p}\ln\|x\| + B\|x\|^{2p}$, $0 \le p \le m+1$, are conditionally positive-definite of order $m+1$. The second term can be excluded from the basis for interpolation if $p < m+1$ because it contributes a polynomial of degree $p < m+1$, annihilated by the projector $P - PQ$. The Duchon spline with even n can then be obtained.

Referring to Ex. 1, we have $n = 1$, $m = 1$, so $p = 1$, and as indicated in the example, $\varphi_i(x) = |x-x_i|^3$, $i = 1,\ldots,N$.

The use of conditionally positive-definite functions can be seen to result in translation-invariant interpolants, while the spherically-symmetric nature of the functions in the examples above yields what may be called isotropic interpolants.

Because the basis for the interpolating projector consists of translates, and depends continuously on the location of the data, Haar's theorem is not a limiting factor in the existence of unique interpolants. It is, however, necessary that W be conditionally positive-definite -- a property not immediately assured by the use of conditionally positive-definite functions.

We conclude with an indication of how the kriging interpolant relates to the Duchon thin-plate spline. Detailed discussion can be found in the recent papers of MEINGUET. Referring to Ex. 2, Sec. 3, we write $X = P_{m-1} \oplus X$, P_{m-1} being the kernel of the semi-norm defined on X. This decomposition may be carried out by using a projector $P: X \to P_{m-1}$; MEINGUET [1] uses a Lagrangean interpolant on ω (cf. Sec. 4). It is then necessary to determine representers in X_0 of the functionals $\delta_{(x_i)}$. These are constructed from a fundamental solution Φ of the partial differential equation $\Delta^m \Phi = \delta$, where Δ^m is the m-th iterated Laplacian arising from the semi-inner-product of X. The required solutions are, for $m > \frac{n}{2}$, constant multiples of those given in Ex. 2 and are assembled into an interpolating projector P. Then the required interpolant has the form $\Pi f = P(f-Qf) + Qf$, a Boolean sum, $\Pi f \in C^k(\Omega)$, $k = 2m-n-1$, and Πf has least semi-norm. When $n = m = 2$, this last property can be interpreted as corresponding to a minimum energy configuration of the interpolant.

During the preparation of this paper, a recent work by MATHERON [5] came to the author's attention. It demonstrates the formal equivalence of kriging and splines, but in a Hilbert space context.

References

[1] Duchon, J. (1976) Splines minimizing rotation-invariant semi-norms in Sobolev spaces. Constructive theory of functions of several variables, Oberwolfach, W. Schempp and K. Zeller eds., Springer, Berlin-Heidelberg.

[2] Gel'fand, I. M. & Vilenkin, N. Ya. (1964) Generalized functions, vol. 4, Academic Press, New York-London.

[3] Krige, D. G. (1951) A statistical approach to some mine evaluation and allied problems on the Witwatersrand, University of Witwatersrand.

[4] Matheron, G. (1973) The intrinsic random functions and their applications. Advances in Appl. Probability 5, 439-468.

[5] Matheron, G. (1981) Splines and kriging: their formal equivalence. Syracuse University Geology Contribution 8, D. F. Merriam ed., Department of Geology, Syracuse University, Syracuse, New York.

[6] Meinguet, J. (1978) Multivariate interpolation at arbitrary points made simple. Rapport No. 118, Seminaire de mathématique appliquée et mechanique, Institut de Mathématique Pure et Appliquée, Université Catholique de Louvain, Louvain-La-Neuve.

[7] Meinguet, J. (1981) On surface spline interpolation: basic theory and computational aspects. Res. Report No. 81-05, Seminar fuer angewandte Mathematik, Eidgenoessische Technische Hochschule, Zuerich.

[8] Schagen, I. P. (1979) Interpolation in two dimensions -- a new technique. J. Inst. Math. Appl., 23, 53-59.

A VIEW OF MATHEMATICS

Arthur Sard †

1. Future Axioms

For me, mathematical concepts are real. Mathematics is the logical study of mathematical concepts. Logic is rational reasoning, the reasoning that we use in making compelling deductions (even in our discussions of alternative "logics".)

Consider the euclidean line E. The evidence that E is real is this: We have deep, coherent, elegant theorems about E and about the things that we construct out of E. (These constructs include E^n, n = 1, 2, ..., and the noneuclidean geometries, and more.) The theorems have been confirmed again and again. New, unanticipated theorems, have been discovered. So we conclude, inductively, that E is real.*

We need not take E as a fundamental concept. Instead, we may construct E in the theory of sets and classes. Then the concepts of set and class are fundamental, and the evidence that E is real engenders evidence that sets and classes are real.

The evidence that sets and classes are real is to me fully as convincing as the physical evidence that the electric and magnetic fields are real.

I take the position that sets and classes are real, and that the axioms of von Neumann, Bernays, and Gödel [1], or some modification thereof [6], are truths about reality, discovered and confirmed by experience.

*Induction is present in any case, since mathematics is immersed in language.

Now, reality is consistent. It follows, from Gödel's incompleteness theorem [3], that the axioms are only a partial description of sets and classes. In the course of time, we may expect that we will recognize new axioms: new truths about sets and classes.

The axiom of choice is an instance. It was dislodged from the unknown by Zermelo [8, 9] and his predecessors. It has been intensively studied. It implies conclusions that were surprising when discovered. It has withstood its tests and is now accepted, I believe, by most mathematicians.

The continuum hypothesis [2, 4, 5] is not yet an axiom. I argue that the generalized continuum hypothesis is true, on the basis of its simplicity and elegance. The question is not whether it or its contrary is consistent with the present axioms of set and class theory, but rather whether it or one of its alternatives will be consistent with future axioms.

Discussions, like the present one, of beliefs and intuitions, should be part of the philosophy of mathematics. I deliberately base my view on the notions of reality, reason, simplicity, and elegance.

2. On Mathematics in all its Aspects

Clearly mathematics is more than its formal description, just as a Bach concerto is more than its printed score. In a sense, a formal proof, printed on paper, is not a proof until it is comprehended by someone, possibly only its author.

Mathematics involves discovery, logical construction of theorems from prior truths, and applications to the sciences and to human activity.

But the formal axioms and proofs are certainly part of mathematics and, indeed, the central part. A mathematical investigation is complete only if its formal counterpart has been or is capable of being put on paper.

3. An Analogy

The strength of analogies is widely recognized. Now there is an analogy between Gödel's incompleteness theorem and our mathematical experience. The theorem assures us that mathematics is inexhaustible. Our experience assures us that mathematics is inexhaustibly interesting.

References

1. Bernays, P.: A system of axiomatic set theory. J. Symbolic Logic, 1937-1954 = the first pages of reference [7] below

2. Cohen, P.J.: The independence of the continuum hypothesis. Proc. Nat. Acad. of Sciences 50 (1963) 1143-1148 and 51 (1964) 105-110

3. Gödel, K.: Über formal unentscheidbare Sätze der Principia Mathematica und verwandter Systeme I. Monatshefte für Math. und Physik 38 (1931) 173-198

4. Gödel, K.: The consistency of the axiom of choice and of the generalized continuum hypothesis with the axioms of set theory. Princeton, N.J., 1940

5. Gödel, K.: What is Cantor's continuum problem? Amer. Math. Monthly 54 (1947) 515-525

6. Lévy, A.: The role of classes in set theory, 1973 = Pages 173-215 of reference [7] below

7. Müller, G.H., ed.: Sets and classes; on the work of Paul Bernays. Amsterdam, 1976

8. Zermelo, E.: Beweis, dass jede Menge wohlgeordnet werden kann. Math. Annalen 59 (1904) 514-516

9. Zermelo, E.: Neuer Beweis für die Möglichkeit einer Wohlordnung. Math. Annalen 65 (1908) 107-128

DREI STATT EINER REELLEN VARIABLEN?

Walter Schempp

Lehrstuhl für Mathematik I der Universität Siegen,
Siegen

The purpose of this paper is twofold: We shall characterize the Čebyšev polynomials of the second kind as the characters of the irreducible unitary linear representations of that sphere among the compact Euclidean n-spheres $S_n (n>1)$ that admits a Lie group structure. Moreover, we shall indicate a geometric idea to prove the existence and uniqueness theorem of cardinal spline interpolation by an application of harmonic analysis of the three-dimensional real Heisenberg nilpotent Lie group $\tilde{A}(\mathbb{R})$. Finally, we point out some applications of nilpotent harmonic analysis to information theory.

1. Einleitung

In der konstruktiven Funktionentheorie mehrerer Variablen wird häufig das Verfahren angewandt, mehrdimensionale Approximationsprobleme auf eindimensionale Probleme zurückzuführen. Als Beispiele seien Kubaturformeln genannt, die (etwa mit Hilfe von Booleschen Methoden) aus Quadraturformeln zusammengesetzt sind, oder Diskretisierungsverfahren zur numerischen Behandlung partieller Differentialgleichungen, welche aus eindimensionalen

Gittern aufgebaute Punktgitter benutzen.

Der umgekehrte Weg, daß sich Eigenschaften von Funktionen einer reellen Variablen auf natürliche Weise aus der Sicht höherer Dimension ergeben, ist weitaus weniger geläufig. Ziel der vorliegenden Arbeit ist es, an Hand zweier Beispiele diesen umgekehrten Weg aufzuzeigen: Wir charakterisieren die Čebyšev-Polynome $(\check{C}_m)_{m\geq 0}$ zweiter Art als Charaktere der irreduziblen unitären linearen Darstellungen genau derjenigen Sphäre unter den kompakten euklidischen n-Sphären S_n in den Vektorräumen $\mathbb{R}^{n+1}$ $(n>1)$, welche eine Liegruppen-Struktur aufweist. Außerdem weisen wir auf eine geometrische Beweisidee für den Existenz- und Eindeutigkeitssatz der kardinalen Spline-Interpolation mit Hilfe der harmonischen Analyse der reellen drei-dimensionalen nilpotenten Heisenberg-Gruppe $\tilde{A}(\mathbb{R})$ hin. Die Überlegungen im Sphärenfall dienen dazu gewissermaßen als konstrastierende Leitlinie. Weitere Anwendungen, z.B. auf die Signalübertragung und die Radarortung bewegter Zielobjekte, werden kurz angedeutet.

2. Die Čebyšev-Polynome $(\check{C}_m)_{m\geq 0}$ zweiter Art

Die kompakte euklidische n-Sphäre

$$S_n = \{x \in \mathbb{R}^{n+1} \mid \|x\| = 1\}, \qquad (n\geq 1)$$

ist auf natürliche Weise zur homogenen Mannigfaltigkeit $SO(n,\mathbb{R})\backslash SO(n+1,\mathbb{R})$ diffeomorph, denn die kompakte Gruppe $SO(n+1,\mathbb{R})$ der orientierungserhaltenden Rotationen des $\mathbb{R}^{n+1}$ um den Nullpunkt operiert stetig und transitiv auf der S_n und besitzt bezüglich eines beliebigen Punktes der S_n, z.B. bezüglich des "Nordpoles" $\mathbb{1}=(0,\ldots,0,1)$, den Stabilisator $SO(n,\mathbb{R})$ [15]. Wir beschränken unsere Betrachtungen auf den mehrdimensionalen Fall $(n>1)$, schließen also den eindimensionalen Fall (die 1-Sphäre S_1 ist diffeomorph zur kompakten Torusgruppe $\mathbb{T}=\mathbb{R}/\mathbb{Z}$) aus. Auf die Frage, welche der euklidischen Einheitssphären $(S_n)_{n>1}$ eine Liegruppen-Struktur tragen, gibt der Satz von SAMELSON [7] Auskunft: Genau die 3-Sphäre S_3 hat diese Eigenschaft. Der Beweis wird mit Methoden der algebraischen Topologie geführt. Weil die (2n+1)-Sphäre S_{2n+1} des $\mathbb{C}^{n+1}$ auf natürliche Weise auch zur kompakten komplexen homogenen Mannigfaltigkeit $SU(n,\mathbb{C})\backslash SU(n+1,\mathbb{C})$ diffeomorph ist, kann die kompakte 3-Sphäre S_3 mit der Lie-Gruppe $SU(2,\mathbb{C})$ identifiziert werden. Ordnet man jedem Punkt $x=(x_j)_{1\leq j\leq 4} \in S_3$ das Paar komplexer Zahlen $z=(z_1,z_2) \in \mathbb{C}^2$ mit $z_1=x_1+ix_2, z_2=x_3+ix_4$ zu, so wird eine solche Identifizierung von S_3 mit $SU(2,\mathbb{C})$ durch den Diffeomorphismus

$$S_3 \ni x \rightsquigarrow -i \begin{pmatrix} -\bar{z}_2 & z_1 \\ \bar{z}_1 & z_2 \end{pmatrix} = u_x \in SU(2,\mathbb{C})$$

realisiert. Der Nordpol $\mathbb{1}=(0,0,0,1)$ wird als mit dem Punkt $(0,i) \in \mathbb{C}^2$ identifiziert betrachtet. Dem standardisierten Lebesgueschen Oberflächenmaß der S_3 entspricht unter dieser Korrespondenz das Haar-Maß der kompakten Gruppe $SU(2,\mathbb{C})$, so daß aufgrund des Satzes von Peter-Weyl eine vollständige Familie von irreduziblen unitären Darstellungen der $SU(2,\mathbb{C})$ eine natürliche Orthonormalbasis des komplexen Hilbert-Raumes $L^2_{\mathbb{C}}(S_3)$ liefert.

Zur Konstruktion einer vollständigen Familie von irreduziblen unitären Darstellungen der $SU(2,\mathbb{C})$ kann man ein von der Gruppe $GL(2,\mathbb{C})$ her bekanntes Standardverfahren anwenden. Es bezeichne $\mathcal{P}_m$ den aus den homogenen Polynomen vom Grad $m\geq 0$ bestehenden Untervektorraum von $\mathbb{C}[z] = \mathbb{C}[z_1,z_2]$. Dann wird für jedes $m\geq 0$ durch die Abbildung

$$\mathfrak{T}'_m(u) : \quad \mathcal{P}_m \ni p(z) \rightsquigarrow p({}^t uz) \in \mathcal{P}_m \qquad (u \in GL(2,\mathbb{C}))$$

eine irreduzible lineare Darstellung $\mathfrak{T}'_m$ von $GL(2,\mathbb{C})$ in $\mathcal{P}_m$ definiert. Es gilt $\dim_{\mathbb{C}} \mathcal{P}_m = m+1$ und die Monome $\{z_1^k z_2^{m-k}\}_{0\leq k\leq m}\}$ bilden eine $\mathbb{C}$-Basis von $\mathcal{P}_m$. Durch die Festsetzung

$$\left\langle \sum_{0\leq k\leq m} a_k z_1^k z_2^{m-k} \,\middle|\, \sum_{0\leq l\leq m} b_l z_1^l z_2^{m-l} \right\rangle = \sum_{0\leq j\leq m} j!\,(m-j)!\,a_j \bar{b}_j$$

bei komplexen Koeffizienten $(a_k)_{0 \leq k \leq m}$, $(b_l)_{0 \leq l \leq m}$ wird ein Skalarprodukt $\langle . | . \rangle$ auf $\mathcal{P}_m$ definiert, bezüglich dessen die Darstellung $\mathfrak{T}'_m$ von $GL(2,\mathbb{C})$ in $\mathcal{P}_m$ eine irreduzible unitäre Darstellung $\mathfrak{T}_m$ von $SU(2,\mathbb{C})$ im komplexen Hilbert-Raum $\mathcal{P}_m$ subduziert. Die Familie $(\mathfrak{T}_m)_{m \geq 0}$ kann als eine Repräsentantenfamilie des unitären Duals der Gruppe $SU(2,\mathbb{C})$ angesehen werden.

Wir betrachten den Charakter

$$\chi_m: SU(2,\mathbb{C}) \ni u \rightsquigarrow \operatorname{Tr} \mathfrak{T}_m(u) \qquad (m \geq 0)$$

von $\mathfrak{T}_m$. Aus den Eigenschaften der Spur (=Tr) linearer Abbildungen folgt, daß χ_m konstant ist auf den Klassen konjugierter Elemente von $SU(2,\mathbb{C})$. Weil aber jede Matrix $u \in SU(2,\mathbb{C})$ unitär äquivalent zu einer Diagonalmatrix ist, wird χ_m durch seine Werte auf der Untergruppe

$$H = \left\{ \begin{bmatrix} t & 0 \\ 0 & \bar{t} \end{bmatrix} \middle| \, t \in \mathbb{T} \right\}$$

von $SU(2,\mathbb{C})$ bereits eindeutig festgelegt. Offenbar ist H zur eindimensionalen Torusgruppe $\mathbb{T}$ topologisch isomorph. Demnach ist H insbesondere kompakt und abelsch und sogar eine maximale abelsche Untergruppe von $SU(2,\mathbb{C})$.

Für jedes Element $h = \begin{bmatrix} t & 0 \\ 0 & \bar{t} \end{bmatrix} \in H$ gilt wegen $\bar{t} = t^{-1}$ offenbar

$$\chi_m(h)(z_1^k z_2^{m-k}) = t^{2k-m} z_1^k z_2^{m-k} \qquad (0 \leq k \leq m).$$

Also nimmt der Charakter χ_m im Gruppenelement h (für $t = e^{2\pi i\theta}$, $\theta \in \,]0,1[\, - \frac{1}{2}$) den Wert

$$\chi_m(h) = \sum_{0 \leq k \leq m} t^{2k-m} = \frac{1-t^{2m+2}}{t^m(1-t^2)} = \frac{t^{m+1}-t^{-(m+1)}}{t-t^{-1}} = \frac{\sin 2\pi(m+1)\theta}{\sin 2\pi\theta}$$

$$= \check{C}_m(\cos 2\pi\Theta) = \frac{1}{m+1} C'_{m+1}(\cos 2\pi\Theta) \qquad (m \geq 0)$$

an, wobei $(C_m)_{m\geq 0}$ und $(\check{C}_m)_{m\geq 0}$ die Čebyšev-Polynome erster bzw. zweiter Art bezeichnen (vgl. z.B. RIVLIN [6]). Also gilt folgender

Satz 1. Die Čebyšev-Polynome $(\check{C}_m)_{m\geq 0}$ zweiter Art sind genau die Charaktere der irreduziblen unitären Darstellungen jener Sphäre S_3 unter den euklidischen n-Sphären $S_n (n>1)$, welche eine Liegruppen-Struktur trägt.

Die Orthogonalitätsrelationen für die Charaktere $(\chi_m)_{m\geq 0}$ der kompakten Gruppe $SU(2,\mathbb{C})$ liefern sofort die bekannten Orthogonalitätsrelationen für die Polynome $(\check{C}_m)_{m\geq 0}$ auf dem kompakten Einheitsintervall $[-1,+1]$ bezüglich des Maßes $\sin^2 2\pi\Theta d\Theta$. Aus diesen folgt wiederum leicht die bekannte dreigliedrige Rekursion für beide Familien $(\check{C}_m)_{m\geq 0}$ und $(C_m)_{m\geq 0}$.

3. Kardinale Spline-Interpolation

Für jede natürliche Zahl $m \geq 1$ bezeichne $\mathfrak{S}_{m-1}$ den komplexen Vektorraum der kardinalen Spline-Funktionen vom Grad m-1 auf $\mathbb{R}$. Sei $(y_k)_{k\in\mathbb{Z}}$ eine beliebig vorgegebene Folge komplexer Zahlen. Das Problem der kardinalen Spline-Interpolation besteht darin, zu dieser Folge ein Element $s \in \mathfrak{S}_{m-1} (m=1)$ so zu konstruieren, daß die Bedingungen

$$s(k) = y_k \qquad (k \in \mathbb{Z})$$

erfüllt sind. Dem Problem der kardinalen Spline-Interpolation läßt sich ein entsprechendes Problem für periodische Splines zur Seite stellen [8]. Man hat dazu an $N \geq 1$ äquidistant auf der eindimensionalen Torusgruppe $\mathbb{T}$ liegenden Knoten komplexe Werte vorzugeben. Als Bindeglied zwischen beiden kann die Theorie der Abminderungsfaktoren angesehen werden [11].

Während bei der im vorausgegangenen Abschnitt behandelten gruppentheoretischen Charakterisierung der Čebyšev-Polynome $(\check{C}_m)_{m\geq 0}$ zweiter Art das Auftreten einer Gewichtsfunktion (Dichte) in den zugehörenden Orthogonalitätsrelationen auf dem kompakten Einheitsintervall $[-1,+1]$ ein Hinweis darauf sein kann, daß eine mehrdimensionale Sicht, nämlich die der nicht-kommutativen harmonischen Analyse der Gruppe $SU(2,\mathbb{C})$, tiefere Einblicke gewähren kann, scheint dagegen auf den ersten Blick die nicht-kommutative harmonische Analyse weder auf das Problem der kardinalen noch auf das der periodischen Spline-Interpolation anwendbar zu sein. Denn der erste Fall legt die Betrachtung der aus den Knoten bestehenden Untergruppe $\mathbb{Z}$ von $\mathbb{R}$ nahe, während im zweiten Fall die zyklische Untergruppe $\mathbb{Z}/N\mathbb{Z}$ von $\mathbb{T}$ diese Rolle übernimmt. Man kommt jedoch zu einer neuen Idee, wenn man bedenkt, daß die Theorie der kardinalen Spline-Funktionen das klassische Whittaker-Shannonsche Abtasttheorem der Signalübertragung weiterentwickelt. In der Signalübertragungstechnik hat man jedoch stets <u>zwei</u> Bereiche, nämlich den Zeit- <u>und</u> den Frequenzbereich, simultan zu berücksichtigen (GABOR [4]). Eine solche Simultanbehandlung erlaubt aber gerade die harmonische Analyse der reellen nilpotenten Heisenberg-Gruppe $\tilde{A}(\mathbb{R})$ [12], [13].

Die reelle nilpotente <u>Heisenberg-Gruppe</u> $\tilde{A}(\mathbb{R})$ ist die aus allen unipotenten Matrizen

$$\begin{pmatrix} 1 & x & z \\ 0 & 1 & y \\ 0 & 0 & 1 \end{pmatrix} = (x,y,z) \in \mathbb{R}^3$$

bestehende Untergruppe von $SL(3,\mathbb{R})$. Ihr Zentrum wird durch $\{(0,0,z) \mid z \in \mathbb{R}\}$ gegeben. Die harmonische Analyse der lokalkompakten Gruppe $\tilde{A}(\mathbb{R})$ unterscheidet sich von der harmonischen Analyse der in Abschnitt 2 betrachteten kompakten Gruppe $SU(2,\mathbb{C})$ in den beiden folgenden Punkten auf entscheidende Weise:

(i) Es existieren unendlichdimensionale irreduzible unitäre Darstellungen von $\tilde{A}(\mathbb{R})$ und gerade diese sind die eigentlich interessanten irreduziblen unitären Darstellungen von $\tilde{A}(\mathbb{R})$.

(ii) Es existiert nicht nur eine diskrete Schar von unendlichdimensionalen irreduziblen unitären Darstellungen von $\tilde{A}(\mathbb{R})$, sondern sogar zu jedem Parameterwert $\lambda \in \mathbb{R}^{\times} = \mathbb{R} - \{0\}$ eine (bis auf unitäre Isomorphie) eindeutig bestimmte Darstellung U_λ dieser Art. Auf dem dicht liegenden Schwartzschen Untervektorraum $\mathcal{S}(\mathbb{R})$ des komplexen Hilbert-Raumes $L^2_{\mathbb{C}}(\mathbb{R})$ nimmt U_λ die Form

$$U_\lambda(x,y,z)f(t) = e^{2\pi i\lambda(z+ty)}f(t+x) \qquad (\lambda \neq 0)$$

an. Demnach setzt sich U_λ $(\lambda \neq 0)$ aus einer Translation und einer Drehung zusammen. Der Prototyp U_1 der Familie $(U_\lambda)_{\lambda\in\mathbb{R}^\times}$ irreduzibler unitärer Darstellungen von $\tilde{A}(\mathbb{R})$ in $L^2_{\mathbb{C}}(\mathbb{R})$ heißt <u>Schrödinger-Darstellung</u> von $\tilde{A}(\mathbb{R})$.

Für einen gruppentheoretischen Zugang zum Subbotin-Schoenbergschen Existenz- und Eindeutigkeitssatz der kardinalen Spline-Interpolation sind zwei weitere Realisierungen der Schrödinger-Darstellung U_1 von $\tilde{A}(\mathbb{R})$ wichtig. Durch die das Zentrum von $\tilde{A}(\mathbb{R})$ punktweise festlassende Abbildung

$$\tau : (x,y,z) \rightsquigarrow (y,-x,z-xy)$$

wird ein Automorphismus von $\tilde{A}(\mathbb{R})$ definiert. Der Satz von Stone-von Neumann impliziert, daß $U_1^\tau = U_1 \circ \tau$ und U_1 unitär isomorph sind. Die Fourier-Kotransformation $\bar{\mathcal{F}}_{\mathbb{R}}$ liefert einen unitären Isomorphismus von U_1 auf U_1^τ:

$$\bar{\mathcal{F}}_{\mathbb{R}} \circ U_1 = U_1^\tau \circ \bar{\mathcal{F}}_{\mathbb{R}}$$

Ähnlich wie die in natürlicher Weise eingebettete Untergruppe $SO(n,\mathbb{R})$ von $SO(n+1,\mathbb{R})$ im Sphärenfall gibt die diskrete Untergruppe $\mathbb{Z}^3$ von $\tilde{A}(\mathbb{R})$ Anlaß zu einer kompakten homogenen Mannigfaltigkeit. Man nennt $\mathbb{Z}^3\backslash\tilde{A}(\mathbb{R})$ die <u>Heisenbergsche</u>

Nilmannigfaltigkeit. Zwar ist diese kompakte Mannigfaltigkeit in global topologischer Hinsicht ziemlich kompliziert - es handelt sich um ein Kreisbündel über dem zweidimensionalen Torus $\mathbb{T}^2$ als kompakter Basis-Mannigfaltigkeit - aber ein Fundamentalbereich läßt sich leicht angeben: $\{(x,y,z) \in \tilde{A}(\mathbb{R}) \mid (x,y,z) \in [-\frac{1}{2},+\frac{1}{2}[^3\}$. Im komplexen Hilbert-Raum $L^2_{\mathbb{C}}(\mathbb{Z}^3\backslash\tilde{A}(\mathbb{R}))$ läßt sich ein τ-stabiler abgeschlossener Untervektorraum ausmachen, auf dem die Schrödinger-Darstellung U_1 von $\tilde{A}(\mathbb{R})$ als rechtsreguläre Darstellung δ_1 operiert. Der Weil-Brezin-Isomorphismus

$$W: f \rightsquigarrow ((x,y,z) \rightsquigarrow \sum_{n \in \mathbb{Z}} e^{2\pi i(z+ny)} f(x+n))$$

liefert einen unitären Isomorphismus von U_1 auf δ_1, d.h. es gilt

$$W \circ U_1 = \delta_1 \circ W.$$

Zwischen den Verflechtungsoperatoren $\bar{\mathcal{F}}_{\mathbb{R}}$ und W besteht die wichtige Identität

$$\tau \circ W = W \circ \bar{\mathcal{F}}_{\mathbb{R}}.$$

Kombiniert man diese mit der Faltungseigenschaft der zentralen Basis-Splines $M_m \in \mathcal{G}_{m-1}$ $(m \equiv 0 \bmod 2)$, so erhält man mit Hilfe des Inversionssatzes für Toeplitz-Matrizen von KREIN [5] und CALDERÓN-SPITZER-WIDOM [3] den nachstehenden, auf Subbotin und Schoenberg zurückgehenden Existenz- und Eindeutigkeitssatz für die kardinale Spline-Interpolation (vgl. SCHOENBERG [16]):

Satz 2. Sei $(y_k)_{k \in \mathbb{Z}} \in \ell^2_{\mathbb{C}}(\mathbb{Z})$ eine vorgegebene Folge komplexer Zahlen. Falls $m \equiv 0 \bmod 2$ besitzt das Problem der kardinalen Spline-Interpolation $s(k)=y_k$ $(k \in \mathbb{Z})$ genau eine Lösung $s \in \mathcal{G}_{m-1} \cap L^2_{\mathbb{C}}(\mathbb{R})$. Im Falle $m \equiv 1 \bmod 2$ hat dieses Problem bei gleichen Knoten, aber um $\frac{1}{2}$ verschobenenen Interpolationspunkten, ebenfalls genau eine Lösung

4. Weitere Anwendungen

Die harmonische Analyse der nilpotenten Heisenberg-Gruppen besitzt, neben der hier nur kurz skizzierten Anwendung auf das Problem der kardinalen Spline-Interpolation, eine Vielzahl verwandter Anwendungen. Aus der numerischen Mathematik sei nur die schnelle Fourier-Transformation (FFT-Algorithmus; vgl. AUSLANDER-TOLIMIERI [1] und AUSLANDER-TOLIMIERI-WINOGRAD [2]), aus der Informationstheorie nur die Untersuchung des Wigner-Woodward-Reliefs genannt, welches für die Sicherheit der Radarortung bewegter Zielobjekte mit Hilfe des Impulsradars von entscheidender Bedeutung ist [9], [10], [14]. Tatsächlich hängt der FFT-Algorithmus zur Berechnung der endlichen Fourier-Kotransformation $\overline{\mathfrak{F}}_N$ eng mit dem Weil-Brezin-Isomorphismus der endlichen nilpotenten Heisenberg-Gruppe $A(\mathbb{Z}/N\mathbb{Z})$ zusammen und die als Untergruppe der Automorphismengruppe von $\tilde{A}(\mathbb{R})$ auffaßbare Gruppe $Sp(1,\mathbb{R})=SL(2,\mathbb{R})$ besteht, bis auf Phasenfaktoren, gerade aus den energieerhaltenden, linearen Invarianten der Wigner-Woodward Reliefs vorgegebener Pulse endlicher Energie. Das führt zu interessanten zahlentheoretischen Folgerungen und hat Bedeutung für die praktische Radar-Synthese.

Literatur

1. Auslander, L., Tolimieri, R.: Is computing with the finite Fourier transform pure or applied mathematics? Bull. (New Series) Amer. Math. Soc. 1, 847-897 (1979)

2. Auslander, L., Tolimieri, R., Winograd, S.: Hecke's theorem in quadratic reciprocity, finite nilpotent groups and the Cooley-Tukey algorithm. Adv. in Math. 43, 122-172 (1982)

3. Calderón, A., Spitzer, F., Widom, H.: Inversion of Toeplitz matrices. Illinois J. Math. 3, 490-498 (1959)

4. Gabor, D.: Theory of communication. J. Inst. Elec. Engrs. (London) 93 429-457 (1946)

5. Krein, M.G.: Integral equations on a half-line with kernel depending upon the difference of the arguments. Uspehi Mat. Nauk (N.S.) 13, 3-120 (1958). Amer. Math. Soc. Transl. Ser. 2, Vol. 22, pp. 163-288 (1962)

6. Rivlin, T.J.: The Chebyshev polynomials. New York-London-Sydney-Toronto: Wiley 1974

7. Samelson, H.: Über die Sphären, die als Gruppenräume auftreten. Comment. Math. Helv. 13, 144-155 (1940)

8. Schempp, W.: Approximation und Transformationsmethoden III. In: Functional Analysis and Approximation, pp. 409-420. ISNM 60. Basel-Boston-Stuttgart: Birkhäuser 1981

9. Schempp, W.: Radar reception and nilpotent harmonic analysis I. C.R. Math. Rep. Acad. Sci. Canada 4, 43-48 (1982)

10. Schempp, W.: Radar reception and nilpotent harmonic analysis II. C.R. Math. Rep. Acad. Sci. Canada (to appear)

11. Schempp, W.: Attenuation factors and nilpotent harmonic analysis. Resultate Math. (in print)

12. Schempp, W.: On the rôle of the Heisenberg group in information theory (to appear)

13. Schempp, W.: Gruppentheoretische Aspekte der Signalübertragung und der kardinalen Interpolationssplines I. Math. Meth. in the Appl. Sci. (erscheint demnächst)

14. Schempp, W.: Gruppentheoretische Aspekte der Signalübertragung und der kardinalen Interpolationssplines II. (In Vorbereitung)

15. Schempp, W., Dreseler B.: Einführung in die harmonische Analyse. Stuttgart: Teubner 1980

16. Schoenberg, I.J.: Cardinal spline interpolation. Regional Conference Series in Applied Mathematics. SIAM, Philadelphia, Pennsylvania 1973

Prof. Dr. Walter Schempp
Lehrstuhl für Mathematik I
Universität Siegen
Hölderlinstraße 3
D-5900 Siegen, W. Germany

EINE METHODE ZUR KONSTRUKTION VON C^1-FLÄCHEN ZUR INTERPOLATION UNREGELMÄSSIG VERTEILTER DATEN

Rita Schmidt

Hahn-Meitner-Institut für Kernforschung GmbH
Berlin, FRG

A method for the construction of an interpolating C^1-surface through scattered data is described. It consists of triangulating the set of nodes and connecting piecewise cubic polynomials, defined over each triangle, interpolating the given data in the corners and matching the cubic polynomials and their normal derivatives on common edges. The method is global and needs no further information beyond the data.

1. Problemstellung

Gegeben ist eine endliche Menge von Tripeln $D:=\{(x_j,y_j,z_j), j=1,\dots,M\} \subset R^3$. Gesucht ist eine C^1-Funktion in $G \subset R^2$ $z=f(x,y)$, $f \in C^1(G)$, die die Daten interpoliert: $z_j=f(x_j,y_j)$ für $j=1,\dots,M$. Die Stützstellen sollen disjunkt sein, d.h. es gelte $(x_j,y_j)=(x_\ell,y_\ell)$ nur für $j=\ell$. Sie dürfen beliebig unregelmäßig in der (x,y)-Ebene verteilt sein. Das Definitionsgebiet G ist ein polygonal berandetes Gebiet, das alle Stützstellen enthält. Oft wird das Rechteck verwendet, das durch die Extremwerte der Stützstellenmenge bestimmt ist.

2. Anwendungen

Im Hahn-Meitner-Institut Berlin ist die Hauptanwendung der Flächeninterpolation die Darstellung von Isolinien, die aus einer diskreten Datenmenge gewonnen werden. In vielen Fällen läßt sich die Meßapparatur so einrichten, daß die Stützstellen auf einem regelmäßigen Gitter liegen. Dann lassen sich C^1-Flächen mit einem von BIRKHOFF und GARABEDIAN [2] angegebenen Verfahren direkt konstruieren. Für den Fall, daß das Gitter in jeder der beiden Variablen eine äquidistante Einteilung hat, wird diese Methode im Programm ISOFIX [7] als eine der möglichen Varianten zum Zeichnen von Isolinien angeboten. Vielfach sind die Stützstellen jedoch unregelmäßig verteilt, wie die nachfolgenden Bilder zeigen, die aus konkreten Anwendungen stammen. Den gesuchten Flächen ist gemeinsam, daß für ihre Darstellung kein analytisches Modell bekannt ist.

Stützstellen für

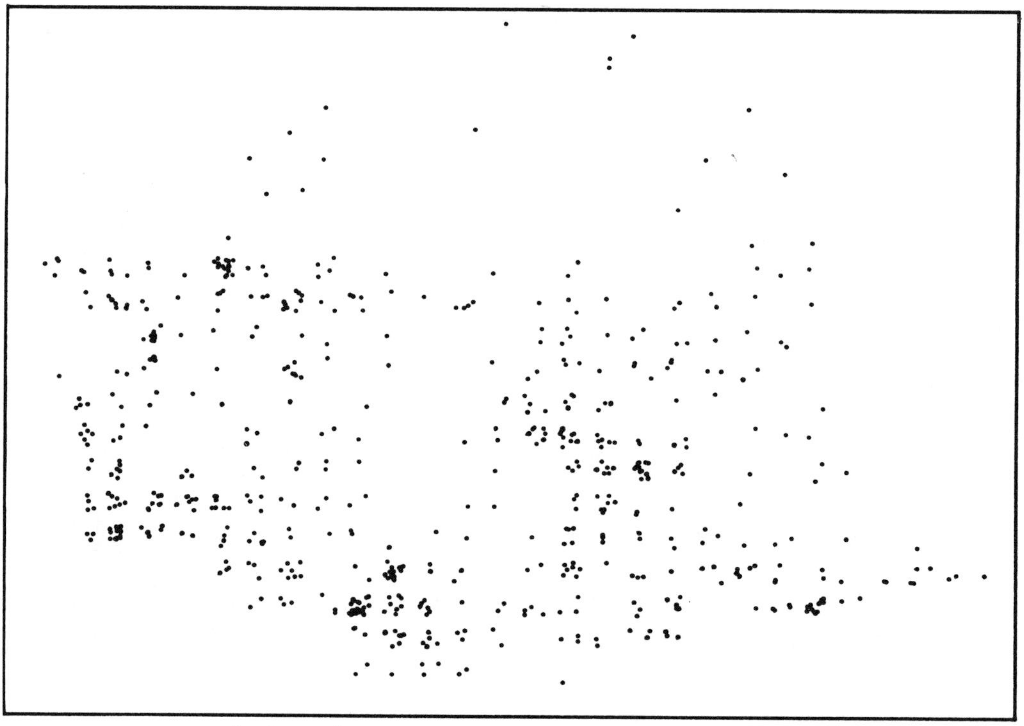

- Niederschlagsmessungen in der Sahel-Zone

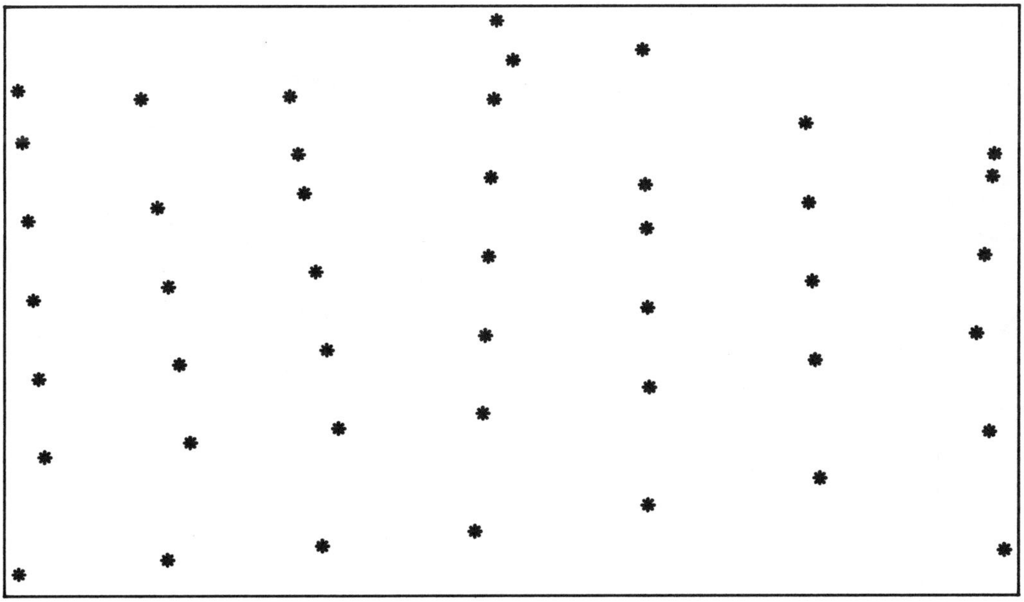

- Schwermetallkonzentrationen im Sediment des Bodensees

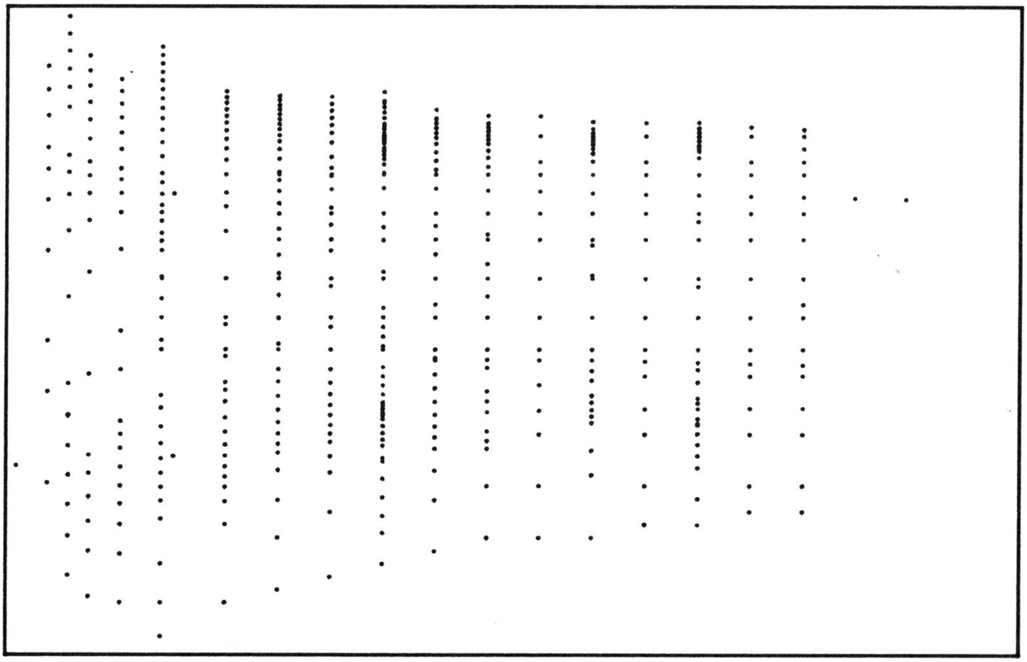

- Energieflächen bei Ion-Atom-Kollisionen

3. Das Verfahren

Es ist das Ziel des entwickelten Verfahrens, zur Herstellung der C^1-Fläche möglichst nur die vorhandene Information zu benutzen, also nur die gegebenen Funktionswerte in den Stützstellen. Andere Vorgehensweisen sind in [1] und [4] beschrieben worden. Das Verfahren besteht aus den folgenden Schritten:

- Triangulierung der Stützstellenmenge; hierdurch wird der entstehenden Fläche ein Dreiecksmuster aufgeprägt
- Konstruktion der Fläche aus stückweise kubischen Polynomen derart, daß
  - über jedem Dreieck ein Polynom liegt, das
  - die gegebenen Funktionswerte in den Eckpunkten interpoliert
  - mit jedem Nachbarpolynom stetig differenzierbar verbunden ist.

Die kubischen Flächenstücke gewährleisten eine ausgewogene Anzahl von Freiheitsgraden und die Behandlung des Problems mit Methoden der linearen Algebra.

3.1 Das Triangulierungsverfahren

Die Triangulierung der Stützstellenmenge (siehe [3]) entsteht nach einem Vorschlag von MC LAINE [5] . Das Prinzip werde am folgenden Bild erläutert. Gegeben seien die Punkte

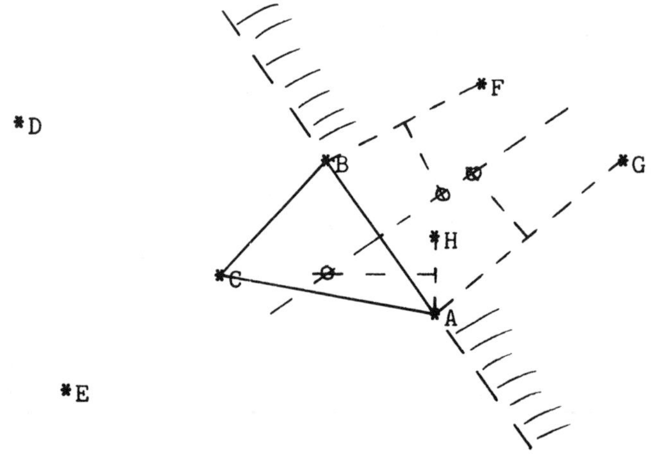

und das Dreieck ABC. Gesucht werde ein Nachbardreieck ABX_o, also ein an die Seite AB anschließendes Dreieck. Für X_o kommen alle die Punkte in Betracht, die in der durch AB erzeugten, dem Punkt C gegenüberliegenden Halbebene H_{AB}^C enthalten sind, im Beispiel also $\Phi_{AB} := \{F,G,H\}$. Die Auswahl geschieht nach folgendem Verfahren: Für alle Tripel ABX mit $X \in \Phi_{AB}$ werden die Umkreismittelpunkte m_x bestimmt. Sie liegen auf der zur Strecke AB gehörenden Mittelsenkrechten. Wird diese in Richtung der Halbebene H_{AB}^C orientiert, so ist $X_o \in \Phi_{AB}$ derjenige Punkt, für den m_{x_o} am weitesten links liegt. Es wird bewiesen [3], daß die so entstehende Triangulierung überschneidungsfrei ist. Der Algorithmus startet mit der kürzesten Strecke, die in der Stützstellenmenge enthalten ist, und wählt eine der beiden Halbebenen willkürlich aus. Das entstehende Gebiet ist die konvexe Hülle der Stützstellenmenge. Ist für die weitere Bearbeitung ein Rechteckgebiet erforderlich, so müssen vor Beginn des Verfahrens die fehlenden Eckpunkte des Rechtecks hinzugenommen werden. Die ebenfalls nicht vorhandenen Funktionswerte werden durch Extrapolation aus den Nachbarwerten bestimmt, was auch gegen das Ziel verstößt, nur Bekanntes zu verwenden.

3.2 Die Konstruktion der Fläche

Die Forderungen an f sind

- die Interpolationsforderungen $z_j = f(x_j, y_j)$ für $j=1,\dots,M$ und
- die Glattheitsforderung $f \in C^1(G)$.

3.2.1 Die Ansatzfunktionen

Die C^1-Fläche wird aus kubischen Polynomstücken stetig und stetig differenzierbar zusammengesetzt. Sei ℓ die Anzahl der Dreiecke. Über jedem Dreieck $D^{(\lambda)}$, $\lambda = 1,\dots,\ell$, liegt das Flächenstück

$$p^{(\lambda)}(x,y) := a_{oo}^{(\lambda)} + a_{1o}^{(\lambda)} x + a_{o1}^{(\lambda)} y + a_{2o}^{(\lambda)} x^2 + a_{11}^{(\lambda)} xy + a_{o2}^{(\lambda)} y^2$$

$$+ a_{3o}^{(\lambda)} x^3 + a_{21}^{(\lambda)} x^2 y + a_{12}^{(\lambda)} xy^2 + a_{o3}^{(\lambda)} y^3 .$$

Es hat zehn freie Parameter, die zum Vektor $a_\lambda = (a_{oo}^{(\lambda)}, \dots, a_{o3}^{(\lambda)})^T$, $a_\lambda \in R^{10}$, zusammengefaßt werden.

3.2.2 Die Interpolationsbedingungen

Die Interpolationsbedingungen lassen sich dreiecksweise erfüllen: Auf jedes Dreieck $D^{(\lambda)}$ mit den Eckpunkten $(x_{\lambda i}, y_{\lambda i})$ entfallen die drei Bedingungen

$$z_{\lambda_i} = P^{(\lambda)}(x_{\lambda_i}, y_{\lambda_i}),\quad i=1,2,3,\quad \lambda=1,\dots,\ell.$$

Es gibt also 3ℓ Interpolationsbedingungen.

3.2.3 Die Glattheitsbedingungen

Durch die Interpolationsforderungen werden die kubischen Flächenstücke i.a. noch nicht einmal stetig miteinander verheftet, da sie auf gemeinsamen Kanten nur in den beiden Eckpunkten übereinstimmen. Damit $f \in C^1(G)$ ist, müssen also die Stetigkeit und die stetige Differenzierbarkeit auf den inneren Dreieckskanten gefordert werden. Diese Kantenbedingungen lassen sich durch folgende Forderungen erfüllen:

- die Flächenstücke stimmen auf aneinandergrenzenden Dreiecksseiten überein
- die Ableitung der Flächenstücke auf gemeinsamen Kanten in Normalenrichtung ist stetig.

Die erste Forderung gewährleistet die Stetigkeit und die stetige Differenzierbarkeit in Kantenrichtung, die zweite erlaubt es, die Ableitung in jeder beliebigen Richtung durch Linearkombination zu gewinnen. Beide zusammen lassen sich, weil die Interpolation in den Eckpunkten schon gefordert wird, durch die fünf Übergangsbedingungen (siehe [6])

$$P_x^{(1)}(x_j,y_j)=P_x^{(2)}(x_j,y_j),\quad P_y^{(1)}(x_j,y_j)=P_y^{(2)}(x_j,y_j) \text{ für } j=1,2 \text{ und}$$

$$\frac{\partial P^{(1)}}{\partial n}\left(\frac{x_1+x_2}{2}, \frac{y_1+y_2}{2}\right) = \frac{\partial P^{(1)}}{\partial n}\left(\frac{x_1+x_2}{2}, \frac{y_1+y_2}{2}\right)$$

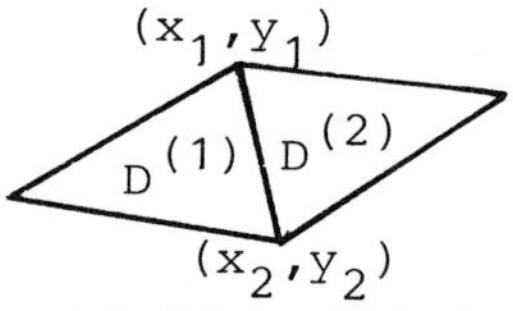

erfüllen. Es werden also stetige Übergange der partiellen Ableitungen P_x und P_y in den Eckpunkten der Dreiecke und der Normalableitungen in den Kantenmitten gefordert, d.h. aber, daß ein globales Verfahren vorliegt.

4. Die Lösungsmannigfaltigkeit

Für die Berechnung der Polynomkoeffizienten der Fläche liegen Flächen- und Kantenbedingungen vor. Damit Aussagen über die Lösbarkeit des Problems gemacht werden können, ist es erforderlich, die Zahl der bei der Triangulierung einer ebenen Punktmenge entstehenden Flächen und Kanten zu kennen. Weiterhin muß die Zahl der Übergangsbedingungen berechenbar sein.

4.1 Die Invarianten einer Triangulierung

Sei m_1 die Zahl der inneren, m_2 die der Randpunkte einer Triangulierung, k_1 die Zahl der inneren, k_2 die der Außenkanten und ℓ die Zahl der Flächen. Gesucht sind für gegebene Zahlen m_1 und m_2 die Größen k_1, k_2 und ℓ. Es ist (siehe [4] und [6])

$$\begin{aligned} k_2 &= m_2 \\ \ell &= m_2 + 2(m_1 - 1) \\ k_1 &= m_2 + 3(m_1 - 1). \end{aligned}$$

Hieraus ist ersichtlich, daß in einer konvexen Triangulierung die Zahl der Dreiecke und die Zahl der Kanten nur durch die Anordnung der Punkte, nicht aber durch die Art der Triangulierung bestimmt ist.

4.2 Die Anzahl der Übergangsbedingungen in den Eckpunkten der Dreiecke

Für die Übergangsbedingungen in den Eckpunkten der Dreiecke gilt folgendes: Wird als Ordnung π_μ eines Punktes P_μ die Zahl der Kanten definiert, die von diesem Punkt ausgehen, so ist die Summe der Ordnungen aller Punkte einer Triangulierung

$$S_P := \sum_{\mu=1}^{m} \pi_\mu = 2k,$$

denn jede Kante des Dreiecksnetzes leistet für genau zwei Punkte einen Beitrag zur Ordnung.

Es ist nun zwischen inneren und Randpunkten zu unterscheiden. In inneren Punkten P_{μ_1} stoßen π_{μ_1} Dreiecke zusammen,

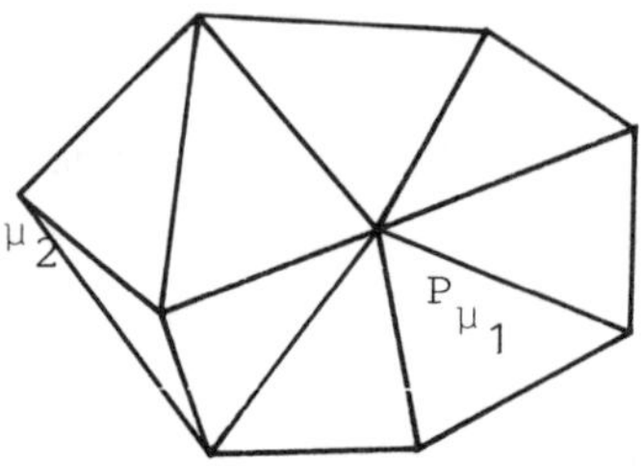

gerade so viele, wie Kanten von diesem Punkt ausgehen. Wegen der Transitivität ist eine der Übergangsbedingungen automatisch erfüllt. Deshalb können in den inneren Punkten $\pi_{\mu_1}-1$ linear unabhängige Übergangsbedingungen gestellt werden. In den Randpunkten P_{μ_2} gehen zwei Kanten nach außen, für die keine Übergangsbedingungen gestellt werden können. Pro Randpunkt gibt es deshalb $\pi_{\mu_2}-2$ Bedingungen. Die Gesamtzahl der Übergangsbedingungen in den Punkten eines Dreiecksnetzes beträgt also

$$\sum_{\mu_1=1}^{m_1}(\pi_{\mu_1}-1)+\sum_{\mu_2=1}^{m_2}(\pi_{\mu_2}-2)=\sum_{\mu=1}^{m}\pi_\mu-(m_1+2m_2)=2k-(m_1+2m_2).$$

4.3 Die Anzahl der Gleichungen

Sei s die Anzahl der Gleichungen, die die geforderte Eigenschaft der Fläche, nämlich $f\in C^1(G)$, garantiert. Sie besteht aus den $s_1=3\ell$ Interpolationsbedingungen in den Eckpunkten der Dreiecke, den $s_2=k_1$ Übergangsbedingungen für die Normalableitungen in den Kantenmitten und den $s_3=2(2k-(m_1+2m_2))$ Übergangsbedingungen für f_x und f_y in den Eckpunkten der Triangulierung. Wegen $k=k_1+k_2$, $k_2=m_2$ ist $s_3=4k_1-2m_1$ und $s=3\ell+5k_1-2m_1$.

4.4 Das lineare Gleichungssystem

Sei r die Zahl der Parameter, die die Fläche $f\in C^1(G)$ definieren. Da pro Dreieck ein kubisches Polynom mit zehn Parametern vorliegt, ist $r=10\ell$. Zur Bestimmung der interpolierenden C^1-Fläche liegt also ein lineares Gleichungssystem mit

$r=10m_2+20m_1-20$ Unbekannten und
$s=\ 8m_2+19m_1-21$ Gleichungen

vor. Die Lösungsmannigfaltigkeit hat also mindestens die Dimension $q:=r-s=2m_2+m_1+1$.

5. Die Zusatzbedingungen

Es gibt viele Möglichkeiten, für die q freien Parameter Zusatzbedingungen aufzustellen. Die Zahl s der bisher vorliegenden Gleichungen könnte zum Beispiel durch die Hinzunahme je einer linearen Bedingung in allen inneren, je zweier in den Randpunkte und einer willkürlich gewählten auf ein quadratisches System erweitert werden. Charakteristisch für die Gesamtfläche sind jedoch die kubischen Flächenstücke über den Dreiecken und deren Verheftung über gemeinsamen Kanten. Deshalb erscheint es sinnvoll, die Zusatzbedingungen in Form von Flächen- und Kantenbedingungen zu stellen. Diese lassen sich jedoch für allgemeine Punkteverteilungen mit $m=m_1+m_2$ Punkten nicht mehr durch genau q lineare Bedingungen formulieren, denn damit $q=\alpha\ell+\beta k_1$ ist, müßte das System

$$\begin{aligned}(\alpha+\beta)\ m_2 &= m_2\\ (2\alpha+3\beta)m_1 &= m_1\\ 2\alpha+3\beta\quad &= -1\end{aligned}$$

für beliebige m_1 und m_2 nichtnegative Lösungen α und β haben. Offensichtlich liegt jedoch für $m_1\neq 0$ ein Widerspruch vor, und auch der Fall $m_1=0$ führt auf die unzulässige Lösung $\alpha=7$, $\beta=-5$. Demnach kann es höchstens in Spezialfällen genau q Zusatzbedingungen geben. Sollen also zusätzliche Bedingungen auf alle Flächen oder Kanten gleichmäßig angewendet werden, so läßt sich das Problem der Parameterbestimmung für die Interpolationsfläche mit der geforderten Glattheitseigenschaft nicht mehr durch ein lineares Gleichungssystem lösen. Die Zusatzforderungen werden deshalb als Ausgleichsforderungen nach der Methode der kleinsten Abweichungsquadrate berücksichtigt. Dadurch verbleibt das Problem innerhalb der linearen Algebra und wird als Ausgleichsproblem mit linearen Nebenbedingungen behandelt. Bisher wurden zwei Arten von Nebenbedingungen verwendet: Flächen- und Kantenbedingungen. Flächenbedingungen sind Forderungen, die das Flächenstück über jedem Dreieck möglichst gut erfüllen soll, während die Kantenbedingungen zusätzliche Forderungen an die Güte der Verheftung der Flächenstücke stellen. Alle diese Bedingungen betreffen das Verhalten der zweiten Ableitungen der Fläche.

5.1 Die Flächenbedingung

Eine sinnvolle Zusatzbedingung ist die Minimierung von

$$F:=\iint_G (\Delta f)^2 dxdy,$$

da dieses Funktional in erster Näherung die Biegeenergie einer Platte darstellt. Das absolute Minimum dieses Funktionals wird für $\Delta f=0$, also für harmonische Flächenstücke, angenommen. Die harmonischen Polynome dritten Grades haben die allgemeine Darstellung

$$h(x,y;\gamma):=\gamma_1+\gamma_2 x+\gamma_3 y+\gamma_4(x^2-y^2)+\gamma_5 xy+\gamma_6 x(x^2-3y^2)$$
$$+\gamma_7 y(y^2-3x^2),$$

also sieben freie Parameter. Der Ansatz, die Gesamtfläche aus harmonischen Flächenstücken zusammenzusetzen, liefert für die s globalen Bedingungen nicht genügend Freiheiten. Sei $a=(a_1,\dots,a_\ell)^T$ mit $a_\lambda=(a_{00}^{(\lambda)},\dots,a_{03}^{(\lambda)})^T$, $a_\lambda\in R^{10}$, $a\in R^r$, der Parametervektor der Fläche $f\in C^1(G)$ mit $f(x,y;a):=\{P^{(\lambda)}(x,y;a_\lambda),\ \lambda=1,\dots,\ell\}$. Es ist

$$f_{/D^{(\lambda)}}=P^{(\lambda)},\quad \Delta f_{/D^{(\lambda)}}=\Delta P^{(\lambda)}$$

mit

$$\Delta P^{(\lambda)}(x,y;a_\lambda)=2(a_{20}^{(\lambda)}+a_{02}^{(\lambda)}+(3a_{30}^{(\lambda)}+a_{12}^{(\lambda)})x+(a_{21}^{(\lambda)}+3a_{03}^{(\lambda)})y)$$

eine lineare Funktion der drei Parameter

$$\alpha_0^{(\lambda)} := a_{20}^{(\lambda)} + a_{02}^{(\lambda)}$$
$$\alpha_1^{(\lambda)} :=3a_{30}^{(\lambda)} + a_{12}^{(\lambda)}$$
$$\alpha_2^{(\lambda)} := a_{21}^{(\lambda)} +3a_{03}^{(\lambda)} .$$

Damit die Flächenstücke $\{P^{(\lambda)}, \lambda=1,\ldots,\ell\}$ möglichst harmonisch sind, wird das Funktional

$$Z(\alpha):=\sum_{\lambda=1}^{\ell}\{(\alpha_o^{(\lambda)})^2+(\alpha_1^{(\lambda)})^2+(\alpha_2^{(\lambda)})^2\}$$

für $\alpha\in R^3$ minimiert. Das führt im Gleichungsystem auf 3ℓ lineare Zusatzbedingungen. Diese reichen zur eindeutigen Lösbarkeit des Ausgleichsproblems für die r Parameter mit den s Nebenbedingungen aus, wenn mindestens ein innerer oder mindestens sieben Randpunkte in einer Triangulierung enthalten sind, denn dann ist $3\ell-q=m_2+5m_1-7\geq 0$. $m_2\geq 3$ ist ja immer erfüllt.

5.2 Die Kantenbedingungen

Es werden Übergangsbedingungen für die zweiten Ableitungen auf den Innenkanten und Linearitätsforderungen auf den Randkanten gestellt.

5.2.1 Die Stetigkeit der zweiten Ableitungen

Diese Bedingung ist geometrisch anschaulich: Über jedem Dreieck entsprechen die partiellen Ableitungen zweiter Ordnung, nämlich $P_{xx}^{(\lambda)}$, $P_{xy}^{(\lambda)}$ und $P_{yy}^{(\lambda)}$, Ebenenstücken. Diese Ebenenstücke haben über die Dreieckskanten hinweg keinen Zusammenhang. Es wird nun gefordert, daß diese unstetigen Flächen möglichst zu Facettenflächen werden, d.h. daß die Sprünge in den zweiten Ableitungen so gering wie unter den Nebenbedingungen möglich gehalten werden. Da über jeder Dreieckskante für jede der angestrebten Facettenflächen f_{xx}, f_{xy} und f_{yy} ein Paar sich nicht notwendig schneidender Strecken liegt, lautet die Forderung, den Abstand dieser Streckenpaare möglichst klein zu machen. Das läßt sich durch die Bedingung erreichen, den Abstand der zweiten Ableitungen in den Eckpunkten der Dreiecke zum Minimum zu machen. Pro Innenkante ergeben sich so die sechs Forderungen

$$P_{xx}^{(1)}(x_j,y_j) \approx P_{xx}^{(2)}(x_j,y_j)$$

$$P_{xy}^{(1)}(x_j,y_j) \approx P_{xy}^{(2)}(x_j,y_j)$$

$$P_{yy}^{(1)}(x_j,y_j) \approx P_{yy}^{(2)}(x_j,y_j), \quad j=1,2,$$

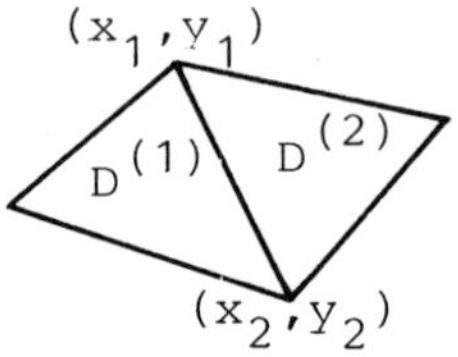

die nach der Methode der kleinsten Abweichungsquadratsumme für alle k_1 Kanten erfüllt sein sollen. Es braucht hierbei kein Unterschied zwischen inneren und Randpunkten gemacht zu werden, da alle Beziehungen nur näherungsweise erfüllt werden und deshalb Abhängigkeiten keine Rolle spielen. Im nebenstehenden Bild werden für den Punkt B alle vier Kantenbedingungen aufgestellt. Es

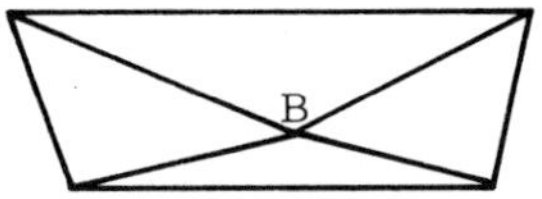

ist $6k_1-q=4m_2+17m_1-19 \geq 0$, wenn $m_2 \geq 5$ oder $m_1 \geq 1$ ist. Die beiden Skizzen zeigen die Mindestforderungen dafür, daß diese Zusatzbedingungen zur eindeutigen Lösung des Problems ausreichen.

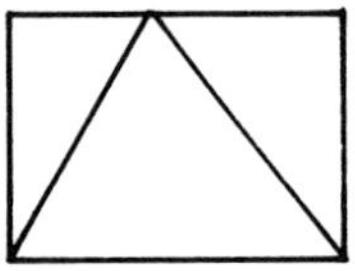

$q = 11$

$6k_1 = 12$

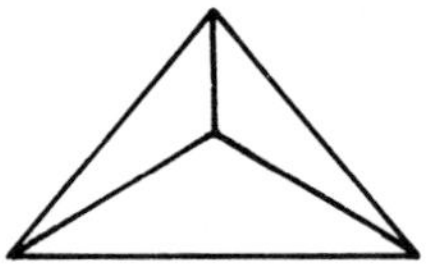

$q = 8$

$6k_1 = 18$

5.2.2 Die Linearität auf dem Rande

Die nach dem vorigen Abschnitt konstruierten Flächen besitzen relativ hohe Extremwerte, die i.a. keinen realen Hintergrund haben. Das wirkt sich am Bildrand besonders störend aus. Deshalb wird die Möglichkeit eingeräumt, zusätzlich zu den unter 5.2.1 angegebenen Bedingungen zu verlangen, daß die Fläche auf den Randkanten möglichst linear verläuft. Da diese Forderung im Widerspruch zur stetigen Differenzierbarkeit der Gesamtfläche steht, darf auch diese Zusatzforderung nur ausgleichsweise gestellt werden. (Die Forderung nach einer C^1-Fläche im Inneren und einer C^0-Fläche auf dem Rande würde für die numerische Behandlung sehr viele Komplikationen bringen, s. Abschnitt 6).

Beispiel:

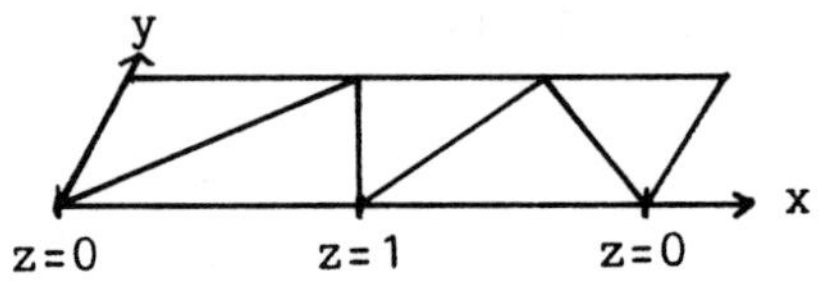

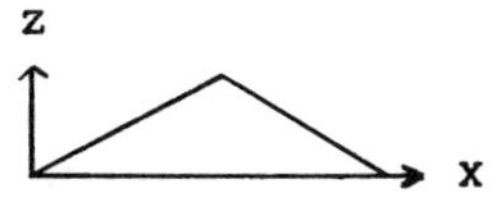

Die Dachfunktion ist nicht aus C^1.

Die Forderung nach einem linearen Verlauf der Fläche über einer Randkante läßt sich folgendermaßen formulieren: Sei AB Randkante einer Triangulierung mit den Koordinaten $A:=(x_1,y_1)$, $B:=(x_2,y_2)$, die dem Dreieck $D^{(\lambda)}$ angehöre. Die Parametrisierung der Randkante durch

$$x(\sigma):= x_1+\frac{x_2-x_1}{\|AB\|}\,\sigma, \quad y(\sigma):=y_1+\frac{y_2-y_1}{\|AB\|}\,\sigma$$

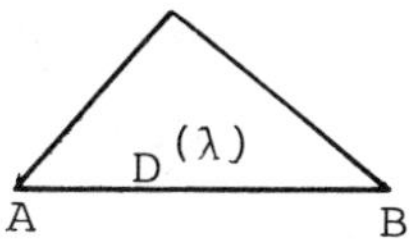

erlaubt die Darstellung der Fläche in der Form

$$P^{(\lambda)}(x(\sigma),y(\sigma)):=g(\sigma)=\alpha_0+\alpha_1\sigma+\alpha_2\sigma^2+\alpha_3\sigma^3 ,$$

die linear über der Kante AB verläuft, wenn $\alpha_2=\alpha_3=0$ ist. Pro Randkante treten also zwei Zusatzbedingungen auf, die sich durch die Parameter der Flächenstücke und deren Ortskoordinaten darstellen lassen (siehe [6]). Damit die $2m_2$ Forderungen gegenüber den $6k_1=6m_2+18m_1-18$ Kantenbedingungen überhaupt berücksichtigt werden, müssen sie mit einem großen Gewicht versehen werden.

6. Die numerische Behandlung

Das lineare Ausgleichsproblem für die r Unbekannten mit den s linearen Nebenbedingungen hat eine sehr dünn besetzte Matrix. Die Nebenbedingungen bestehen ja aus den Interpolationsbedingungen, die an jedes kubische Flächenstück über einem Dreieck gestellt werden und deshalb genau 10 von Null verschiedene Elemente pro Matrixzeile enthalten, und den Übergangsbedingungen, die genau 18 Parameter miteinander verknüpfen, da sie sich auf die ersten Ableitungen je zweier Flächenstücke beziehen. Die Zusatzbedingungen berücksichtigen nur zweite Ableitungen und haben deshalb pro Matrixzeile höchstens 14 von Null verschiedene Elemente. Da es $r=10m_2+20m_1-20$ Unbekannte gibt, ist der Besetztheitsgrad der Matrix etwa der Zahl der inneren Punkte umgekehrt proportional. Deshalb ist es erforderlich, ein spezielles Verfahren zur numerischen Lösung dieses Problems zu verwenden.

6.1 Die Elimination der Interpolationsbedingungen

Sei $a_\lambda^T:=(a_{oo}^{(\lambda)},a_{1o}^{(\lambda)},a_{o1}^{(\lambda)},a_{2o}^{(\lambda)},a_{11}^{(\lambda)},a_{o2}^{(\lambda)},a_{3o}^{(\lambda)},a_{21}^{(\lambda)},$ $a_{12}^{(\lambda)},a_{o3}^{(\lambda)})$ der transponierte Parametervektor für das kubische Flächenstück über dem Dreieck $D^{(\lambda)}$. Aus den 10 Komponenten werden drei Teilvektoren gebildet, die gesondert behandelt werden: $\alpha_\lambda:=(a_{oo}^{(\lambda)},a_{1o}^{(\lambda)},a_{o1}^{(\lambda)})^T$, $\beta_\lambda:=(a_{2o}^{(\lambda)},a_{11}^{(\lambda)},a_{o2}^{(\lambda)},a_{3o}^{(\lambda)},a_{o2}^{(\lambda)})^T$ und

$\gamma_\lambda:=(a_{21}^{(\lambda)},a_{12}^{(\lambda)})^T$. Aus den Interpolationsbedingungen

$$p^{(\lambda)}(x_{\lambda_i},y_{\lambda_i})=z_{\lambda_i} \quad \text{für } i=1,2,3 \text{ und } \lambda=1,\ldots,\ell$$

lassen sich die drei ersten Parameter jeder Teilfläche formal eliminieren, da die Determinante

$$\Delta^{(\lambda)}:=\begin{vmatrix} 1 & x_{\lambda_1} & y_{\lambda_1} \\ 1 & x_{\lambda_2} & y_{\lambda_2} \\ 1 & x_{\lambda_3} & y_{\lambda_3} \end{vmatrix}$$

dem von Null verschiedenen Flächeninhalt von $D^{(\lambda)}$ direkt proportional ist. Das System der Nebenbedingungen kann also um die 3ℓ Interpolationsbedingungen reduziert werden, indem die Vektoren α_λ eindeutig durch lineare Funktionen der restlichen Parameter β_λ und γ_λ dargestellt werden. Es ist

$$\alpha_\lambda = l^{(\lambda)}(\beta_\lambda,\gamma_\lambda) \text{ mit } l^{(\lambda)}:\mathcal{L}[R^7 \longrightarrow R^3]$$

6.2 Die Verkettung der Dreiecke

Die zweite Reduktion des Systems der Nebenbedingungen geschieht durch formale Elimination von Teilen des Vektors β, $\beta^T:=(\beta_1,\ldots,\beta_\ell)$, $\beta_\lambda\in R^5$. Der Restvektor wird in fünf und zwei Komponenten aufgespalten, weil auf den Innenkanten höchstens fünf Bedingungen an zwei aneinandergrenzende Flächenstücke gestellt werden können. Die Dreiecke werden so miteinander verkettet, daß in einer Folge von Dreiecken, einer Kette, je zwei eine gemeinsame Kante haben und die Anzahl der Ketten minimal ist. Dadurch bekommt das System der Nebenbedingungen die folgende Struktur (für eine Kette, sonst siehe [6] und das folgende Beispiel):

$$\begin{array}{rl}
A_{11}\beta_1+A_{12}\beta_2 & =C_{11}\gamma_1+C_{12}\gamma_2+C_1 \\
A_{22}\beta_2+A_{23}\beta_3 & =C_{22}\gamma_2+C_{23}\gamma_3+C_2 \\
\vdots & \vdots \\
A_{\ell-2,\ell-2}\beta_{\ell-2}+A_{\ell-2,\ell-1}\beta_{\ell-1} & =C_{\ell-2,\ell-2}\gamma_{\ell-2}+C_{\ell-2,\ell-1}\gamma_{\ell-1}+C_{\ell-2} \\
A_{\ell-1,\ell-1}\beta_{\ell-1} & =C_{\ell-1,\ell-1}\gamma_{\ell-1}+C_{\ell-1,\ell}\gamma_\ell+C_{\ell-1} \\
 & \quad -A_{\ell-1,\ell}\beta_\ell \\
B_{\ell 1}\beta_1+B_{\ell 2}\beta_2+\dots\dots\dots+B_{\ell,\ell-1}\beta_{\ell-1} & =D_{\ell 1}\gamma_1+\dots+D_{\ell\ell}\gamma_\ell+D_\ell-B_{\ell\ell}\beta_\ell
\end{array}$$

Der obere Teil des Systems besteht aus den (5x5)-Matrizen $A_{\lambda\lambda}$ und $A_{\lambda,\lambda+1}$, den (5x2)-Matrizen $C_{\lambda\lambda}$ und $C_{\lambda,\lambda+1}$ sowie dem Vektor $C_\lambda \in R^5$. Jeder Block enthält eine volle Kantenbedingung, d.h. die fünf Übergangsbedingungen auf einer Innenkante zwischen benachbarten Dreiecken. Der untere Teil besteht aus allen durch die Verkettung nicht erfaßten Übergangsbedingungen sowie den Zusatzbedingungen. Die Matrizen $B_{\ell\lambda}$ für $\lambda=1,\dots,\ell$ haben also fünf Spalten, während ihre Zeilenanzahl durch die gewählte Zusatzbedingung bestimmt wird. Entsprechendes gilt für die zweispaltigen Matrizen $D_{\ell\lambda}$ für $\lambda=1,\dots,\ell$ sowie den Vektor D_λ. Die Verkettung der Dreiecke und der Aufbau des Gleichungssystems werde an folgendem Beispiel erläutert: Werden die Dreiecke $D^{(1)}$ bis $D^{(6)}$ zu einer Kette zusammengefaßt, so entstehen drei Ketten, da die Dreiecke $D^{(7)}$ und $D^{(8)}$ unverkettet bleiben. Durch den Algorithmus entstehen die beiden Ketten $(D^{(1)},D^{(2)},D^{(5)},D^{(6)})$ und $(D^{(7)},D^{(3)},D^{(4)},D^{(8)})$. Im System der Ne-

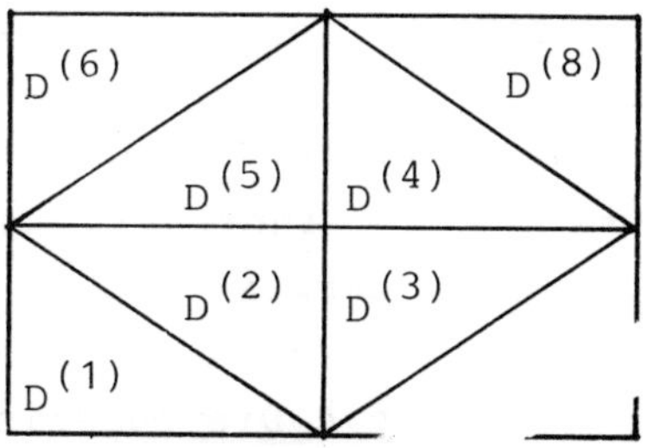

benbedingungen enthält der obere Teil für jede der Ketten drei volle Kantenbedingungen. Die Vektoren β_6 und β_8 kommen als Unbekannte auf die rechte Seite. In der Originalnumerierung hat das System die Form

$$\begin{aligned}
A_{11}\beta_1+A_{12}\beta_2 &= C_{11}\gamma_1+C_{12}\gamma_2+C_1\\
A_{22}\beta_2+A_{25}\beta_5 &= C_{22}\gamma_2+C_{25}\gamma_5+C_2\\
A_{55}\beta_5 &= C_{55}\gamma_5+C_{56}\gamma_6+C_5-A_{56}\beta_6\\
A_{77}\beta_7+A_{73}\beta_3 &= C_{77}\gamma_7+C_{73}\gamma_3+C_7\\
A_{33}\beta_3+A_{34}\beta_4 &= C_{33}\gamma_3+C_{34}\gamma_4+C_3\\
A_{44}\beta_4 &= C_{44}\gamma_4+C_{48}\gamma_8+C_4-A_{48}\beta_8
\end{aligned}$$

Der untere Teil enthält fünf Übergangsbedingungen zwischen $D^{(2)}$ und $D^{(3)}$, jedoch nur noch drei zwischen $D^{(4)}$ und $D^{(5)}$, da die Verheftung der Dreiecke im Mittelpunkt schon vollständig ist.

6.3 Die zweite Reduktion des Systems

Der obere Teil des Gleichungssystems läßt sich formal auflösen. Es ist ja

$$\beta_\lambda = A_{\lambda\lambda}^{-1}(C_{\lambda\lambda}\gamma_\lambda+C_{\lambda,\lambda+1}\gamma_{\lambda+1}+C_\lambda-A_{\lambda,\lambda+1}\beta_{\lambda+1})$$

eine lineare Funktion der Parameter $\gamma_\lambda,\gamma_{\lambda+1}$ und $\beta_{\lambda+1}$. Rekursives Ersetzen von β_λ im Restsystem führt auf ein lineares System, das nur noch die Parameter $\{\beta_{\lambda_1},\dots,\beta_{\lambda_d},\gamma_1,\dots,\gamma_\ell\}$ enthält. Die für das rekursive Ersetzen notwendigen Größen

$$Z_\lambda := \{A_{\lambda\lambda}^{-1}C_{\lambda\lambda}, A_{\lambda\lambda}^{-1}C_{\lambda,\lambda+1}, A_{\lambda\lambda}^{-1}C_\lambda, A_{\lambda\lambda}^{-1}A_{\lambda,\lambda+1}\}$$

werden durch numerisches Auflösen der höchstens $\ell-1$ (5x5) Gleichungssyteme $A_{\lambda\lambda}Z_\lambda=U_\lambda$ mit den (5x10)-Matrizen $U_\lambda:=(C_{\lambda\lambda},C_{\lambda,\lambda+1},$

$C_\lambda,A_{\lambda,\lambda+1})$ gewonnen. Dagegen werden die Parameter $\{(a_{10}^{(\lambda)},a_{01}^{(\lambda)}),$ $\lambda=1,\dots,\ell\}$ innerhalb der Übergangsbedingungen analytisch durch β_λ und γ_λ dargestellt.

6.4 Die numerische Lösung des Restsystems

Durch diese beiden formalen Eliminationen kann das System der Nebenbedingungen bis auf 4% des Originalumfangs reduziert werden. Für die numerische Behandlung verbleiben ja $2\ell+5d$ statt $r=10\ell$ Unbekannte, wenn d die Anzahl der bei der Verkettung entstehenden Ketten bezeichnet, und die Zahl der Nebenbedingungen beträgt $3m_1+5(d-1)$ statt $s=8m_2+19m_1-21$. Die Zahl der Zusatzbedingungen bleibt erhalten. Das reduzierte Ausgleichsproblem mit den linearen Nebenbedingungen wird durch die Anwendung von Householder-Transformationen gelöst, wobei die Nebenbedingungen über Lagrange-Faktoren an das Zielfunktional angekoppelt werden. Das Problem lautet dann: Minimiere

$$H(v) := \lim_{\omega\to\infty} \left\| \begin{matrix} \omega(Cv-d) \\ Av-b \end{matrix} \right\|^2 .$$

A,b,C,d und v stehen für die nach der Reduktion verbliebenen Größen des Originalproblems. Der Grenzübergang $\omega\to\infty$ wird analytisch ausgeführt. Dadurch ändert sich die Transformationsvorschrift im Householderverfahren (siehe [6]).

7. Ein Beispiel

Um wenigstens einen optischen Gütetest zu haben, wird das beschriebene Verfahren auf einen Datensatz angewandt, dem eine Funktion zugrundeliegt. Die folgenden Bilder zeigen die Isolinien 0.2(0.2)1.2, 1.5(0.5)6.0 der Funktion

$$f(x,y) = \frac{1}{|\Gamma((x-0.5),\ i(y-0.5))|} \quad \text{für } 0\le x\le 3,\ 0\le y\le 2,\ (x,y)\ne(0.5,0.5),$$

die ausgewählten 16 Stützstellen (*) mit der Triangulierung, die Isolinien der Facettenfläche, der möglichst harmonischen Polynome, der möglichst stetigen Facettenflächen der zweiten Ableitungen und zusätzlich der möglichst linearen Funktionen auf dem Rande.

Literaturverzeichnis

Akima, H. (1977) A method of bivariate interpolation and smooth surface fitting for irregularly distributed data points. ACM TRANS. MATH. SOFTWARE 4, 148-159

Birkhoff, G. and H.L. Garabedian (1960) Smooth surface interpolation. J. Math. and Phys. 39, 258-268

Gleue, J. (1981) Triangulierung und Interpolation von im R^3 unregelmäßig verteilten Daten. HMI-B 357, Berlin

Lawson, C.L. (1977) Software for C^1-surface interpolation, in Rice, J.R.(ed.) Mathematical Software III, Academic Press, 159-192

McLain, D.H. (1976) Two dimensional interpolation from random data. Comp.J. 19, 178-181 and 384

Schmidt, R.M. und M. Steinfeld (1982) Die Konstruktion von C^1-Flächen zur Interpolation unregelmäßig verteilter Daten. HMI-B 371, Berlin

Stühler, M. (1980) ISOFIX - Ein System zur grafischen Darstellung von Isolinien. HMI-B 330, Berlin

Dr. R. Schmidt, Bereich Datenverarbeitung und Elektronik, Hahn-Meitner Institut für Kernforschung GmbH, Glienicker Str. 100, D-1000 Berlin 39, Germany

APPROXIMATION THEORY AND "DOMAIN OF DEPENDENCE" FOR P.D.E. OF HYPERBOLIC TYPE

Harold S. Shapiro

Mathematics Institute, Royal Institute of Technology, Stockholm, Sweden

1. Introduction

This is a preliminary report on work in progress, and a fuller account will appear elsewhere. We illustrate our theme mainly in terms of the classical wave equation, and (purely for simplicity of notation!) in two space dimensions:

$$(1.1) \quad Lu \equiv \left(\frac{\partial^2}{\partial x_1^2} + \frac{\partial^2}{\partial x_2^2} - \frac{\partial^2}{\partial x_3^2}\right)u = 0$$

(here x_3 denotes "time", and the propagation speed is 1).

Consider a solution u of (1.1) which is of class C^2 in a neighborhood of the half-space $x_3 \geq 0$. Then it is uniquely determined by its "Cauchy data" on the plane $x_3 = 0$, i.e. if u and its partial derivatives of first order vanish on this plane, u is zero for all $x_3 \geq 0$. A refinement of this (see [2, p. 643]) is, <u>if the Cauchy data vanish on the disc</u> $D : x_1^2 + x_2^2 \leq a^2,\ x_3 = 0$ <u>then</u> u <u>vanishes for all</u> (x_1,x_2,x_3) <u>in the solid cone</u> K <u>where</u>

$$K = \{x : 0 \leq x_3 \leq a\ ,\ x_1^2 + x_2^2 \leq (a - x_3)^2\} .$$

Physically this is interpreted so, that if the "signal" whose propagation in the (x_1,x_2) plane is represented by u has at time $x_3 = 0$ not penetrated into the disc D, then it cannot reach the point (x_1,x_2) of D before the time $x_3 = a - (x_1^2 + x_2^2)^{1/2}$ which is what one expects of a signal travelling with speed 1. In other language, the (precise) "domain of dependence" in the plane $x_3 = 0$ for the point (0,0,a) of space-time is D. A slightly more general formulation [2, l.c.] is

Theorem A Let $\Omega \subset \mathbb{R}^3$ be a bounded open set and for some fixed $a \geq 0$ denote by S_a the conical ("characteristic") surface

$$(1.2) \quad S_a = \{x \in \mathbb{R}^3 : x_1^2 + x_2^2 - (x_3 - a)^2 = 0\} .$$

If $u \in C^2(\bar{\Omega})$ satisfies (1.1) and u and its first-order partial derivatives vanish on $(\partial\Omega) \setminus S_a$ then $u = 0$ in Ω.

This theorem is usually proven by the use of "energy integrals". The main purpose of this paper is to deduce Theorem A (and its extension when L includes lower-order derivatives) from results on polynomial approximation of an unconventional type, that does not seem to have been studied before. So far we can only treat second-order hyperbolic p.d.e. with constant coefficients, and some very special higher order equations, however it is hoped that further progress with the approximation problem (i.e. extending Theorem 1 below to other operators L and polynomials P) may increase the scope of these applications. The method can to some extent be adapted to mixed initial-boundary problems, as indicated in § 4.

2. An approximation theorem

Theorem 1. Let B denote any closed ball in $\mathbb{R}^3$. For each integer $m \geq 2$ the closure in C(B) of the polynomials of the form

(2.1) $f = L(P^m g)$

where L *is given by (1.1),* g *is an arbitrary polynomial in* $x = (x_1, x_2, x_3)$ *and*

(2.2) $P(x) = x_1^2 + x_2^2 - x_3^2$,

consists precisely of those functions in C(B) *which vanish on* $\{x \in B : P(x) = 0\}$.

Remark. For $m = 0$, the closure in question is C(B); this follows from a simple modification of the following proof which (since the result is not required in the sequel) we leave to the reader.

Before giving the proof, let us show how this theorem (with $m = 2$) implies Theorem A. So, assume u,Ω satisfy the hypotheses of Theorem A. Since the set of solutions of (1.1) is invariant with respect to translation of the independent variables we may assume $a = 0$. Thus P^2g and its first order partial derivatives vanish on

$$S_o = \{x \in \mathbb{R}^3 : x_1^2 + x_2^2 - x_3^2 = 0\}$$

and by a standard argument based on partial integration [2,p.239] we get

$$\int_\Omega uL(P^2g)dx = 0 \ .$$

By Theorem 1, $\int_\Omega uh\,dx = 0$ for all $h \in C(\bar{\Omega})$ which vanish on $S_o \cap \bar{\Omega}$; this set being of measure zero in $\mathbb{R}^3$ we conclude that $u = 0$ in Ω, as was to be shown. (This argument is of course modelled on the standard proof [2, p. 237 ff.] of Holmgren's uniqueness theorem; the point is that the use of Theorem 1 in place of the Cauchy-Kovalevska theorem enables us to get the *maximal* domain of uniqueness.)

We deduce Theorem 1 from the purely algebraic

Lemma 2.1 Let H_k $(k = 0,1,\ldots)$ denote the set of homogeneous polynomials in $x = (x_1,x_2,x_3)$ of degree k. Then, for each $m \geq 1$ the range of the map

(2.3) $g \longmapsto L(P^m g)$

from H_k to $H_{k+2(m-1)}$ is precisely the set $P^{m-1}H_k$.

Indeed, from this lemma we see that the set of f defined by (2.1) definitely contains all polynomial multiples of P^{m-1}, which easily implies Theorem 1. To prove Lemma 2.1 we require

Lemma 2.2 For any $u \in C^\infty$, and $m \geq 1$, $L(P^m u) = P^{m-1}v$ where v is a finite linear combination of partial derivatives of u with polynomial coefficients.

Proof. By the generalized "Leibniz rule" (see [4, p. 10]), $L(P^m u) = L(P^m)u + \ldots$ where each omitted term contains as a factor P^m or $(\frac{\partial}{\partial x_i})P^m$ for some i, and hence the factor P^{m-1}. We have therefore to show that $L(P^m)$ is divisible by P^{m-1}. This is trivially true for $m = 1$, and the rest is a simple induction based on the Leibniz rule which we omit.

Proof of Lemma 2.1. We introduce into each of the vector spaces H_k (homogeneous polynomials of degree k with complex coefficients) the "Fischer inner product" whereby (with usual multi-index notations as in [4]) $\langle x^\alpha, x^\beta \rangle$ is 0 for $\alpha \neq \beta$, and $\alpha!$ for $\alpha = \beta$. (For discussion of this formalism see [1,5,6,7].) The main point is the theorem of E. FISCHER [3] that for any $Q \in H_r$, the operator "multiplication by Q" from $H_k \to H_{k+r}$ and the differential operator $Q^*(D)$ from $H_{k+r} \to H_k$ are mutually adjoint. (Q^* denotes the polynomial whose coefficients are complex-conjugate to those of Q.)

Now, by Lemma 2.2 the range of the map (2.3) is contained in $P^{m-1}H_k$. To show that the map covers this set we have to show that if

$$(2.4) \quad \langle P(D)(P^m g) \; , \; P^{m-1}h \rangle = 0$$

for some $h \in H_k$ and all $g \in H_k$, then $h = 0$. But the inner product in (2.4), using Fisher's theorem, equals $\langle P^m g, P^m h \rangle$, hence taking $g = h$ gives $\| P^m h \|^2 = 0$ whence $h = 0$, completing the proof of Lemma 2.1, and hence of Theorem 1.

A variant of Theorem 1 that is more natural from the standpoint of distribution theory and p.d.e. is

Theorem 1' The restrictions to any closed ball $B \subset \mathbb{R}^3$ of the set of functions

$$\{L\varphi : \varphi \in C_0^\infty(\mathbb{R}^3 \setminus S_0)\}$$

span, in the norm of $C(B)$, all functions in $C(B)$ vanishing on $B \cap S_0$.

Proof. It is easy to see that, for a sufficiently large m, every polynomial multiple of P^m can be approximated with an arbitrarily small error in the norm of $C^2(B)$ by (the restriction to B of) a function in $C_0^\infty(\mathbb{R}^3)$ vanishing on a neighborhood of S_0 so the result follows from Theorem 1.

A dual form of Theorem 1' reads: if u is any measure of compact support in $\mathbb{R}^3$ which is a (distributional) solution of (1.1) in $\mathbb{R}^3 \setminus S_0$ then supp $u \subset S_0$. (The analogous assertion where u is a distribution of finite order can be proven similarly.) If one establishes this result by traditional methods of p.d.e. one gets an alternative proof of Theorem 1.

In particular, a solution of (1.1) on $\mathbb{R}^3 \setminus S_0$ which is in $L^1(\mathbb{R}^3)$ and has compact support vanishes identically; this is a kind of "removable singularities" theorem insofar as any distributional solution on all of $\mathbb{R}^3$ with compact support is 0.

3. Other partial differential operators

3.1 As is well known, Theorem A remains valid when L is replaced by the more general operator

$$L_1 = L + \sum_{j=1}^{3} a_j \frac{\partial}{\partial x_j} + b$$

where a_j, b are real constants (with S_a remaining unchanged). Let us denote as Theorem A' this generalization. It is obtainable by the method of the preceding section; one has only to show that Theorem 1 remains valid when L is replaced by L_1, and this we now do. Introducing the polynomial

$$P_1(x) = P(x) + \sum_{j=1}^{3} a_j x_j + b$$

so that $L_1 = P_1(D)$ we can again check that for any polynomial g, $L_1(P^m g)$ is a multiple of P^{m-1}. Because P_1 is not homogeneous we cannot however proceed as before with the finite-dimensional algebraic theory of Fischer, but require its "transcendental" extension [5,6,7] to the Hilbert space F_3 of entire functions f on $\mathbb{C}^3$ satisfying $\int |f|^2 e^{-|z|^2} d\sigma < \infty$ (here $z = (z_1, z_2, z_3)$ and σ is Lebesgue measure in $\mathbb{R}^6$), and some familiarity with this theory must be assumed here.

The unbounded linear operator "multiplication by P^{m-1}" on its natural domain of definition in F_3 has a closed range R and we shall first show that R is spanned by the family $\{L_1(P^m g) : g \text{ polynomial}\}$, i.e. for every $h \in F_3$ in the domain of this multiplication operator the relation

$$\langle P_1(D)(P^m g) \ , \ P^{m-1}h\rangle = 0$$

for all polynomials g, implies h = 0. We can rewrite this relation as

$$(3.1) \quad \langle P_1(D)P(D)^{m-1}(P^m g), h\rangle = 0 \ .$$

Now, as shown in the proof of Theorem 1, as g ranges over all polynomials, $P(D)(P^m g)$ ranges over all multiples of P^{m-1} and hence, proceeding inductively, $P(D)^s(P^m g)$ ranges over all multiples of P^{m-s} for $s = 1,2,\dots m-1$. The choice $s = m-1$ shows that (3.1) is equivalent to

$$\langle P_1(D)Pk,h\rangle = 0$$

for all polynomials k. This implies $P(D)(P_1 h) = 0$. Therefore, the desired conclusion $h = 0$ follows if we can establish: <u>an entire function</u> f <u>on</u> $\mathbb{C}^3$ <u>satisfying</u> $P(D)f = 0$ <u>and vanishing on the set</u> $P_1(z) = 0$ <u>vanishes identically</u>.

Since we are now working with <u>complex</u> variables, we might as well (for greater symmetry) prove this in the following equivalent form:

<u>Lemma 3.1</u> <u>Let</u> $Q(z) = z_1^2 + z_2^2 + z_3^2$, <u>and let</u> Q_1 <u>be any quadratic polynomial of the form</u> Q + (lower degree terms). <u>Let</u> f <u>be holomorphic on</u> $\mathbb{C}^3$, <u>satisfy</u> $Q(D)f = 0$ <u>and vanish on</u> $\{z : Q_1(z) = 0\}$. <u>Then</u> $f = 0$.

<u>Proof</u>. For any $\zeta \in \mathbb{C}^3$, $f_0(z) = f(z+\zeta)$ satisfies $Q(D)f_0 = 0$ and vanishes wherever $Q_2(z) = Q_1(z+\zeta)$ does. Choosing ζ suitably we can arrange that $Q_2(z) = Q(z) - c$ where c is a complex constant. If $c = 0$, f_0 is an entire multiple of the homogeneous polynomial Q. In this case writing $f_0 = p_0 + p_1 + \dots$ where p_j is a homogeneous polynomial of degree j, each p_j is a multiple of Q and satisfies $Q(D)p_j = 0$ which, by Fischer's theorem, implies $p_j = 0$, hence $f_0 = 0$.

The case $c \neq 0$ may be reduced to $c = 1$ by considering $f_1(z) = f_0(\lambda z)$ for a suitably chosen complex $\lambda \neq 0$. So, we have only to show that <u>an entire solution</u> f_1 of $Q(D)f_1 = 0$ <u>vanishing wherever</u> $z_1^2 + z_2^2 + z_3^2 = 1$ <u>vanishes identically</u>; but then Re f_1 and Im f_1 restricted to $z \in \mathbb{R}^3$ are real-valued harmonic functions vanishing on the real unit sphere, and so identically, whence $f_1 = 0$. This concludes the proof.

The proof of Theorem A' is now completed by the remark that convergence in the norm of F_3 implies uniform convergence on compact subsets of $\mathbb{C}^3$. (Actually, convergence in F_3 implies more, namely convergence on $\mathbb{C}^3$ in the norm

$$|||f||| \mathrel{\hat{=}} \max_{z\in\mathbb{C}^3} e^{-\frac{|z|^2}{2}} |f(z)|$$

which allows extensions of the theorems of this paper to unbounded domains, however we shall not enter into this here.)

3.2 One other p.d. operator to which the methods of this paper apply readily is $L_2 = \frac{\partial^3}{\partial x_1 \partial x_2 \partial x_3}$. The associated "characteristic" manifolds are

$$T_\xi = \{x : (x_1 - \xi_1)(x_2 - \xi_2)(x_3 - \xi_3) = 0\} , \quad \xi \in \mathbb{R}^3 .$$

Theorems A and 1 hold with L, S_a, P replaced by L_2, T_ξ and $R = x_1\, x_2\, x_3$, respectively. In this case Theorem 1 has an almost trivial direct proof: for, choosing $g = x_1^{n_1} x_2^{n_2} x_3^{n_3}$ with $n_i \geq 0$,

$$L_2(R^m g) = \prod_{j=1}^{3} (m + n_j) x_j^{m+n_j-1}$$

so the range of the operator $g \longmapsto L_2(R^m g)$ for polynomial g contains all polynomial multiples of $(x_1 x_2 x_3)^{m-1}$, and for each ball B these span the functions in $C(B)$ vanishing on $x_1\, x_2\, x_3 = 0$. (The two-dimensional version yields Theorem A for the one-dimensional wave equation, apart from a 45° rotation of axes.)

3.3 We indicate <u>very</u> briefly a different argument by which the results of this paper could have been obtained. For the purpose of approximation on compact subsets of $\mathbb{R}^n$ by functions of the form $P(D)(Qg)$, where P and Q are given polynomials, we can let g run over all finite linear combinations of <u>exponential functions</u> E_ζ, where

$$E_\zeta(x) = \exp(\Sigma\ \zeta_j x_j) \ , \qquad \zeta \in \mathbb{C}^n$$

rather than polynomials. Using the easily proved symbolic identity

$$P(D)(QE_\zeta) = \big(P(D+\zeta)Q\big)E_\zeta$$

the question whether the restrictions to $B \subset \mathbb{R}^n$ of these latter functions, as ζ ranges over $\mathbb{C}^n$, span (say) $L^2(B)$ takes the form: for $f \in L^2(B)$, does

$$\int_B f(x)\big(P(D+\zeta)Q\big)E_\zeta(x)\ dx = 0 \ , \quad \forall\ \zeta \in \mathbb{C}^n$$

imply $f \equiv 0$? This formulation leads to a <u>linear partial differential equation with polynomial coefficients for the (entire) Laplace transform</u> F <u>of</u> f, that is, for $F(\zeta) = \int f(x)E_\zeta(x)\ dx$, which we can write in the concise form

$$\big(P(D+\zeta)Q(p)\big)F = 0$$

where now $D \equiv (\frac{\partial}{\partial p_1}, \ldots \frac{\partial}{\partial p_n})$ and after evaluation of $P(D+\zeta)Q(p)$ we are to replace p by $(\frac{\partial}{\partial \zeta_1}, \ldots \frac{\partial}{\partial \zeta_n})$. For example, when $P(D) = D_1D_2$ and $Q(p) = p_1^2p_2^2$ this analysis leads to

$$\frac{\partial^2}{\partial\zeta_1\partial\zeta_2}\left(\zeta_1^2\ \zeta_2^2\ \frac{\partial^2 F}{\partial\zeta_1\partial\zeta_2}\right) = 0$$

whence "Fischer space" considerations yield $\frac{\partial^2 F}{\partial\zeta_1\partial\zeta_2} = 0$, and this together with

$$\int_{\mathbb{R}^2} |F(i\eta_1, i\eta_2)|^2\ d\eta < \infty$$

implies $F \equiv 0$.

4. Mixed boundary value problem

The method of this paper can be modified to study some mixed initial-boundary value problems. We illustrate only with one very simple example. Consider a C^2 solution u of $D_1D_2u = 0$ with vanishing Cauchy data on a curve Γ in the first quadrant of the (x_1,x_2) plane that joins the x_1 and x_2 axes (e.g. for simplicity of description, imagine for Γ a segment of $x_1 + x_2 = 1$). Then (cf. § 3.2 above) u vanishes in the triangular domain bounded by Γ and the positive x_1 and x_2 axes. Suppose now however that the Cauchy data vanishes only on the part of Γ between its points of intersection with the positive x_1 axis and with the line $x_2 = x_1$. Now if u also vanishes (but not necessarily its normal derivative) on the line $x_1 = x_2$, between 0 and the intersection with Γ, then $u = 0$ in the triangle Ω bounded by the positive x_1 axis, Γ and the line $x_1 = x_2$. This is a well-known result for the one-dimensional wave equation, typical of "mixed" problems.

These hypotheses imply, by a simple computation based on partial integration that

$$\int_\Omega u \frac{\partial^2}{\partial x_1 \partial x_2} \left(x_2^2(x_1 - x_2)g\right) dx_1dx_2 = 0$$

for all polynomials g. (The extra factor $x_1 - x_2$ is now needed to compensate that the full Cauchy data of u does not vanish on the line $x_1 = x_2$.) To prove by our methods the desired vanishing of u in Ω it is now sufficient to show (following the pattern of our proof of Theorem A): <u>the range of the map</u>

$$(4.1) \quad g \longmapsto D_1D_2\left(x_2^2(x_1 - x_2)g\right) \quad , \quad g \in \text{polynomials}$$

<u>consists of all multiples of</u> x_2. (Note that every polynomial in the range <u>is</u> a multiple of x_2.)

It suffices to show that (4.1) maps H_k onto $\{x_2h : h \in H_k\}$, i.e. that

$$\langle D_1D_2\; x_2^2(x_1 - x_2)g \;,\; x_2h\rangle = 0$$

for some $h \in H_k$ and all $g \in H_k$ implies $h = 0$. Using Fischer's Theorem as before, this reduces to showing: _if_ $h \in H_k$ _satisfies the differential equation_

$$(4.2) \quad (D_1 - D_2)D_2^2(x_1x_2^2\ h) = 0$$

then $h = 0$.

But this is immediate, since (4.2) implies that $D_2^2(x_1x_2^2h)$ is a function of $x_1 + x_2$, say $p(x_1 + x_2)$, but also it contains the factor x_1 so $p(x_2) = 0$. Thus $D_2^2(x_2^2(x_1h)) = 0$ and a trivial application of Fischer's theorem now shows $x_1h = 0$.

5. Concluding remarks

The fundamental problem of this paper is, for which polynomials P,Q does the map $g \longmapsto P(D)(Qg)$ have dense range in (say) the space $L^2(\Omega)$ for bounded open Ω as g ranges over all polynomials (and more generally, to determine the closure of this range in various topologies). This problem embodies a very "finely tuned" relationship between P and Q; its delicacy can be seen from the following considerations. For $P = x_1^2 - x_2^2$, $Q = (x_1^2 - x_2^2)^2$ we got an affirmative answer. However, if Q is changed to $(cx_1^2 - x_2^2)^2$ for some positive $c \neq 1$ the answer is _negative_. This is perhaps not so easy to see directly, but follows from the relation to p.d.e.: if the "dense range" persisted when $c \neq 1$ then we could deduce a corresponding uniqueness theorem for the wave equation $(D_1^2 - D_2^2)u = 0$ à la Theorem A, but with the _false_ characteristics $cx_1^2 - x_2^2 = 0$ in place of the usual ones; and the simplest examples show that any such assertion is untrue.

References

1. Bargmann, V. (1961) On a Hilbert space of analytic functions and an associated integral transform, Comm. Pure Appl. Math. 14, 187-214.

2. Courant, R. and D. Hilbert (1962) Methods of mathematical physics, vol. II (Interscience, New York).

3. Fischer, E. (1911) Über algebraische Modulsysteme und lineare homogene partielle Differentialgleichungen mit konstanten Koefizienten, J. für Math 140, 48-81.

4. Hörmander, L. (1969) Linear partial differential operators, third revised printing (Springer-Verlag, New York).

5. Newman, D.J. and H.S. Shapiro (1964) A Hilbert space of entire functions related to the operational calculus (mimeographed notes, Ann Arbor, 92 pp.)

6. ——— (1966) Certain Hilbert spaces of entire functions, Bull A.M.S. 72, 971-977.

7. ——— (1968) Fischer spaces of entire functions, in Proc. Symposia Pure Math. vol 11 (Amer. Math. Society).

Prof. Harold S. Shapiro, Mathematics Institute, Royal Institute of Technology, S-10044 Stockholm, Sweden.

PROPERTIES OF SPLINE PROJECTIONS

Boris Shekhtman

Department of Mathematics
University of Southern California
Los Angeles, CA 90007

I. Splines in Hilbert Spaces

Let X,Y be Hilbert spaces, T be a linear bounded operator that maps X into Y and $\Lambda \subset X$ be a subset of X.

For an arbitrary subspace $M \subset X$, P_M denotes the orthogonal projection on M.

For every $x \in X$ an element σ is an interpolating spline if

$$(1) \qquad \sigma = x - (T|T(\Lambda^{\perp}))^{-1} P_{T(\Lambda^{\perp})} Tx$$

This definition makes sense iff $(T|T(\Lambda^{\perp}))^{-1}$ exists which is equivalent to anyone of the following conditions

$$\Lambda \cap \ker T = \{0\} \text{ and } T(\Lambda^{\perp}) \text{ is closed}$$

or

$$\cos(\ker\Lambda, \ker T) < 1.$$

In this paper we assume that these properties are satisfied. Then formula (1) defines an interpolation projection

(2) $$p(T,\Lambda) = I - (T|T(\Lambda^{\perp}))^{-1}P_{T(\Lambda^{\perp})}T$$

and the range of this projection defines a space of splines $Sl(T,\Lambda) \subset X$.

The projection $p(T,\Lambda)$ has "the minimal norm property":

(3) $$\text{for every } x \in X: \ ||Tp(T,\Lambda)\ x|| \leq ||Ty||: (y-x) \in \Lambda^{\perp}.$$

The purpose of this talk is to discuss some of the properties and open problems for $p(T,\Lambda)$ as a function of two variables: T and Λ.

We will need the following notation: Let (A,t) be a topological space; A_n A then

$$\operatorname{Lim} A_n := \{a \in A: a = t\text{-}\lim a_n;\ a_n \in A_n\}$$

II. P(T, Λ) as a function of Λ.

It is natural to expect that if $\operatorname{Lim} sp(\Lambda_n) = X$ then

$$p(T,\Lambda_n) \to I_X.$$

Theorem 1. (cf[7]). Let $\dim \ker T = q < \infty$, and $\operatorname{Lim} sp\ \Lambda_n = X$ then

$$||p(T,\Lambda_n)\ x - x|| \to 0 \quad \forall x \in X.$$

The proof of this theorem is based on the "minimal norm property":

$$||P_{(\ker^{\perp} T)}\ p(T,\Lambda_n)x|| = ||(T|(\ker^{\perp} T)\ )^{-1} Tp(T,\Lambda_n)x|| \leq$$

$$||(T|\ker^{\perp}T)^{-1}|| \, ||Tp(T,\Lambda_n)x|| \leq ||(T|\ker^{\perp}T)^{-1}|| \, ||Tx||.$$

To prove the uniform boundedness of $||p(T,\Lambda_n)||$ we now have to prove that it's finite-dimensional part is bounded:

$$||P_{\ker T}\, p(T,\Lambda_n)x|| = \mathcal{O}(1)$$

which was done in [7]. So the property (2) takes care of the large portion of $p(T,\Lambda_n)$. This idea will dominate most of the proofs in this paper.

Theorem 1 was improved by C. de Boor [2] who gave sharp necessary and sufficient conditions in

Theorem 2 $p(T,\Lambda_n) \overset{s}{\to} 1$ iff sup cos (ker T, $\Lambda_n^{\perp}$) < 1 and (ker $T)^{\perp} \subseteq$ Lim Λ_n.

His proof was based directly on the formula (1). The same theorem could be proved using generalized inverses of operators (cf [5]).

Theorems 1 and 2 could be viewed as an abstract analog of the question raised by Marsden [6] of uniform boundedness of spline-interpolation projections.

Back to back with this question goes a question of C. de Boor [3] of the uniform boundedness of L_2- orthogonal spline projections in C-norm.

To state an abstract analog of this problem let T be a closed linear operator from Y into Y. Let X be the closure of D(T) in a norm

$$||\cdot||_X^2 = ||\cdot||_Y^2 + ||T\cdot||_Y^2$$

and let $\Lambda_n \subset X$. Let $Q(T,\Lambda_n)$ be Y-orthogonal projections on the spaces $Sl(T, \Lambda_n) \subset Y$.

Problem 1. Under what conditions on T and Λ_n

$$||Q(T,\Lambda_n)|| = O(1)?$$

Assume for simplicity that $T(X) = Y$. Then under a rather unsatisfactory condition (4) (below) we can prove the result. We reserve the symbol $\perp$ for the orthogonality in X.

Proposition 3 Let Λ_n be finite-dimensional subspaces of X; $\dim \ker T = q < \infty$ and

(4) there exists $\rho > 0$ such that
$$||(T^*T)^{-1}z||_Y \geq \rho||z||_X \quad z \in \Lambda_n \cap \ker^{\perp} T.$$

Then $||Q(T,\Lambda_n)||_X = O(1)$

Proof: To prove the uniform boundedness of $||Q(T,\Lambda_n)||_X$ it is sufficient to prove the uniform boundedness of $||TQ(T,\Lambda_n)x||_Y$ for all $x \in X$. Let $\Lambda_n = \mathrm{sp}\,\{\lambda_1^{(n)},\ldots,\lambda_n^{(n)}\}$

where $\lambda_j^{(n)}$ are X-orthogonal bases for Λ_n. Then by the "minimal norm property" it is sufficient to prove that

$$< T\lambda_j^{(n)},\ TQ(T,\Lambda_n)\,x > = O(1) \text{ independent of } j \text{ and } n.$$

We can ignore the part of $\lambda_j^{(n)}$-s that belong to $\ker T$ and prove that

$$<T\lambda_j^{(n)},\ TQ(T,\Lambda_n)x > = \mathcal{O}(1), \quad j = 1,\ldots,n-q$$

where $\lambda_j^{(n)} \in (\ker^{\perp} T)$. Then (cf [1])

$$TQ(T,\Lambda_n)\,x = \sum_{j=1}^{n-q} \alpha_j^{(n)}(T^*)^{-1}\lambda_j^{(n)} \text{ for some } \alpha_j^{(n)} \in R.$$

Hence

$$< T\lambda_j^{(n)},\ TQ(T,\Lambda_n)x>_Y = < \lambda_j^{(n)},\ \Sigma\alpha_j^{(n)}\ \lambda_j^{(n)}\ >_X = \alpha_j^{(n)}.$$

So we have reduced our problem to proving the uniform boundedness of $|\alpha_j^{(n)}|$ individually.

Introducing operator $T^{-1} = (T|(\ker^{\perp}T)\ )^{-1}$ we have

$$||Q(T,\Lambda_n)x||_Y = ||\Sigma\ \alpha_j^{(n)}T^{-1}(T^*)^{-1}\ \lambda_j^{(n)}||_Y = O(1)$$

therefore

$$O(1) = |\alpha_{max}^{(n)}|||\ \Sigma\ \mu_j^{(n)}(T^*T)^{-1}\ \lambda_j^{(n)}||_Y$$

where $\mu_j^{(n)} = \alpha_j^{(n)}/\alpha_{max}^{(n)}$ and $|\mu_j^{(n)}| \leq 1$. So

$$|\alpha_{max}^{(n)}| \leq ||x||_Y/||\ \Sigma\ \mu_j^{(n)}(T^*T)^{-1}\lambda_j^{(n)}||_Y$$

and condition (4) implies

$$||(T^*T)^{-1}\ \Sigma\ \mu_j^{(n)}\lambda_j^{(n)}||_Y \geq \rho||\mu_j^{(n)}\lambda_j^{(n)}||_X \geq \rho$$

thus proves the proposition.

III. $p\ (T,\Lambda)$ as a function of T.

In this section we consider projection

$$p(T,\Lambda) = I-(T|\Lambda^{\perp})^{-1}P_{T(\Lambda^{\perp})}T$$

as a function of T. For simplicity we will assume that $\dim \ker T = q < \infty$.

Let $P(\Lambda) \subset L(X,Y)$ be the set of operators T so that $p(T,\Lambda)$ exists.

Theorem 4. The set $P(\Lambda)$ is open in $L(X,Y)$.

Proof: We indicate the proof of one part of this statement. Let $T \in P(\Lambda)$. Then $\ker T \cap \Lambda^{\perp} = \{0\}$. We have to prove that for any B in some neighborhood of T

$$\ker B \cap \Lambda^{\perp} = \{0\}.$$

By the existence condition, $T(\Lambda^{\perp})$ is closed and so

$$Y = T(\Lambda) \dot{+} M; \text{ M is a subspace of Y.}$$

For every $B \in L(X,Y)$, define $\phi_B \in L(S+M,Y)$ by

$$\phi_B(s,m) = Bs + m$$

Then the map $\phi: B \to \phi_B$ is continuous. To finish the proof we have to mention that the set of all invertable operators in $L(S+M,Y)$ is open and the inverse image of an open set is open.

Theorem 5. Let T_n, $T \in P(\Lambda)$ and $T_n \to T$ uniformly. Then

$$p(T_n, \Lambda) \to p(T,\Lambda) \text{ pointwise.}$$

Proof Since $T_n \to T$ the usual argument $(T_n, T \in P(\Lambda))$ implies that $(T_n|\Lambda^{\perp})^{-1} \to (T|\Lambda^{\perp})^{-1}$ pointwise. Finally the angle between spaces $T(\Lambda^{\perp})$ and $T_n(\Lambda^{\perp})$ tends to zero and so $P_{T_n(\Lambda^{\perp})} \to P_{T(\Lambda^{\perp})}$

This theorem implies that the exponential splines converge to polynomial splines as the frequences tend to zero.

Another obvious corollary of Thereom 5 is the following

Proposition 6. Let $K \subset L(X,Y)$ be compact and $x \in X$. Then there exists $T' \in K$ such that

$$||x-p(T', \Lambda)x|| \leq ||x-p(T,\Lambda)x|| \ \forall T \in K.$$

Of course such an operator T' is not unique. Nevertheless.

Problem 2. Characterize an optimal operator T'. In particular let Π_k be a uniformly bounded set of monic polynomials of degree k. Let

$$K = \left\{ s(\frac{d}{dt}); \ s \in \Pi_k \right\}.$$

What is the optimal polynomial $s' \in \Pi_k$: $s'(\frac{d}{dt}) = T'$.

I would like to finish this section with a remark that the Theorem 4 has a different set of applications to solvability of interpolation problems. For instance, the solvability of the Hermite-Birkhoff interpolation problem in the class of polynomials implies the solvability of the same problem in the class of exponential functions with small frequences and the solvability of this problem with perturb mesh ([9]).

IV. $p(T^m, \Lambda)$ as a function of m.

We start with stating the result that was proved by A. Cavaretta and D. Newman, W. fon Golitchek and I. Schoenberg independently (cf. [4]):

Let $\Delta = \{t_1,...,t_{2n+1}\} \subset [-\Pi,\Pi]$. Let σ_m be a sequence of periodic splines that interpolate given function x on a set Δ, and are generated by $\frac{d^m}{dt^m}$. Then

$$\lim \sigma_m = s \text{ as } m\to\infty$$

where s is the trigonometric polynomial of degree n which interpolates x on Δ.

In this section, we present a generalization of this theorem (cf [8]).

Let as before $T:Y\to Y$ and $X_m = \overline{D(T^m)}$ where the closure is taken with respect to the norm

$$||\cdot||_m^2 = ||\cdot||_Y^2 + ||T^m||_Y^2$$

Let X_m dense in Y and singular values of T - $\{\mu_j\}$ are discrete and increasing. Let $\Lambda = \{\lambda_1,\ldots,\lambda_n\}$ be a set of functionals on Y such that $\Lambda|X_m \subset X_m$ for all m. Assume further that $\{s_j\}$ is the corresponding set of singular vectors of operator T and

$$\operatorname{span}\{s_j\}_{j=1}^{n} \cap \Lambda^{\perp} = \{0\}.$$

Theorem 7. If $|\mu_n| < |\mu_{n+1}|$ then $\lim p(T^m, \Lambda)x = s(x)$ where $s(x) = \sum_{j=1}^{n} \alpha_j s_j$ interpolates x.

(b) If $|\mu_n| = |\mu_{n+1}|$ let p, r be such that

$$|\mu_{p-1}| < |\mu_p| = \ldots = |\mu_n| = \ldots = |\mu_r| < |\mu_{r+1}|.$$

Then $\lim p(T^m,\Lambda)x = s(x)$ where $s(x) = \sum_{j=1}^{r} \alpha_j s_j$ interpolates x and $\sum_{j=p}^{r} \alpha_j^2$ is minimal among all interpolants of x of this form.

This theorem explains the appearance of trigonometric

polynomials in case of periodic spline functions.

The theorem does not apply to polynomial spline functions since in this case $\ker T^m \Lambda \neq \{0\}$ for large m. So it is natural to ask

Problem 3 What happens to $\lim p(T^m, \Lambda_m)$ as Λ_m increasing?

In particular let f be an analytic function of finite order. Let $\Delta_m = \{t_1^{(m)},\ldots,t_m^{(m)}\}$ and let σ_m be splines generated by $\frac{d^m}{dt^m}$ that interpolate f on the set Δ_m. Is it true that $\lim \sigma_m = f$?

References

[1] P.M. Anselone, P.T. Laurent, A general method for the construction of interpolating and smoothing spline-functions, Numer. Math. 12 (1968) 66-82.

[2] C. de Boor, Convergence of abstract splines, J. Approx. Theory 31 (1981) 80-89.

[3] C. de Boor, The quasiinterpolant as a tool in elementary polynomial spline theory, in "Approximation Theory", G.G. Lorentz ed., Academic Press (1973), 269-276.

[4] A.S. Cavaretta, Jr. and D.J. Newman, Periodic interpolating splines and their limits, Proceeding of the Koninklijke Nederlandse Academic Series A, V81 (4), (1978) 515-526.

[5] S. Izumino, Convergence of generalized inverses and abstract splines, submitted to J. Approx. Theory.

[6] M. Marsden, Cubic spline interpolation of continuous functions. J.A.T. 10, (1974) 103-111.

[7] B. Shekhtman, Unconditional convergence of abstract splines, J. Approx. Theory 30 (1980) 237-246.

[8] B. Shekhtman, The limits of abstract splines, Numer. Funct. Anal. and Optimiz. 2(5) (1980) 375-385.

[9] B. Shekhtman, Interpoltion in abstract spaces, Ph.D. Dissertion, Kent, Ohio 1980.

Prof. Boris Shekhtman, Department of Mathematics, University of Southern California, Los Angeles, California 90007, USA.

A SURVEY OF RECENT RESULTS ON APPROXIMATION THEORY IN CHINA

Shen Xie-chang

Peking University, Department of Mathematics
Beijing, China

In this paper some results obtained in last years on approximation theory of functions of real variables as well as complex variables are introduced.

1. Approximation of functions of a real variable

1.1. Approximation by linear operators

For simplicity, X is applied to denote all three spaces $C_{2\pi}, M_{2\pi}, L_{2\pi}$. For given modulus of continuity w(t) we introduce some notation:

$$H_X^0 = \{\varphi(x):\ \varphi \in X,\ \|\varphi\|_X \leq 1,\ \int_0^{2\pi} \varphi(t)dt = 0\},$$

$$H_X^w = \{\varphi(x):\ \varphi \in X,\ \|\varphi(.+t) - \varphi(.)\|_X \leq w(t),\ \int_0^{2\pi} \varphi(t)dt = 0\}.$$

Let the kernel $K(t) \in L_{2\pi}$ satisfy the following conditions:

1. $K(t)=K(-t)$,
2. $\frac{1}{\pi}\int_{-\pi}^{\pi} K(t)dt=1.$

Suppose

$$\mathfrak{M}_X = \{f(x): f = \frac{a_0}{2} + \frac{1}{\pi}K * \varphi,\ \varphi \in H_X^0\},$$

$$\mathfrak{M}_X^w = \{f(x): f = \frac{a_0}{2} + \frac{1}{\pi}K * \varphi,\ \varphi \in H_X^w\},$$

where a_0 is a constant.

If we take

$$K(t) = K_r(t) = \sum_{n=1}^{\infty} \frac{1}{n^r} \cos(nt + \frac{\pi}{2}r),\ r > 0,$$

then $\mathfrak{M}_X$ and $\mathfrak{M}_X^w$ are denoted by W_X^r and $W^r H_X^w$, respectively; and if

$$K(t) = \tilde{K}_r(t) = \sum_{n=1}^{\infty} \frac{1}{n^r} \sin(nt + \frac{\pi}{2}r),\ r > 0,$$

then $\mathfrak{M}_X$ and $\mathfrak{M}_X^w$ are denoted by $\tilde{W}_X^r$ and $\tilde{W}^r H_X^w$.

If we take

$$K(t) = K_r^{\alpha}(t) = \sum_{n=1}^{\infty} \frac{1}{n^r} \cos(nt + \frac{\pi}{2}\alpha),\ r > 0, \alpha > 0,$$

or

$$K(t) = \tilde{K}_r^{\alpha}(t) = \sum_{n=1}^{\infty} \frac{1}{n^r} \sin(nt + \frac{\pi}{2}\alpha),\ r > 0, \alpha > 0,$$

then we have the spaces $W_X^{r,\alpha}$, $W^{r,\alpha} H_X^w$ and $\tilde{W}_X^{r,\alpha}$, $\tilde{W}^{r,\alpha} H_X^w$.

Let $L_n(t) = \frac{1}{\pi} K_n * f$ be a sequence of convolution operators where the kernel K_n satisfies the preceding two conditions. We write

$$\mathcal{E}[\mathfrak{M}_X, L_n]_X = \sup_{f \in \mathfrak{M}_X} \| f - L_n(f) \|_X,$$

and

$$\mathcal{E}[\mathfrak{M}_X^w, L_n]_X = \sup_{f \in \mathfrak{M}_X^w} \| f - L_n(f) \|_X.$$

Two problems could be raised as follows:

1. Do these inequalities

$$(1)\quad \mathcal{E}[\mathfrak{M}_L, L_n]_L \leqq \mathcal{E}[\mathfrak{M}_C, L_n]_C,$$
$$(2)\quad \mathcal{E}[\mathfrak{M}_L^W, L_n] \leqq \mathcal{E}[\mathfrak{M}_L^W, L_n]_C$$

hold? Under what conditions do these inequalities become equalities or asymptotic equalities as $n \to +\infty$?

2. For some concrete classes of functions and operators what are the main terms in its asymptotic expansion?

For arbitrary trigonometric polynomials $K_n(t)$, Nikol'ski S.M [1] established (1) for $K_r(t)$ and pointed out that (1) becomes asymptotic equality for Fejér operator; Stechkin S.B. and Teliakovski S.A. [2] obtained (1) for $K_r^{\alpha}(t)$ and pointed out that in many cases these inequalities become equalities or asymptotic equalities (also see Motoruye V.P. [3]).

For $K(t) \in L_{2\pi}, K_n(t) \in M_{2\pi}, K_n^*(t) = K * K_n - K, x(t,c) = K_n^*(t) - c$, where c is a constant. We say $K_n^*(t) \in N$, if there exist c^*, β and a positive integer p such that

$$(3)\quad \sigma\, x(t,c^*)\,\mathrm{sgn}\,\sin(pt-\beta\pi) \geqq 0 \text{ a.e.}, \quad \sigma = \pm 1.$$

In particular, when $K_n^*(t) - c^* \neq 0$ a.e., then (3) takes the form

$$\sigma x(t,c^*) = \mathrm{sgn}\,\sin(pt-\beta\pi) \text{ a.e.}, \ \sigma = \pm 1.$$

<u>Theorem</u> (Sun Y.S. [4]). If $K_n^*(t) \in N$, then

$$\mathcal{E}[\mathfrak{M}_L, L_n]_L = \mathcal{E}[\mathfrak{M}_C, L_n]_C = \frac{1}{\pi} E_1(K_n^*)_L$$
$$= \frac{1}{\pi} \left| \int_0^{2\pi} K_n^*(t)\,\mathrm{sgn}\,\sin(pt-\beta\pi)\,dt \right|.$$

In particular, if $K(t) = K_r(t)$, then

$$K_n^*(t) = F_{n,r}(t) = \sum_{k=1}^{\infty} \frac{1-\lambda_k^{(n)}}{k^r} \cos(kt + \frac{\pi}{2} r),$$

if $K(t)=\tilde{K}_r(t)$, then

$$K_n^*(t)=\tilde{F}_{n,r}(t)=\sum_{k=1}^{\infty}\frac{1-\lambda_r^{(n)}}{k^r}\sin(kt+\frac{\pi}{2}r),$$

where

$$\lambda_r^{(n)}=\frac{1}{\pi}\int_{-\pi}^{\pi}K_n(t)\cos ktdt.$$

<u>Theorem</u> (Sun Y.S. [4]). If the kernels $K_n(t)$ satisfy the condition

$$(4)\quad G_n(t) = \int_t^{\pi}K_n(u)du\geq 0,\quad 0\leq t\leq\pi, n=1,2,...,$$

then

1. $F_{n,r}(t)\in N$,

2. For odd r,

$$\sigma_r \operatorname{sgn}\{F_{n,r}(t)\}\operatorname{sgn}\sin t\geq 0,\ |\sigma_r|=1,$$

3. For even r,

$$\sigma_r \operatorname{sgn}\{F_{n,r}(t)-F_{n,r}(\tfrac{\pi}{2})\}\operatorname{sgn}\cos t\geq 0,\ |\sigma_r|=1,$$

4. $$\mathcal{E}[W_L^r,L_n]_L = \mathcal{E}[W_C^r,L_n]_C$$
$$=\frac{1}{\pi}\left|\int_0^{2\pi}F_{n,r}(t)\operatorname{sgn}\sin(t-\beta\pi)dt\right|$$
$$=\frac{4}{\pi}\left|\sum_{\nu=0}^{\infty}\frac{(-1)^{\nu(r+1)}(1-\lambda_{2\nu+1}^{(n)})}{(2\nu+1)^{r+1}}\right|,\quad n,r=1,2,...,$$

where

$$\beta=\begin{cases}0 & \text{for } r \text{ odd}\\ \frac{1}{2} & \text{for } r \text{ even.}\end{cases}$$

There are analogous results for $\tilde{W}_L^r$ and $\tilde{W}_C^r$. However, for the cases $K(t)=K_r^{\alpha}(t)$ or $\tilde{K}_r^{\alpha}(t)$ the analogous estimates have not been obtained yet.

Theorem (Sun Y.S. [4]). Suppose that the kernels $K_n(t)$ satisfy the condition (4) and that $w(t)$ is a given upward convex modulus of continuity, then

1. For odd r,

$$(5)\quad \mathcal{E}[W^r H_L^w, L_n]_L = \mathcal{E}[W^r H_C^w, L_n]_C = \frac{2}{\pi}\left|\int_0^{\pi} F_{n,r}(t)\tilde{w}(t)\,dt\right| = \left|\sum_{\nu=0}^{\infty} \frac{(-1)^{\nu(r+1)}(1-\lambda_{2\nu+1}^{(n)})}{(2\nu+1)^{r+1}} b_{2\nu+1}\right|,$$

2. For even r,

$$(6)\quad \mathcal{E}[\tilde{W}^r H_L^w, L_n]_L = \mathcal{E}[\tilde{W} H_C^w, L_n]_C = \frac{2}{\pi}\left|\int_0^{\pi} \tilde{F}_{n,r}(t)\tilde{w}(t)\,dt\right| = \left|\sum_{\nu=0}^{\infty} \frac{(-1)^{\nu r}(1-\lambda_{2\nu+1}^{(n)})}{(2\nu+1)^{r+1}} b_{2\nu+1}\right|,$$

where

$$\tilde{w}(t) = \begin{cases} \frac{1}{2}\, w(2t), & 0\le t\le \frac{\pi}{2}, \\ \frac{1}{2}\, w(2\pi-2t), & \frac{\pi}{2}\le t\le \pi, \end{cases}$$

$$\tilde{w}(-t) = \tilde{w}(t),\quad \tilde{w}(t+2\pi) = \tilde{w}(t)$$

and

$$b_{2\nu+1} = \frac{2}{\pi}\int_0^{\pi} w(2t)\sin(2\nu+1)t\,dt.$$

These results generalized the corresponding results obtained by Motoruye V.P. [3], Baskakov V.A. [5] for $K_n(t)\ge 0$.

The corresponding results for even r in (5) and for odd r in (6) are not known.

1.2. Approximation by concrete linear operators

If $f(x) \in L_{2\pi}$ we denote by $S_n(f;x)$ the n-th partial sum of the Fourier series of f(x).

Let f(x) be a locally integrable function on the interval (a,b), $(-\infty \leq a < b \leq +\infty)$ and

$$u(f;h)=\sup_{|I|=h} \frac{1}{|I|}\int_I |f(x)-f_I|dx,$$

where

$$f_I = \frac{1}{|I|}\int_I f(x)dx$$

and the supremum is taken over all closed intervals $I \subset (a,b)$ with the length h. We say that a function f(x) is a bounded mean oscillation function, if u(f;h) is bounded for all h. In 1981, Wang S. L. [6] obtained the following result.

Suppose that $f \in C_{2\pi}$ and

$$u(f;t)=o(\ln \frac{1}{t}),$$

then the Fourier series of f(x) uniformly converges on $[0,2\pi]$. If we replace "o" by "O" in (1), or $f(x) \in L_{2\pi}$, then the above conclusion can not be valid.

Let

$$V_{n,\ell}(f;x)=\frac{1}{\ell+1}\sum_{\nu=n}^{n+\ell} S_\nu(f;x),$$

$$\mathfrak{M}_\varepsilon = \{f : E_\nu(f) \leq \varepsilon_\nu,\ \nu=1,2,...\},$$

where $\varepsilon_\nu \downarrow 0$, $E_\nu(f)$ is the best value of approximation to f(x) by the trigonometric polynomials of degree $\leq \nu$. In 1981, Xie T.F. [7] obtained the following results.

Suppose $\ell=O(n)$, then there exist two constants $C_1 > C_2 > 0$ such that

$$C_2 \sum_{\nu=n}^{2(n+\ell)} \frac{\varepsilon_\nu}{(\nu+1)^r(\nu+\ell+1-n)} \leqq \sup_{f^{(r)}\in \mathfrak{M}_\varepsilon} \|f-V_{n,\ell}(f;x)\|_C$$
$$\leqq C_1 \sum_{\nu=n}^{2(n+\ell)} \frac{\varepsilon_\nu}{(\nu+1)^r(\nu+\ell+1-n)},$$

and for every $r=0,1,\ldots,$ one can construct $f_\varepsilon \in C_{2\pi}, f^{(r)} \in \mathfrak{M}_\varepsilon$ such that

$$\overline{\lim_{n\to+\infty}} \frac{\|f_\varepsilon - V_{n,\ell}(f_\varepsilon;x)\|_C}{\sum_{\nu=n}^{2(n+\ell)} \frac{\varepsilon_\nu}{(\nu+1)^r(\nu+\ell+1-n)}} \geqq C_3 > 0.$$

This kind of inequalities for $r = 0$, $\ell = 0$ (i.e., the case of $S_n(f;x)$) was earlier obtained by Oskalkov K.I. [8] and then, in 1978, Damen V. [9] has generalized it to the case of $r=0$, $\ell \geqq 0$. We should indicate that the function f_ε constructed by them, depended on n and ℓ, and in Xie Ting-fan's paper it was not. Sun Yong-sheng [10] considered the case for $r \geqq 0$ and $\ell=0$ also.

If the condition $\ell=o(n)$ is not satisfied, but an additional condition on $\{\varepsilon_\nu\}$ is imposed, then an analogous theorem is valid.

We denote by

$$\sigma_n^\alpha(f;x) = \frac{1}{(\alpha)_n} \sum_{\nu=0}^{n} (\alpha-1)_{n-\nu} S_\nu(f;x)$$

the α-th Cesàro summation (C,α), where $\alpha > -1$ and

$$(\alpha)_\nu = \frac{\Gamma(\nu+\alpha+1)}{\Gamma(\nu+1)\Gamma(\alpha+1)}.$$

Shi Y.L. [11] obtained the following results.

Suppose that $X = C_{2\pi}$ or $L_{2\pi}$ and that a subsequence $\{m_n\}$ of integers satisfies the condition

$$m_n = o(n),$$

then for every $f \in X$ it holds

$$\left\| \frac{1}{(\alpha)_n} \sum_{\nu=0}^{n} (\alpha-1)_{n-\nu} S_{m_\nu}(f;x) - f(x) \right\|_X$$

$$\leqq C_1 \left\{ \frac{1}{(\alpha)_n} \sum_{\nu=0}^{n} (\alpha-1)_{n-\nu} E_{m_\nu}(f)_X \right\}, \quad \alpha > 0,$$

where the constant C_1 depends on $\{m_n\}$ and α only.

He also considered the case where $\{m_n\}$ is a convex sequence, satisfying the condition

$$\ln m_n = O(\sqrt{n}).$$

The case $\alpha = 1$ and $m_n = n$, $n=1,2,\ldots$, was considered by Stechkin S.B. [12].

<u>1.3 The exact constant in Jackson's type inequalities</u>

Let

$$C^*_{n,X} = \sup_{\substack{f \in X \\ f \neq c}} \frac{\|f - L_n(f;x)\|_X}{\omega(f;\frac{\pi}{n+1})_X}$$

where $L_n(f;x) = \frac{1}{\pi} K_n * f$ and let $K_n(t)$ be a trigonometric polynomial of degree $\leqq n$.

It is easy to prove that (see Huang X.H. [13])

$$C^*_{n,L} \leqq C^*_{n,C}, \quad n = 1,2,\ldots,$$

under the condition $K_n(t) \geqq 0$, $n=1,2,\ldots$.

Sun Y.S. [14, 15] obtained the following results:

Suppose that the continuous function $\varphi(t)$ has its bounded variation on $[0,1]$ and satisfies

$$\varphi^2(0) + \varphi^2(1) > 0.$$

Suppose

$$K_n(t) = K_n^{\varphi}(t) = \frac{1}{2A_n} \left| \sum_{s=0}^{n} \varphi\left(\frac{s}{n}\right) e^{ist} \right|^2 = \frac{1}{2} + \sum_{k=1}^{n} \lambda_k^{(n)} \cos kt,$$

where

$$A_n = \sum_{s=0}^{n} \varphi^2(\tfrac{s}{n}).$$

If

$$L_n(f) = \frac{1}{\pi} K_n^{\varphi} * f,$$

then

$$C^*_{n,L} \sim C^*_{n,C} \sim \frac{\varphi^2(0)+\varphi^2(1)}{\pi^2\int_0^1 \varphi^2(t)dt} \ln n.$$

Moreover for α-th Cesàro operator $\sigma_n^{\alpha}(f;x)$, $\alpha>0$,

$$C^*_{n,L} \sim C^*_{n,C} \sim \frac{2\alpha}{\pi^2} \ln n.$$

In 1976, Dudas V.O. [16] obtained for $\alpha=1$:

$$\sup_{\{n\}} \frac{C^*_{n,C}}{\ln n} = \frac{1}{\ln 2}.$$

2. Multiple interpolation and approximation harmonic analysis

1.2. Bivariate interpolation

Let D denote a bounded and closed domain in the plane. Suppose that H is linear subspace of C(D), and that the linearly independent polynomials $P_i(x,y)$, $1\leq i\leq \ell$, constitute a basis for H. If Q_i, $1\leq i\leq \ell$, are distinct points in D such that the determinant

$$(1) \quad \| P_j(Q_i) \| \neq 0$$

then for any function $f \in C(D)$, there is an interpolation polynomial $P \in H$ such that

$$P(Q_i) = f(Q_i), \quad 1\leq i\leq \ell.$$

For this reason we call a set of points $\{Q_i\}$, $1 \leq i \leq \ell$, satisfying (1) the properly posed nodes for H. Let H_n denote the space of all the real bivariate polynomials of degree $\leq n$. Liang X.Z. [17, 18] obtained the following result:

If $\{Q_i\}$, $1 \leq i \leq \ell$, is the properly posed set of nodes for H_n, and if none of these points is on the i-th irreducible curve $f(x,y)=0$ (either i=1 or i=2), then the set $\{Q_i\}$ combining with the $(2n+3)i-1$ distinct points selected freely on the irreducible curve must constitute a set of properly posed nodes for H_{n+i}.

The theorem was used to construct many schemes of the bivariate interpolation for irregularly spaced data.

2.2. Approximation of multiple variables

Liang X. Z. [19] discussed the polynomials with the least deviation from zero in the domain

$$A=\{-1\leq x,y\leq 1\},\quad B=\{x\geq -1, y\geq -1, x+y\leq 0\},\quad C=\{x^2+y^2\leq 1\}$$

and obtained the polynomials

$$A_{m,n}(x,y)=\prod_{i=1}^{m}\left(x+\cos\frac{2i-1}{2m}\pi\right)\prod_{j=1}^{n}\left(y+\cos\frac{2j-1}{2n}\pi\right),$$

$$B_{m,n}(x,y)=\prod_{i=1}^{m}\left(x+\cos\frac{2i-1}{2(m+n)}\pi\right)\prod_{j=1}^{n}\left(y+\cos\frac{2j-1}{2(m+n)}\pi\right),$$

$$C_{m,n}(x,y)=\prod_{i=1}^{m}\left(x+\sin\frac{2i-m-1}{2(m+n)}\pi\right)\prod_{j=1}^{n}\left(y+\sin\frac{2j-n-1}{2(m+n)}\pi\right)$$

with the least deviations 2^{-m-n+2}, 2^{-m-n+1} and 2^{-m-n+1} from zero of $x^m y^n$ in the domains A, B, and C, respectively.

Let $f \in L(R^k)$ and let $S_{\nu_1\nu_2\cdots\nu_k}(f;,x)$ be the $(\nu_1,\nu_2,\ldots,\nu_k)$ partial sum of the Fourier series of f. The Vallee-Poussin summation is

$$V_n^{n+\ell}(f;x) = \frac{1}{(\ell+1)^k}\sum_{\nu_1=n}^{n+\ell}\cdots\sum_{\nu_k=n}^{n+\ell} S_{\nu_1\nu_2\cdots\nu_k}(f;x).$$

Huang Kun-Yang [20] obtained the following results: Let X=C or L, then

1. $$\|f-V_n^{n+\ell}(f;x)\|_X \leq C_k \sum_{\nu=0}^{n+\ell} \frac{E_{n+\nu}(f)_X \ln^{k-1}(3+\frac{\nu}{\ell+1})}{\nu+\ell+1}$$

where C_k is a constant and $E_\nu(f)_X=E_{\nu,\dots,\nu}(f)_X$ is the rectangular best approximation to f.

2. Let $\varepsilon_\nu \downarrow 0$ and

$$X(\varepsilon) = \{f \in X,\ E_k(f)_X \leq \varepsilon_k,\ k=0,1,\dots\},$$

then

$$\sup_{\substack{f\in X \\ f\neq c}} \|f-V_n^{n+\ell}(f;x)\|_X \geq C_k' \sum_{\nu=0}^{n+\ell} \frac{E_{n+\nu}(f)_X \ln^{k-1}(3+\frac{\nu}{\ell+1})}{\nu+\ell+1},$$

where C_k' is a constant.

These results generalized the corresponding theorems of Oskolkov K.I. [21, 8], Stechkin S.B. [22] and Damen V. [9] in the case of one variable and the theorem of Sun Y.S. [23] in the case of two variables.

He also obtained that it holds

$$(2)\ \left\| \frac{1}{(\alpha)_n} \sum_{j=0}^{n} (\alpha-1)_{n-j} |S_{k,k}(f;x,y)-f(x,y)| \right\|_C$$
$$\leq C_\alpha \frac{1}{(\alpha)_n} \sum_{j=0}^{n} (\alpha-1)_{n-j} E_{jj}(f)_C, \qquad \alpha>0.$$

This result improved the result of Timan M.F. and Ponomarenko V.G. [24], where an analogous estimation was obtained only for summability (not for strong summability as in (2)).

2.3. Harmonic analysis

Let $Q^k = \{x=(x_1,x_2,\dots,x_k) \in R^k, -\pi \leq x_i \leq \pi, 1\leq i\leq k\}$ and let $f\in L[Q;^k]$ be a 2π-periodic function in each variable. The Fourier series of f is defined by $f(x)\sim\sum a_n e^{i\langle n,x\rangle}$, where $n=(n_1,\dots,n_k)$, $\langle n,x\rangle = \sum_{i=1}^{k} n_i x_i$. Bochner S. introduced the Riesz spherical mean summation of order δ :

$$S_R^{\delta}(f;x) = \sum_{|n|<R} (1-\frac{|n|^2}{R^2}) a_n e^{i\langle n,x\rangle},$$

where $|n| = \sqrt{n_1^2+\ldots+n_k^2}$. The case $\delta = \frac{k-1}{2}$ is called the critical case and $\delta = \frac{k-1}{2}$ - the critical index. The problem of convergence in the case $\delta > \frac{k-1}{2}$ has been solved, so we are interested only in the case $\delta \leqq \frac{k-1}{2}$.

Let

$$f_{x_0}(t) = \frac{1}{|S(x_0,t)|} \int_{S(x_0,t)} f(x)\,dx$$

where $S(x_0,t)$ is the sphere with center x_0 and radius t and $|S(x_0,t)|$ is its volume.

Lu S.Z. [24, 26] obtained the following results.

Suppose $f(x) \in L\ \ln^+ L[Q^k]$ and

$$(3) \quad \frac{1}{h^k} \int_0^h t^{k-1}\{f_{x_0}(t)-f(x_0)\}dt = o(1) \qquad (h\to 0),$$

$$\int_h^{\delta_0} \frac{|f_{x_0}(t+h)-f_{x_0}(t)|}{t}dt = o(1) \qquad (h\to 0),$$

then

$$\lim_{R\to\infty} S_R^{\frac{k-1}{2}}(f;x_0) = f(x_0).$$

This result improved a result of Stein [27], in which the condition (3) is placed by an absolute value in the integral.

<u>Theorem</u> (Lu X.Z. [28]). Suppose $f \in L\ \ln^+ L[Q^k]$ and $f_{x_0}(t)$ is HBV in some $[0,\delta]$, i.e.,

$$\sup_{[a_n,b_n]\subset[0,\delta]} \sum \frac{|f(b_n) - f(a_n)|}{n} < +\infty,$$

where the supremum is taken over all nonoverlapping intervals

$[a_n, b_n] \subset [0, \delta]$, then

$$\lim_{R\to\infty} S_R^{\frac{k-1}{2}}(f;x_0) = f_{x_0}(+0).$$

Lu S.Z. investigated some other kinds of summation [29, 30]. Pan W.J. [31] considered the case $\delta < \frac{k-1}{2}$.

Suppose that $\Phi \in L[R^k]$ satisfies the following conditions:

1. $\Phi(x) = \Phi_0(|x|)$, so as the Fourier transform

$$\hat{\Phi}(y) = \varphi(y) = \psi(|y|),$$

2. $$\lim_{T\to\infty} \int_{|y|\leq T} \varphi(y)\,dy = \lim_{T\to\infty} \frac{(2\pi)^{\frac{k}{2}}}{\Gamma(\frac{k}{2})} \int_0^T r^{k-1}\psi(r)\,dr = 1.$$

Let $f \in L[R^k]$ and

$$G_\varepsilon(f)(x) = \int_{R^k} \Phi(\varepsilon y)\hat{f}(y)\, e^{2\pi i\langle x,y\rangle} dy, \qquad \varepsilon > 0.$$

The integral

$$\int_{R^k} \hat{f}(y) e^{2\pi i\langle x,y\rangle} dy$$

is called Φ-summable to $f(x)$ if

$$\lim_{\varepsilon\to 0} G_\varepsilon(f)(x) = f(x).$$

The Riesz spherical means are equivalent to the Φ-means with

$$\Phi(x) = \begin{cases} (1-|x|^2)^\delta & , |x| < 1, \\ 0 & , |x| \geq 1, \end{cases} \qquad \delta > \frac{k-3}{2}.$$

<u>Theorem</u> (Pan W.J. [31]). Suppose that $f \in L[R^k]$ and the total variation of $f_{x_0}(t)$ on $[0, \delta]$ is bounded for all $\delta \in [0, \infty)$, then

$$\lim_{\varepsilon \to 0} G_\varepsilon(f)(x) = f_{x_0}(+0).$$

Theorem (Pan W.J. [32]). Suppose that $k \geq 2m-1$, $f \in W^m[R^k]$,

$$\int_0^1 |f_{x_0}(t)|\, dt < +\infty,$$

$$\int_0^1 |f_{x_0}^{(j)}(t)|\, dt < +\infty, \quad j=1,2,\ldots,m,$$

where

$$f_{x_0}^{(j)}(t) = \frac{1}{|S(0,1)|} \int_{S(0,1)} (D_t^{(j)} f)(x+tu)\,du,$$

and

$$(D_t^{(j)} f)(x+tu) = \sum_{i_1=1}^{k} \cdots \sum_{i_j=1}^{k} \frac{\partial^j f}{\partial x_{i_1} \cdots \partial x_{i_j}}(x+tu)\, u_{i_1} \cdots u_{i_j},$$

then the limit

$$\lim_{R \to \infty} \sigma_R^\alpha(f;x_0) = A_{x_0}$$

exists for $\alpha > \frac{k-(2m+1)}{2}$, where A_{x_0} is a constant such that

$$f_{x_0}(t) = \int_0^t f'_{x_0}(r)\,dr + A_{x_0}, \text{ a.e., } t \in (0,+\infty).$$

In particular, if

$$\int_0^h t^{k-1} |f_{x_0}(t) - f(x_0)|\, dt = o(h^k), \qquad (h \to 0),$$

then $A_{x_0} = f_{x_0}(+0)$; if

$$\int_0^h t^{k-1} |f_{x_0}(t) - f(x_0)|\, dt = o(h^k), \qquad (h \to 0),$$

then $A_{x_0} = f(x_0)$.

3. Complex approximation

Let $f(z) \in C(|z|=1)$ and let $\{\alpha_i\}, |\alpha_i| < 1, i=1,2,\ldots,n$, $\{\beta_j\}, |\beta_j| > 1, j=1,2,\ldots,m$, be the poles of a rational function $R_{n+m}(z)$.

We consider the best approximation

$$\rho_{n,m} = \inf_{\{R_{n+m}\}} \| f(z) - R_{n+m}(z) \|_{C(|z|=1)},$$

where the infimum is taken over all the rational functions $R_{n,m}(z)$ with the poles at $\{\alpha_i\}, i=1,2,\ldots,n$, and $\{\beta_j\}, j=1,2,\ldots,m$.

Theorem (Shen X.C. and Lou Y.R. [32])

$$(4) \quad \rho_{n,m} \leq C_k [\varepsilon_n(\alpha)^k \omega(f^{(k)}; \varepsilon_n(\alpha)) + \varepsilon_m(\beta)^k \omega(f^{(k)}; \varepsilon_m(\beta)) + q_1^{\frac{1}{\varepsilon_n(\alpha)}} + q_1^{\frac{1}{\varepsilon_m(\beta)}}],$$

where q_1 is absolute constant, $0 < q_1 < 1$.

This theorem improved the result of Mergeljan S.N. and Dzrbasjan M.M. [33].

It is needed to indicate that Andersson J.R. and Ganelius T. [34], Pekarskii A.A. [35] obtained the equivalent results in 1977.

Moreover the inverse theorem was also obtained by Shen X.C. and Lou Y.R. [32] without any restriction on the limits of $\{\alpha_i\}$ and $\{\beta_j\}$.

Later, the analogous theorems were established in the H_p space by Shen X.C. and Lou Y.R. [36].

There are a few papers concerning the approximation by rational functions in general domains. Elliolt H.M. [37] and Walsh J.L. [38] investigated the case of analytic boundary Γ. In their investigation the poles of rational functions, by which the approximation is considered, have no any limit point on Γ.

In the Al'per's domain (see [39]), i.e., the angle between the tangent of the boundary Γ of the domain G and the positive real axis has a modulus of continuity j(h) satisfying the condition

$$(5)\quad \int_0 \frac{j(h)}{h}|\ln h|\,dh < +\infty,$$

the analogous results were obtained in $C(\Gamma)$ of $C(\bar{G})$ and $E_1(G)$ (see Shen X.C. and Lou Y.R. [40], Shen X.C. [41, 42], Su Z.L. [43]). If the condition (5) is replaced by

$$(6)\quad \int_0 \frac{j(h)}{h}dh < +\infty,$$

then the analogous results can be obtained in $E_p(G)$ (Shen X.C. and Lou Y.R. [44], Shen X.C. [42]), $1 < p < +\infty$.

Recently we studied a new class of domains, the K_q domain, $q > 1$, that is the boundary Γ of domain G is a closed rectifiable Jordan curve and for any $f(\xi) \in L^q(\Gamma)$, the function

$$S(z) = \frac{1}{2\pi i}\int_\Gamma \frac{f(\xi)}{\xi - z}\,d\xi$$

belongs to $E_q(G_\infty)$, where G_∞ is the complement of G. This kind of domains is more general than the domains mentioned above. An estimation analogous to (4) in $E_p(G)$, $p>1$, $\frac{1}{p}+\frac{1}{q}=1$ has been established by Shen X.C. [45].

Theorem (Shen X.C. [45, 46]). Suppose $G \in K_q$, $q > 1$ and $F(z) \in E_p(G)$, $\frac{1}{p}+\frac{1}{q}=1$, then for any positive integer n we have

$$R_n^{(p)}(F,G) \leq C_1 R_n^{(p)}(\tilde{F},|w|<1) \leq C_2[\omega(\tilde{F}, \varepsilon_m(\beta)) + q_1^{\frac{1}{\varepsilon m(\beta)}}]$$

where $R_n^{(p)}(F,G)$ and $R_n^{(p)}(\tilde{F},|w|<1)$ are the best approximations to $F(z)$ and $\tilde{F}(w)$ with the help of rational functions having the poles at $\{b_j\}$, $b_j \in G_\infty$, $1 \leq j \leq n$, and $\{\beta_j\}$, $\beta_j = \varphi(b_j)$, $1 \leq j \leq n$, in the space $E_p(G)$ and H_p, respectively, $w = \varphi(z)$ is a mapping function which maps G_∞ into $|w|>1$, $\varphi(\infty) = \infty$, $\varphi'(\infty) > 0$ and

$$\tilde{F}(w) = \frac{1}{2\pi i} \int_{|\tau|=1} \frac{F[\psi(\tau)]\psi'(\tau)^{\frac{1}{p}}}{\tau - w} d\tau ,$$

where $z = \psi(w)$ is the inverse function of $w = \varphi(z)$.

Besides, the problems concerning the expansion in the series of rational functions with preassigned poles and the estimate of the remainder of this series has been obtained by Shen S.C. [37, 48].

Consider the system of functions

$$\varphi_0(w) = \frac{\sqrt{1-|\alpha_0|^2}}{\sqrt{2\pi}(1-\bar{\alpha}_0 w)} , \quad \varphi_n(w) = \frac{\sqrt{1-|\alpha_n|^2}}{\sqrt{2\pi}(1-\bar{\alpha}_n w)} \prod_{k=0}^{n-1} \frac{\alpha_k - w}{1-\bar{\alpha}_k w} \frac{|\alpha_k|}{\alpha_k}, n \geq 1,$$

where

$$\alpha_k = \frac{1}{\overline{\varphi(b_k)}} , \quad b_k \in G_\infty, \quad k=0,1,\ldots,$$

and set $\frac{|\alpha_k|}{\alpha_k} = -1$, $\alpha_k = 0$. They satisfy the conditions

$$\int_{|w|=1} \varphi_n(w) \overline{\varphi_m(w)} |dw| = \delta_{n,m} = \begin{cases} 1, & m=n \\ 0, & m \neq n. \end{cases}$$

We define $\{M_n^{(\frac{1}{p})}(z)\}$ as the sum of the principal parts of $\varphi_n[\varphi(z)]\varphi'(z)^{\frac{1}{p}}$ at their poles $b_k \in G_\infty$, $k=0,1,\ldots,n$. It is obvious that if $b_k = \infty$, $k=0,1,\ldots,n$, $p = \infty$, then $\{M_n^{(\frac{1}{p})}(z)\}$ is the system of Faber polynomials.

<u>Theorem</u> (Shen X.C. [48]). Suppose $G \in K_q$, $q > 1$, $F(z) \in E_p(G)$, $\frac{1}{p} + \frac{1}{q} = 1$, and

$$S_n(z) = \sum_{k=0}^{n} a_k M_k^{(\frac{1}{p})}(z)$$

is the partial sum of the expansion of $F(z)$ on $\{M_n^{(\frac{1}{p})}(z)\}$ where

a_k is determined by

$$a_k = a_k(f) = \frac{1}{i} \int_{|w|=1} F[\psi(w)]\psi'(w)^{\frac{1}{p}} \frac{1}{w} \overline{\varphi_k(\frac{1}{\bar{w}})}\, dw, \quad k=0,1,\ldots,$$

then

$$(7) \quad \|F(z)-S_n(z)\|_{L^p(\Gamma)} \leq C_1 R_n^{(p)}(\tilde{F}, |w|<1) \leq C_2[\omega(\tilde{F};\varepsilon_n(\alpha))+q_1^{\frac{1}{\varepsilon_n(\alpha)}}].$$

We would like to indicate that the analogous result to (7) under the condition (5) is valid in $C(\bar{G})$ and $E_1(G)$, but in the second inequality of (7) must stand a factor

$$\ln[\sum_{k=1}^{n} (1-|\alpha_k|^{-1})]$$

(see [48, 43]).

References

1. Nikol'skii, S.M., On the approximation of functions in the mean by trigonometric polynomials, Izv. Akad. Nauk SSR Ser. Mat. 10 (1946), 207-256 (Russian)
2. Stechkin, S.B. and Teliakovski, S.A., Approximation of differential functions by trigonometric polynomials in metric L, Trudy Mat. Inst. Steklov 88 (1967), 20-29 (Russian)
3. Motoruye, V.P., Approximation of periodic functions in the mean by trigonometric polynomials, Mat. Zametki 16 (1974), 15-26 (Russian)
4. Sun, Yong-shen, Approximation of periodic functions in the mean by linear positive operators, Acta Math. Sinica (to appear)
5. Baskakov, V.A., The degree of approximation of differential functions by linear polynomial operators, Theory of approximation of functions and its applications, Nauka, 1977, 28-29 (Russian)
6. Wang, Si-lei, Fourier series for the class of BMO functions, Initial issue (1981), 57-59 (Chinese)

7. Xie, Ting-fan, On the approximation of differential functions by Vallee-Poussin sums, Acta Math. Sinica 24:5 (1981), 689-693 (Chinese)
8. Oskolkov K.I., A Lebesgue constant in uniform metric, Mat. Zamatki 25:4 (1979), 551-555 (Russian)
9. Damen, V., On best approximation and Vallee-Poussin sums, Mat. Zamatki 23:5 (1978), 671-651 (Russian)
10. Sun, Yong-sheng, A note on Oskolkov's inequality, Beijing Shifandaxue Xuelao; Nat. Sci. (1979) No. 3, 7-15 (Chinese)
11. Shi, Yan-liang, On the linear approximation of periodic functions by the subsequences of Fourier series (to appear in Advances of Math.) (Chinese)
12. Stechkin S.B., Approximation of periodic functions by Fejér's summation. Trudy Mat. Inst. Steklov 62 (1961), 48-60 (Russian)
13. Huang, Xin-hua, Approximation of continuous functions by linear positive sum of Fourier series, Advances in Math. 8:3 (1965), 322-325 (Chinese)
14. Sun, Yong-sheng, The exact constant of approximation for Cesàro operators, 24:4 (1981), 516-537 (Chinese)
15. Sun, Yong-sheng, On the Jackson's inequality, Beijing Shifandanue Xuebao, Nat. Sci. 1 (1981), 1-6 (Chinese)
16. Dudas, V.U. "Problem of the theory of approximation of functions and its applications", Akad. Nauk Ukrain SSR, Kiev 1976, 109-125
17. Liang, Xue-zhang, Properly posed nodes for bivariate interpolation and the superposed interpolation, Acta Sci. Natur. Univ. Jilinensis 1 (1979), 27-32 (Chinese)
18. Liang, Xue-zhang, Two-dimensional quartic spline interpolation and the finite element method, Journal of Kirin Univ. Natur. Sci. edi. 1 (1978), 85-95 (Chinese)
19. Liang Xue-zhang, The polynomials with least deviation from zero in some multidimensional region; Math. Num. Sinica 1-2 (1977), 189-193 (Chinese)
20. Huang, Kun-yang, Some problems about the approximation of

multiple periodic functions, M.S. Dissertation, Beijing Normal University, 1981 (Chinese)

21. Oskolkov K.I., On the Lebesgue's inequality in the uniform metric and on the set of full measure, Mat. Zamatki 18:4 (1975), 515-526 (Russian)
22. Stechkin S.B., On the approximation of periodic functions by Vallee-Poussin sums, Analysis Math. 4 (1978), 61-74 (Russian)
23. Sun, Young-sheng, The uniform approximation of bivariate periodic Fourier sums, Beijing Shifandaxue Xuebao Nat. Sci. 3 (1979), 16-35 (Chinese)
24. Timan M.F. and Ponomarenko V.G., Approximation of periodic functions with two variables by summation of Marcinkiewicz's type, Izv. Vyss. Ucebn. Zaved. Mathematika (1975), No. 9, 59-67 (Russian)
25. Lu, Shan-zhen, Some problems of Fourier analysis and approximation of functions with several variables, Proceedings of the conference on approximation theory, Hangzhon, 1978, 35-41 (Chinese)
26. Lu, Shan-zhen, Spherical summation of multiple Fourier series and its conjugate Fourier integral, Kexae Tongbao (1980), No. 5, 199-202 (Chinese)
27. Stein E.M., On certain exponential sums arising in multiple Fourier series, Ann. of Math. 73 (1961), 87-109
28. Lu, Shan-zhen, Multiple bounded variation function and spherical summation of its Fourier series, Kexue Tongbao (1981), No. 1, 1-4 (Chinese)
29. Lu, Shan-zhen, The (C,1) summation of multiple conjugate Fourier integral, Kexue Tongbao (1981), No. 19, 865-869 (Chinese)
30. Lu, Shan-zhen, On the convergence of spherical integral and Riesz spherical mean summation, Acta Math. Sinica 23:4 (1980), 609-623 (Chinese)
31. Pan, Wen-jie, On the localization and convergence of multiple Fourier integrals by spherical means, Scienta Sinica (1981), No. 11, 1310-1321 (Chinese)

32. Shen, Xie-chang and Lou, Yuna-ren, The best approximation by rational functions on the unit circle, 20:3 (1977), 232-235 (Chinese)
33. Mergeljan, S.N. and Dzarbasjan M.M.,On best approximation by rational functions, Dokl. Akad. Nauk SSSR, 99:5 (1954), 673-675 (Russian)
34. Andersson, J.E. and Ganelius, T., The degree of approximation by rational functions with fixed poles, Math. Zeit. 153 (1977), 161-166
35. Pekarskii, Q.A., The degree of rational approximation with preassigned poles. Dokl. Akad. Nauk Bararuciea SSR 21:4 (1977), 302-304 (Russian)
36. Shen, Xie-chang and Lou Yuna-ren, On the best approximation by rational functions in the H_p ($p > 1$) space, Acta Sci. Natur. Univ. Pekinensis, (1979), No. 1, 58-72
37. Elliolt, H.M., On approximation to analytic functions by rational functions, Proc. Amer. Math. Soc. 4:1 (1953), 161-167
38. Walsh, J.L., Note on degree of approximation to analytic functions by rational functions with preassigned poles, Proc. Nat. Acad. Sci. U.S.A. 42:12 (1956), 927-930
39. Al'per S.Yu., On the uniform approximation of functions with a complex variable in the closed domain, Izv. Akad. Nauk SSSR, Ser. Mat. 19 (1955), 423-444 (Russian)
40. Shen, Xie-chang and Lou Yuna-ren, The best approximation by rational functions in the domain of complex plane, Acta Math. Sinica 20:4 (1977), 301-303 (Chinese)
41. Shen, Xie-chang, The problem of the best approximation by rational functions with preassigned poles, Acta Math. Sinica 21:1 (1978), 86-90 (Chinese)
42. Shen, Xie-chang, On the best approximation of functions in the E_p ($1 < p \leq +\infty$) space by rational functions with preassigned poles, Annals of Math. 1:1 (1980), 51-62 (Chinese)
43. Su, Zhan-long, On the best approximation of functions in the E_1 space by rational functions with preassigned poles, (to appear in Journal of Math. Res. and Exp.)

44. Shen, Xie-chang and Lou, Yuna-ren, The best approximation by rational functions in space $E_p(p>1)$ to the complex plane, Acta Sci. Natur. Univ. Pekinensis 2 (1979), 1-18 (Chinese)
45. Shen, Xie-chang, Approximation and expansion by rational functions in certain class of domains, A Monthly Journal of Sci. 25:2 (1980), 97-102
46. Shen, Xie-chang, Approximation by rational functions in complex plane, Scienta Sinica 24:8 (1981), 1033-1046
47. Shen, Xie-chang, On the expansion by means of rational functions in a certain class of domains, Scienta Sinica 24:11 (1981) 1489-1496
48. Shen, Xie-chang, The remainder estimation of the expansion by means of rational functions in $E_p(1<p\leq+\infty)$. Annals of Math. 2:3 (1981), 301-320

Prof. Shen Xie-chang, Department of Mathematics, Peking University, Beijing, China

VECTOR SPLINES ON THE SPHERE, WITH APPLICATION TO THE ESTIMATION OF VORTICITY AND DIVERGENCE FROM DISCRETE, NOISY DATA

Grace Wahba

Statistics Department
University of Wisconsin-Madison
USA

Vector smoothing splines on the sphere are defined. Theoretical properties are briefly alluded to. An approach to choosing the appropriate Hilbert space norms to use in a specific meteorological application is described and justified via a duality theorem. Numerical procedures for computing the splines as well as the cross validation estimate of two smoothing parameters are given. A Monte Carlo study is described which suggests the accuracy with which upper air vorticity and divergence can be estimated using measured wind vectors from the North American radiosonde network.

This research was supported by the Office of Naval Research Under Contract N00014-77-G-0675 and by the National Aeronautics and Space Administration under Contract NAG5-128.

1. Introduction

A theory of spline functions on the sphere is rapidly being developed, see WAHBA (1981a), FREEDEN (1981a,b), SHURE, PARKER AND BACKUS (1981). Dr. FREEDEN will be reporting on some of his results elsewhere in this volume. Much of the rich theory surrounding univariate splines and thin plate splines clearly is extendable to the theory of splines on the sphere, via the use of reproducing kernels, n-widths, etc. In particular convergence rates for smoothing splines on the sphere can be obtained from the known rate of decay of the eigenvalues of the relevant reproducing kernels, see e.g. MICCHELLI and WAHBA (1981), WAHBA (1977), UTRERAS (1981).

In this paper we propose a notion of vector splines on the sphere. It is clear that interesting approximation theoretical properties of these splines can be obtained. However, in this paper our focus will be on the solution of certain practical problems which must be solved so that these splines may be usefully applied to the analysis of meteorological data from the upper air radiosonde network.

For the purpose of numerical weather prediction the global radiosonde (weather balloon) network takes measurements every 12 hours of the horizontal wind velocity vectors and other variables, at 9 standardized vertical levels. From this data it is desired to estimate the horizontal wind field and its vorticity and divergence (and other variables) at a regular grid of points, for each level. These estimates on a grid are then merged with estimates of the same variables on the same grid, which have been obtained from a forecast, to provide an estimate of the present state of the atmosphere. This state estimate is then used as the initial conditions to a numerical integration scheme which integrates a set of differential equations describing the dynamics of the atmosphere, to provide a new forecast. Numerical weather forecasts can be quite sensitive to errors in the vorticity and divergence in the initial wind fields. Unfortunately, horizontal wind vectors at, for example the 500 millibar height, of the order of a few tens of meters per second, are measured with an error standard deviation in each component of the order of 2-4 meters per second. Thus, it is not a trivial matter to obtain accurate information concerning the vorticity and divergence from this data, even in areas such as the continental U.S. where the radiosonde network is relatively dense. We believe that the appropriate derivatives of the vector smoothing splines we propose have the potential for doing this relatively well.

Speaking intuitively, the vector smoothing splines we propose will behave like low pass filters. In the splines we propose there will be two regularization or smoothing parameters to be chosen and two (sets of) "shape" parameters. The first smoothing parameter to be chosen, may be thought of as governing the overall half power point of the low pass filter. The second parameter governs the relative distribution of power between vorticity and divergence in the estimate. The choice of the two sets of shape parameters correspond to the choice of Hilbert space norms, but in an important practical sense they govern the rates of decay of the energy spectrum of the solution, one "shape" for vorticity, and one "shape" for divergence. It is well known from the theory and practice of ill posed problems that the appropriate choice of certain of these parameters can affect the practical usefulness of the result.

In this paper we propose the use of generalized cross-validation (GCV) for choosing the two smoothing parameters. GCV can also be used to choose a small number of "shape" parameters (see CRAVEN and WAHBA (1979), WAHBA and WENDELBERGER (1980)). However, in this paper we show how historical meteorological data can be used to choose the "shape" parameters, or Hilbert space norm. We discuss some numerical methods, and we describe the results of some numerical experiments on synthetic data which mimics actual 500 millibar horizontal wind fields over the U.S. In our experiments we have observed that the accuracy in estimating both vorticity and divergence can be quite sensitive to the relative distribution of power allocated between them, (choice of second smoothing parameter) but that GCV can be quite effective in estimating the correct relative power distribution.

For the meteorological experts in the audience we remark that estimating the present state of the atmosphere from current data is not exactly the same problem as estimating the state of the atmosphere from a combination of present data and a forecast of the present. This is so because a data only estimate needs to take account of properties of the atmosphere and measurement system while a data plus forecast estimate needs to take into account the relative error of the data and the forecast. In this paper we are studying the data only problem. However, we believe that this class of techniques can be extended to the data plus forecast problem and hope to do that in a subsequent paper.

In Section 2 we define the vector smoothing splines. In Section 3 we discuss the choice of Hilbert space norms. In Section 4 we describe numerical methods and the cross validation estimate of the smoothing parameter and in Section 5 we describe a Monte Carlo test of the method.

2. Helmholz Theorem and The Definition of Vector Smoothing Splines

We let P be a point on the sphere S, $P = (\lambda,\phi)$, where λ = longitude $(0\le\lambda\le 2\pi)$ and ϕ = latitude $(-\frac{\pi}{2}\le\phi\le\frac{\pi}{2})$. $\underset{\sim}{V} = (U,V)$ is a (sufficiently regular) horizontal vector field on the sphere, where U(P) is the eastward component and V(P) is the northward component at P.

The vorticity ζ and the divergence D of $\underset{\sim}{V}$ are given by

$$\zeta = \frac{1}{a\cos\phi}\left[-\frac{\partial}{\partial\phi}(U\cos\phi) + \frac{\partial V}{\partial\lambda}\right] \tag{2.1}$$

$$D = \frac{1}{a\cos\phi}\left[-\frac{\partial U}{\partial\lambda} + \frac{\partial}{\partial\phi}(V\cos\phi)\right], \tag{2.2}$$

where a is the radius of the sphere. Then there exists (by Helmoltz Theorem) two functions $\Psi(P)$ and $\Phi(P)$, $P\varepsilon S$, called the stream function and the velocity potential respectively, with the following properties:

$$\begin{aligned} U &= \frac{1}{a}\left(-\frac{\partial\Psi}{\partial\phi} + \frac{1}{\cos\phi}\frac{\partial\Phi}{\partial\lambda}\right) \\ V &= \frac{1}{a}\left(\frac{1}{\cos\phi}\frac{\partial\Psi}{\partial\lambda} + \frac{\partial\Phi}{\partial\phi}\right) \end{aligned} \tag{2.3a}$$

$$\begin{aligned} \zeta &= \Delta\Psi \\ D &= \Delta\Phi \end{aligned} \tag{2.3b}$$

where Δ is the (horizontal) Laplacian on the sphere

$$\Delta f = \frac{1}{a^2}\left[\frac{1}{\cos^2\phi} f_{\lambda\lambda} + \frac{1}{\cos\phi}(\cos\phi f_{\phi})_{\phi}\right].$$

Ψ and Φ are uniquely determined up to a constant (which we will take to be determined by $\int_S \Psi(P)dP = \int_S \Phi(P)dP = 0$. We are interested in defining Hilbert spaces of vector fields whose divergence and vorticity exists pointwise. We will do this as follows. Let f(P) be a square integrable function on the sphere which integrates to 0. Then f has an expansion in the normalized spherical harmonics Y_ℓ^s

$$f(P) = \sum_{\ell=1}^{\infty} \sum_{s=-\ell}^{\ell} f_{\ell s} Y_\ell^s(P)$$

where

$$Y_\ell^s(\lambda,\phi) = \begin{cases} \theta_{\ell s} \cos s\lambda P_\ell^s(\sin\phi) & 0 \le s \le \ell \\ \theta_{\ell s} \sin s\lambda P_\ell^{|s|}(\sin\phi) & -\ell \le s < 0, \end{cases} \qquad \ell = 1,2,\ldots$$

$$\theta_{\ell s} = \sqrt{2}\sqrt{\frac{2\ell+1}{4\pi}\frac{(\ell-|s|)!}{(\ell+|s|)!}} \qquad s \neq 0$$

$$= \sqrt{\frac{2\ell+1}{4\pi}} \qquad s = 0$$

and the Fourier Bessel coefficients $f_{\ell s}$ are given by

$$f_{\ell s} = \int f(P) Y_\ell^s(P) dP$$

with

$$\int f^2(P) dP = \sum_{\ell,s} f_{\ell s}^2 .$$

Now Y_ℓ^s are the eigenfunctions of the Laplacian

$$\Delta Y_\ell^s = -\ell(\ell+1) Y_\ell^s .$$

Thus

$$\Delta f = -\sum_{\ell,s}\sum \ell(\ell+1) f_{\ell s} Y_\ell^s .$$

Let $\lambda_{\ell s}$, $\ell = 1,2,\ldots$, $s = -\ell,\ldots,\ell$ be a set of nonnegative numbers with

$\bar{\lambda}_\ell = \max_{s=-\ell,\ldots\ell} \lambda_{\ell s}$ and

$$\sum_{\ell} \ell^2(\ell+1)^2(2\ell+1)\bar{\lambda}_{\ell}<\infty. \tag{2.5}$$

Using the addition formula for spherical harmonics

$$\sum_{s=-\ell}^{\ell} Y_{\ell}^{s}(P)Y_{\ell}^{s}(Q) = \frac{2\ell+1}{4\pi} P_{\ell}(\cos\gamma(P,Q))$$

where γ is the angle between P and Q, the Cauchy-Schwartz inequality and the fact that $P_{\ell}(1) = 1$ gives

$$|\Delta f(P)|=|\sum_{\ell,s} \ell(\ell+1)\sqrt{\lambda_{\ell s}}Y_{\ell}^{s}(P) \frac{f_{\ell s}}{\sqrt{\lambda_{\ell s}}}|$$

$$\leq(\sum_{\ell} \ell^2(\ell+1)^2 \sum_{s=-\ell}^{\ell} \lambda_{\ell s}(Y_{\ell}^{s}(P))^2)^{1/2}(\sum_{\ell,s} \frac{f_{\ell s}^2}{\lambda_{\ell s}})^{1/2}$$

$$\leq(\frac{1}{4\pi} \sum_{\ell} \ell^2(\ell+1)^2(2\ell+1)\bar{\lambda}_{\ell})^{1/2}(\sum_{\ell,s} \frac{f_{\ell s}^2}{\lambda_{\ell s}})^{1/2}$$

Thus $\{\lambda_{\ell s}\}$ satisfying (2.5) and $\sum_{\ell,s} \frac{f_{\ell s}^2}{\lambda_{\ell s}}<\infty$ imply that $\Delta f(P)$ is well defined and finite for all P.

Let H be the collection of all pairs (Ψ,Φ) on the sphere which integrate to zero, are square integrable and

$$J^{(1)}(\Psi) = \sum_{\ell=1}^{\infty} \sum_{s=-\ell}^{\ell} \frac{\Psi_{\ell s}^2}{\lambda_{\ell s}(1)}<\infty \qquad \Psi_{\ell s} = \int\Psi(P)Y_{\ell}^{s}(P)dP$$

$$J^{(2)}(\Phi) = \sum_{\ell=1}^{\infty} \sum_{s=-\ell}^{\ell} \frac{\Phi_{\ell s}^2}{\lambda_{\ell s}(2)}<\infty \qquad \Phi_{\ell s} = \int\Phi(P)Y_{\ell}^{s}(P)dP$$

where $\{\lambda_{\ell s}(1)\}$ and $\{\lambda_{\ell s}(2)\}$ are sequences satisfying

$$\sum_{\ell=1}^{\infty} \ell^2(\ell+1)^2(2\ell+1)\max_{s}\lambda_{\ell s}(i)<\infty, \; i = 1,2.$$

H is clearly a Hilbert space with square norm

$$||(\Psi,\Phi)||^2 = J^{(1)}(\Psi) + \frac{1}{\delta}J^{(2)}(\Phi)$$

for any fixed $\delta > 0$ and both members of each pair possess Laplacians everywhere. It is easy to show that if $\lambda_{\ell s} = [\ell(\ell+1)]^{-m}$, then

$$\begin{aligned} J(f) &= \int(\Delta^{m/2}f)^2 dP && m \text{ even} \\ &= \int\{\frac{(\Delta^{(m-1)/2}f)_\lambda^2}{\sin^2\phi} + (\Delta^{(m-1)/2}f)_\phi^2\}dP && m \text{ odd.} \end{aligned} \quad (2.6)$$

If $\lambda_{\ell s} = [\ell(\ell+1)]^{-m}$, them $m > 3$ guarantees the pointwise existence of the Laplacian.

The observations are assumed to be of the form

$$U_i = U(P_i) + \varepsilon_i^U, \; V_i = V(P_i) + \varepsilon_i^V, \; i = 1,2,\ldots,n \quad (2.7)$$

where $(U(P_i),V(P_i))$ is the true (wind) vector at P_i and $\varepsilon_i^U,\varepsilon_i^V$ are measurement errors. We propose estimating the stream function and velocity potential (Ψ,Φ) associated with U and V by finding $(\Psi,\Phi)\varepsilon H$ to minimize

$$\begin{aligned} &\frac{1}{n}\sum_{i=1}^{n}(-\frac{1}{a}\frac{\partial\Psi}{\partial\phi}(P_i) + \frac{1}{a\cos\phi_i}\frac{\partial\Phi}{\partial\lambda}(P_i)-U_i)^2 \\ &+ \frac{1}{n}\sum_{i=1}^{n}(\frac{1}{a\cos\phi_i}\frac{\partial\Psi}{\partial\lambda}(P_i) + \frac{1}{a}\frac{\partial\Phi}{\partial\phi}(P_i)-V_i)^2 \\ &+ \lambda[J_1(\Psi) + \frac{1}{\delta}J_2(\Phi)] \end{aligned} \quad (2.8)$$

Note that in the residual sum of squares above, $U(P_i)$ and $V(P_i)$ are expressed in terms of Ψ and Φ via (2.3a). A unique minimizer $(\Psi_{\lambda,\delta},\Phi_{\lambda,\delta})$ exists for each $\lambda > 0$, $\delta > 0$ and the resulting wind field $(U_{\lambda,\delta},V_{\lambda,\delta})$ constructed from $(\Psi_{\lambda,\delta},\Phi_{\lambda,\delta})$ may be termed a vector spline field. Its vorticity and divergence will be given by $\zeta_{\lambda,\delta} = \Delta\Psi_{\lambda,\delta}$, $D_{\lambda,\delta} = \Delta\Psi_{\lambda,\delta}$. (Obviously, interpolating splines can be defined as minimizers of $J_1(\Psi) + \frac{1}{\delta}J_2(\Psi)$ subject to the interpolating conditions, we will not discuss these further.) Using WAHBA (1981a) or FREEDEN (1981a) it is straightforward to write an explicit (infinite series) expression for $(U_{\lambda,\delta}, V_{\lambda,\delta})$.

3. On the Choice of $J^{(1)}$ and $J^{(2)}$

Let $\lambda_{\ell s} = |\sum_{j=0}^{m} \alpha_j [(-\ell)(\ell+1)]^j|^{-2}$ and suppose that $\lambda_{\ell s} > 0$ for $\ell = 1,2,\ldots, s = -\ell,\ldots,\ell$. It is not hard to see that

$$J(f) \doteq \sum_{\ell=1}^{\infty} \sum_{s=-\ell}^{\ell} \frac{f_{\ell s}^2}{\lambda_{\ell s}} = \int (\sum_{j=0}^{m} \alpha_j \Delta^j f)^2 dP \tag{3.1}$$

so that the choice of the $\lambda_{\ell s}$ can then be reduced to the choice of m and the $\{\alpha_j\}$. (If $\lambda_{\ell s} = 0$ for one or more s, the minimization problem can be handled by the methods described in KIMELDORF and WAHBA (1971), see also FREEDEN (1981a). In principle m and possibly α_{m-1} (with $\alpha_m = 1$) can be chosen by cross validation (see WAHBA and WENDELBERGER (1980)), but it is undesireable to attempt to choose too many of these parameters from the data, see WAHBA (1981c).

In this section we will use the duality theorem which relates smoothing by splines to Bayesian estimation/Weiner filtering on stochastic processes to suggest how the J's may be chosen based on historical meteorological data.

To give the duality theorem we need some background, which we will give in a univariate context.

Let $X(P)$, $P \in S$ be a (univariate) zero mean Gaussian stochastic process on the sphere with covariance $R(P,Q)$ defined by

$$R(P,Q) = EX(P)X(Q),$$

where E is mathematical expectation. Following PARZEN (1961), CRAMER and LEADBETTER (1967) we can define the Hilbert space $\mathcal{X}$ spanned by $X(P)$, $P \in S$, as all finite linear combinations of random variables (r.v.'s) of the form

$$Z_k = \sum_{j=1}^{n_k} \xi_{kj} X(P_{kj}) \tag{3.2}$$

and their quadratic mean (q.m.) limits. (A sequence $Z_1, Z_2, \ldots,$ of r.v.'s has a q.m. limit if $\lim_{\ell,m\to\infty} E(Z_\ell - Z_m)^2 = 0$). The inner product in $\mathcal{X}$ is

$<X(P),X(Q)> = EX(P)X(Q) = R(P,Q)$, and is extended by linearity to all r.v.'s of the form $Z_k = \sum_{j=1}^{n_k} \xi_{kj} X(P_{kj})$ and their q.m. limits. For example, letting L be a linear functional, the r.v. $LX = \frac{\partial}{\partial\phi}X(P_0)$ will be in X if the sequence of r.v.'s

$$Z_k = \frac{X(\lambda_0,\phi_0+h_k)-X(\lambda_0,\phi_0)}{h_k} \tag{3.3}$$

has a q.m. limit, as $h_k \to 0$, where $(\lambda_0,\phi_0) = P_0$. Then, it is not hard to show that the sequence $\{Z_k\}$ will have a q.m. limit $Z = \frac{\partial}{\partial\phi}X(P_0)$ if and only if

$$\frac{\partial^2}{\partial\phi\partial\phi'} R(P,P')_{P=P'=P_0} \tag{3.4}$$

is well defined and finite. Then the quantity in (3.4) is equal to $E(\frac{\partial}{\partial\phi}X(P_0))^2$, and furthermore

$$E(\frac{\partial}{\partial\phi}X(P_0))X(Q) = \frac{\partial}{\partial\phi_0} R(P_0,Q) .$$

More generally, let H_R be the reproducing kernel Hilbert space with reproducing kernel R. Then each random variable of the form $Z = LX$ can be identified with the bounded linear functional L on H_R, and vice versa. The argument is as follows. If $Z = LX$ is a r.v. in X it can be shown that $EZX(Q) = L_{(P)}R(P,Q) = \eta(Q)$, say, where $L_{(P)}$ means the linear functional L applied to R considered as a function of P. However, by the properties of reproducing kernels, it can be shown that $\eta(\cdot)$ is the representer of L in R, that is $Lf = <\eta,f>_R$, where $<\cdot,\cdot>_R$ is the inner product in H_R. We are now ready to state the

Duality Theorem (KIMELDORF and WAHBA (1970)).

Let $X(P)$, $P \varepsilon S$ be a zero mean Gaussian stochastic process with covariance $bR(P,Q)$, and let H_R be the reproducing kernel Hilbert space with reproducing kernel H_R. Let

$$Y_i = L_i X + \varepsilon_i, \; i = 1,2,\ldots,n,$$

where L_iX, $i = 1,2,\ldots,n$ are n r.v.'s in X, and the $\varepsilon_1,\ldots,\varepsilon_n$ are independent, 0 mean Gaussian r.v.'s, independent of $X(P)$, $P\varepsilon S$, with common variance σ^2. Then the conditional expectation of $X(Q)$, given $Y_i = y_i$, $i = 1,2,\ldots,n$,

$$E\{X(Q)|Y_i = y_i,\ i = 1,2,\ldots,n\} \tag{3.5}$$

is given by $f_\lambda(Q)$, where $f_\lambda(\cdot)$ is the solution to the minimization problem: Find $f\varepsilon H_R$ to minimize

$$\frac{1}{n}\sum_{i=1}^{n}(L_if-y_i)^2 + \lambda||f||_R^2,$$

and $\lambda = \sigma^2/nb$.

Proof: See KIMELDORF and WAHBA (1970,1971), WAHBA (1978). However, the proof proceeds by direct calculation of $f_\lambda(Q)$ and by using the facts that $E(L_iX)X(Q) = \eta_i(Q)$, where $\langle\eta_i,f\rangle_R = L_if$.

Now let f be some atmospheric variable of interest. We will proceed as though the different realizations of f were sample functions from a zero mean Gaussian stochastic process with covariance $R(\cdot,\cdot)$. If repeated (independent!) observations on f were available, then various properties of R could be estimated from this data. We will discuss both "frequency domain" and "space domain" methods for doing this. Using the properties of reproducing kernel spaces (see, e.g. NASHED and WAHBA (1974)) it is not hard to show that if $J(f) = \sum_{\ell,s}\frac{f_{\ell s}^2}{\lambda_{\ell s}}$ is the norm on a reproducing kernel space H, then the reproducing kernel R for H is given by

$$R(P,Q) = \sum_{\ell,s}\lambda_{\ell s}Y_\ell^{\,s}(P)Y_\ell^{\,s}(Q). \tag{3.6}$$

To simplify the discussion, in this paper we are considering only R's whose eigenfunctions are the spherical harmonics. (Other eigenfunctions, i.e. those associated with Laplace's tidal equations, may well be reasonable in certain meteorological applications, see WAHBA (1981b)).

If a stochastic process $X(P)$, $P \varepsilon S$, has covariance

$$\sum_{\ell,s} \lambda_{\ell s} Y_\ell^s(P) Y_\ell^s(Q)$$

then X may be modelled as a random linear combination of the spherical harmonics (Karhumen-Loeve expansion)

$$X(P) = \sum_{\ell s} X_{\ell s} Y_\ell^s(P) \qquad (3.7)$$

where the $X_{\ell s}$ are random variables with

$$EX_{\ell s} X_{\ell' s'} = \lambda_{\ell s}, \ \ell s = \ell' s', = 0, \ \ell s \neq \ell' s'.$$

(To see this, compute $EX(P)X(Q)$ from (3.7) to obtain (3.6).) We have

$$X_{\ell s} = \int X(P) Y_\ell^s(P)$$

and

$$\lambda_{\ell s} = EX_{\ell s}^2 = E(\int X(P) Y_\ell^s(P) dP)^2.$$

If K independent observations, $f^1, \dots, f^K$ of a meteorological variable of interest are available, this suggests choosing $\{\lambda_{\ell s}\}$ based on estimates

$$\hat{\lambda}_{\ell s} = \frac{1}{K} \sum_{k=1}^{K} (f_{\ell s}^k)^2$$

where the sample Fourier-Bessel coefficients $f_{\ell s}^k$, $k = 1,2,\dots,K$ are given by

$$f_{\ell s}^k = \int f^k(P) Y_\ell^s(P) dP.$$

Figure 1 gives a plot of February 1974 monthly averages of some atmospheric mean square sample Fourier Bessel Coefficients collected by STANFORD (1979) from Channels 2 and 4 of the Radiometer on NIMBUS-5. The radiation received by Channels 2 and 4 respectively can be used (crudely) to infer the temperature $T(P)$, $P \varepsilon S$ in the upper and lower stratosphere, respectively below the satellite. By piecing together data

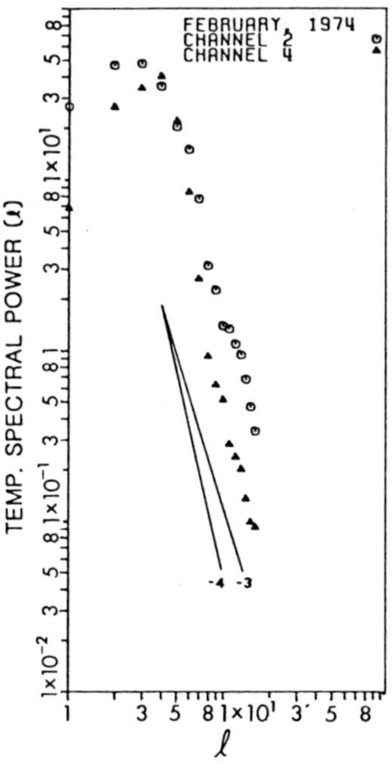

Figure 1: Temperature Spectral Power (ℓ).

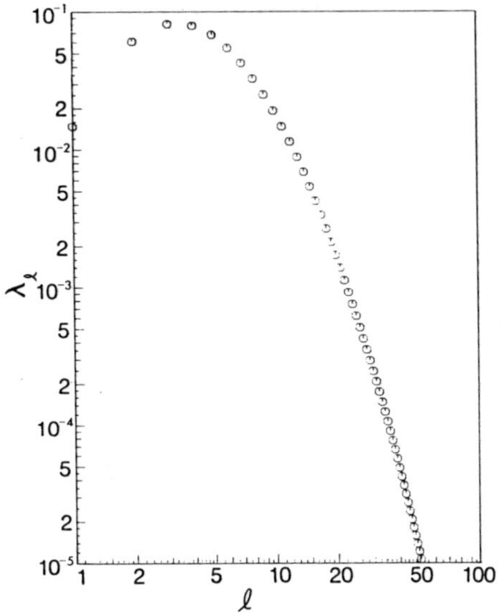

Figure 2: Idealized λ_ℓ.

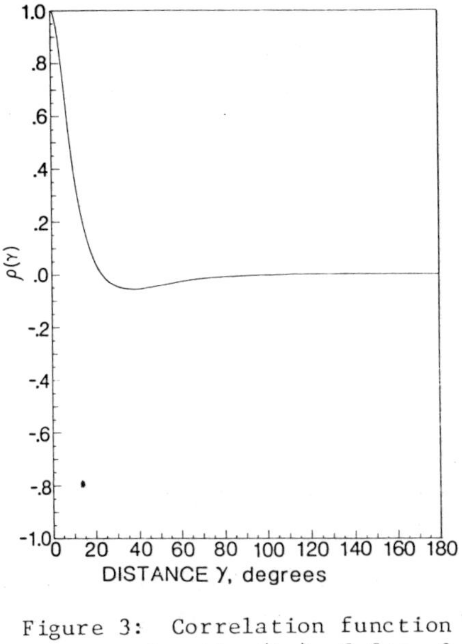

Figure 3: Correlation function for the (λ_ℓ) of fig. 2.

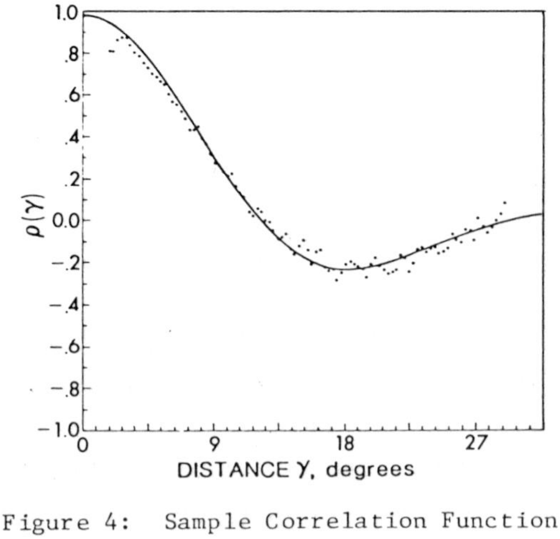

Figure 4: Sample Correlation Function.

from several orbits, (approximations to) $T_{\ell s}^{k} = \int T^{k}(P)Y_{\ell}^{s}(P)dP$ can be obtained. STANFORD has computed monthly mean square values $\bar{T}_{\ell s}^{2}$,

$$\bar{T}_{\ell s}^{2} = \frac{1}{K}\sum_{\ell=1}^{K}(T_{\ell s}^{k})^{2}.$$

What has actually been plotted in Figure 1 is the "TEMP SPECTRAL POWER" defined as

$$\text{TEMP SPECTRAL POWER } (\ell) = \frac{1}{5}\sum_{j=1}^{5}\bar{T}_{\ell+j,j}^{2} .$$

The energy spectrum in Temperature fields is related to the energy spectrum of other meteorological variables, i.e. wind and geopotential. We are not concerned here with the exact details of these pictures but rather that sequences $\{\lambda_{\ell s}\}$ can be fitted to this kind of data to provide meteorologically reasonable Hilbert Space norms. See KASSAHARA (1976) for some plots of sample Fourier-Bessel coefficients with respect to the eigenfunctions of Laplace's Tidal equations for wind and geopotential. Figure 2 gives a plot of an idealized sequence $\lambda_{\ell s} = \lambda_{\ell}$, $\ell = 1,2,\ldots$, where λ_{ℓ} was obtained by fitting (by an ad hoc procedure), a function of the form

$$\lambda_{\ell s} = |\sum_{j=0}^{2}\alpha_{j}[-\ell(\ell+1)]|^{-2}$$

to some of the data behind Figure 1. If $\lambda_{\ell s}$ does not depend on s, $\lambda_{\ell s} = \lambda_{\ell}$, then the covariance

$$R(P,Q) = \sum_{\ell,s}\lambda_{\ell s}Y_{\ell}^{s}(P)Y_{\ell}^{s}(Q)$$

reduces by the addition formula for spherical harmonics, to

$$R(P,Q) = \frac{1}{4\pi}\sum_{\ell=1}^{\infty}(2\ell+1)\lambda_{\ell}\, P_{\ell}(\cos\gamma(P,Q)),$$

where $\gamma(P,Q)$ is the angle between P and Q. Figure 3 gives the function $\rho(\gamma)$ defined by

$$\rho(\gamma) = \sum_{\ell=1}^{\infty}(2\ell+1)\lambda_{\ell}P_{\ell}(\cos\gamma)/\sum_{\ell=1}^{\infty}(2\ell+1)\lambda_{\ell}P_{\ell}(\cos 0)$$

which is associated with the $\{\lambda_\ell\}$ of Figure 2. Figure 4 gives an estimate for $\rho(\gamma)$ for $f(P)$ = the 500 millibar (geopotential) height obtained by JULIAN and THIEBAUX (1975) from sample covariances from data from a network of 51 North American weather stations for the winters of 1966 and 1967. In estimating $\rho(\gamma)$, an isotropic covariance function was assumed. The purpose of providing Figures 1 and 4 here is to convince the reader that historical collected or collectable meteorological data may be used to choose the norm on H, although the particular data sets exhibited here may or may not be the most appropriate. In the numerical experiments to be described we have taken the $\{\lambda_{\ell s}(1)\}$ and $\{\lambda_{\ell s}(2)\}$ both as in Figure 2.

4. Numerical Methods. The Generalized Cross-Validation Estimates of λ and δ.

Given λ,δ, $\{\lambda_{\ell s}(1), \lambda_{\ell s}(2)\}$ and the data $\{(U_i,V_i)\}$, an approximate minimizer (Ψ,Φ) of (2.8) can be obtained in the form

$$\Psi = \sum_{\ell=1}^{N} \sum_{s=-\ell}^{\ell} \alpha_{\ell s} Y_\ell^s \tag{4.1}$$

$$\Phi = \sum_{\ell=1}^{N} \sum_{s=-\ell} \beta_{\ell s} Y_\ell^s \tag{4.2}$$

where N is sufficiently large. For other numerical approaches to the minimization of (2.8) see WAHBA (1980,1981a), WENDELBERGER (1982). Let $\tilde{N} = \sum_{\ell=1}^{N} \sum_{s=-\ell}^{\ell} 1 = N^2-1$ and renumber the indices (ℓ,s), $s = -\ell,\ldots,\ell$, $\ell = 1,\ldots,N$, as $1,2,\ldots,\tilde{N}$. Let X_ϕ be the $n\times\tilde{N}$ matrix with $(i,\ell s)$th entry

$$\frac{1}{a}\frac{\partial}{\partial\phi}Y_\ell^s(P_i)$$

and X_λ be the $n\times\tilde{N}$ matrix with $(i,\ell s)$th entry

$$\frac{1}{a}\frac{1}{\cos\phi_i}\frac{\partial}{\partial\lambda}Y_\ell^s(P_i)$$

and let X be the $2n\times2\tilde{N}$ matrix

$$X = \begin{pmatrix} -X_\phi & X_\lambda \\ X_\lambda & X_\phi \end{pmatrix} \tag{4.3}$$

Let D_δ be the $2\tilde{N}\times 2\tilde{N}$ matrix

$$D_\delta = \begin{pmatrix} D_1 & 0 \\ 0 & \delta D_2 \end{pmatrix} \tag{4.4}$$

where D_i is the $\tilde{N}\times\tilde{N}$ diagonal matrix with $\ell s,\ell s$th entry $\lambda_{\ell s}(i)$, $i = 1,2$. Letting $z = (U_1,\ldots,U_n,V_1,\ldots,V_n)$, $\gamma = (\alpha_1,\ldots,\alpha_{\tilde{N}},\beta_1,\ldots,\beta_{\tilde{N}})$, it is seen by substituting (4.1) into (2.8) that we have to find γ which minimizes

$$\frac{1}{n}||z - X\gamma||^2 + \lambda\gamma' D_\delta^{-1}\gamma.$$

The minimizer is

$$\gamma = (X'X+n\lambda D_\delta^{-1})^{-1}X'z. \tag{4.5}$$

By the use of (2.3a) and (4.3), it follows that the estimated wind field $(U_{\lambda,\delta},V_{\lambda,\delta})$ at the data points satisfies

$$\begin{pmatrix} U_{\lambda,\delta}(P_1) \\ \vdots \\ U_{\lambda,\delta}(P_n) \\ V_{\lambda,\delta}(P_1) \\ \vdots \\ V_{\lambda,\delta}(P_n) \end{pmatrix} = A(\lambda)z \tag{4.6}$$

where $A(\lambda)$ is the $2n\times 2n$ "influence" matrix

$$A(\lambda) = X(X'X+n\lambda D_\delta^{-1})^{-1}X'$$

The generalized cross validation (GCV) estimate of (λ,δ) is the minimizer of the cross validation function $V(\lambda,\delta)$ defined by

$$V(\lambda,\delta) = \frac{\frac{1}{n}||(I-A(\lambda,\delta))z||^2}{\frac{1}{n}[\mathrm{Trace}(I-A(\lambda,\delta))]^2} \quad . \tag{4.7}$$

This method for estimating smoothing parameters in regularization problems was proposed in CRAVEN and WAHBA (1979), GOLUB, HEATH and WAHBA (1979) and WAHBA (1977b) and its numerical and theoretical properties have been studied in various places, see for example UTRERAS (1981). We only note here the useful property of the GCV estimate of λ and δ. Let the predictive mean square error $R(\lambda,\delta)$, when λ and δ are used be defined by

$$R(\lambda,\delta) = \frac{1}{n}\sum_{i=1}^{n}(U_{\lambda,\delta}(P_i)-U(P_i))^2$$

$$+\frac{1}{n}\sum_{i=1}^{n}(V_{\lambda,\delta}(P_i)-V(P_i))^2, \tag{4.8}$$

where $U(P_i)$, $V(P_i)$ is the true (but unknown) wind vector, and suppose the measurement errors ε_i^U and ε_i^V are independent identically distributed zero mean normally distributed random variables. Then under rather general conditions, for large n the minimizer $(\hat{\lambda},\hat{\delta})$ of $V(\lambda,\delta)$ provides a good estimate of the minimizer of $R(\lambda,\delta)$. V is not guaranteed to have a unique, or even a finite minimizer. Practical difficulties in minimizing V though possible appear to be moderately rare when the assumptions are reasonably well satisfied. Various diagnostic tools are available in troublesome cases and will be discussed elsewhere.

The numerical experiment reported in Section 4 was performed on the Amdahl at Goddard Space Flight Center, with $2n = 228$, $N = 15$, $2\tilde{N} = 448$. We outline the calculations used. Let $W_\delta = XD^{1/2}$, and let the singular value decomposition (SVD) of W_δ be

$$W_\delta = UD_WV' \tag{4.9}$$

where $UU' = U'U = I_{2n\times 2n} = V'V$ and D_W is a diagonal matrix with entries $b_1,\dots,b_{2n}$. $U,\{b_i\}$ and V' are computed using LINPACK. Letting

$$w = \begin{pmatrix} w_1 \\ \vdots \\ w_{2n} \end{pmatrix} = U'z,$$

then

$$V(\lambda,\delta) = \frac{\frac{1}{2n}\sum_{i=1}^{2n} z_i^2 - \frac{1}{2n}\sum_{\nu=1}^{2n} w_\nu^2 \left(\frac{b_\nu^2}{b_\nu^2+n\lambda}\right)^2}{\left(1-\frac{1}{2n}\sum_{\nu=1}^{2n}\frac{b_\nu^2}{b_\nu^2+n\lambda}\right)^2} \tag{4.10}$$

$$\gamma = D_\delta^{1/2} V \begin{pmatrix} \frac{b_1}{b_1^2+n\lambda} & & 0 \\ & \ddots & \\ 0 & & \frac{b_{2n}}{b_{2n}^2+n\lambda} \end{pmatrix} w. \tag{4.11}$$

For fixed $\delta, \hat{\lambda}(\delta)$, the minimizer of (4.10), is easily found by a global search in increments of $\log\lambda$. Then $V(\hat{\lambda}(\delta),\delta)$ was plotted for 8 values of δ chosen in powers of 1/6, and the minimum was readily evident. No doubt more efficient and automatic search procedures can be found.

For large n, N, and W_δ poorly conditioned, computing the SVD can be expensive, or it can fail to converge in a reasonable time. Some shortcut methods which alleviate this problem somewhat and use less storage have been developed. (BATES and WAHBA, (1982) in preparation.)

5. A realistic Monte Carlo test of the method

A number of techniques for estimating divergence of the upper atmosphere from radiosonde data have been proposed in the atmospheric sciences literature. For example, see SCHMIDT and JOHNSON (1972). In an attempt to determine how well the proposed method might work in practice a Monte Carlo experiment simulating realistic measured wind data from "model" stream functions and velocity potentials has been coded, and various experiments run. We describe one such experiment.

We obtained a model streamfunction and velocity potential of the form

$$\Psi = C_1 \sum_{\ell=1}^{N} \sum_{s=-\ell}^{\ell} a_{\ell s} Y_\ell^s$$

$$\Phi = C_2 \sum_{\ell=1} \sum_{s=-\ell}^{\ell} b_{\ell s} Y_\ell^s \tag{5.1}$$

by choosing $a_{\ell s}$ and $b_{\ell s}$ as normally distributed pseudo-random numbers with mean 0 and variances $\lambda_{\ell s}(1) = \lambda_{\ell s}(2) = \lambda_{\ell s}$ given in Figure 2. C_1 and C_2 were scale factors chosen so that the simulated $\zeta = \Delta\Psi$ and $D = \Delta\Phi$ had magnitudes typical of real atmospheres.
$(\int \zeta^2 dP)^{1/2} = 6 \times 10^{-5}$/sec., $(\int D^2 dP)^{1/2} = 1 \times 10^{-5}$/sec. Model wind vectors $(U(P_i), V(P_i))$ were computed from the model (Ψ, Φ) of (5.1) for $\{P_i\}$ corresponding to n = 114 North American radiosonde stations. The data $z = (U_1, \ldots, U_n, V_1, \ldots, V_n)$, where $U_i = U(P_i) + \varepsilon_i^U$, $V_i = V(P_i) + \varepsilon_i^V$, were constructed by adding the measurement errors ε_i^U, ε_i^V as normally distributed pseudo random numbers with mean 0 and standard deviation $\sigma = 2.5$ meters/sec., a realistic value for the measurement error standard deviation. Since the ability to estimate divergence will depend on the signal to noise ratio, it is necessary that the values of "signal" and "noise" be chosen realistically. The results reported here can be expected to be rosier than that obtainable in practice, however, primarily to the extent that wave numbers $\ell > N$ occur in practice but are not simulated here, and (secondarily) because in practice $J^{(1)}$ and $J^{(2)}$ cannot be so precisely matched to the "truth" as they are in this experiment.

Figure 5 shows the simulated wind vectors. Figure 6 shows the estimate of the true wind field, plotted on a 5° × 5° grid in latitude and longitude. Figures 7 and 8 show the model and estimated vorticity and divergence, respectively. Figure 9 shows $V(\hat{\lambda}(\delta),\delta)$ and $R(\hat{\lambda}(\delta),\delta)$, (of 4.8) plotted as a function of δ. In Figures 6 - 8 $\hat{\delta} = 1/36$ was used. It can be seen that the minimizer of $V(\hat{\lambda}(\delta),\delta)$ was a good estimate of the minimizer of $R(\hat{\lambda}(\delta),\delta)$. Figure 10 gives $MSE(\zeta_{\hat{\lambda}(\delta),\delta})$ and $MSE(D_{\hat{\lambda}(\delta),\delta})$ and their sum, where

$$MSE(\zeta_{\lambda,\delta}) = \frac{1}{K} \sum_{k=1}^{K} (\zeta_{\lambda,\delta}(P_k) - \zeta(P_k))^2$$

$$MSE(D_{\lambda,\delta}) = \frac{1}{K} \sum_{k=1}^{K} (D_{\lambda,\delta}(P_k) - D(P_k))^2.$$

The $\{P_k\}$ constitute a regular grid inside the U.S. It can be seen from Figure 10 that if δ is taken as too small (i.e. divergence is suppressed), then the mean square error in the estimated vorticity becomes large.

Figure 5: Simulated Wind Data.

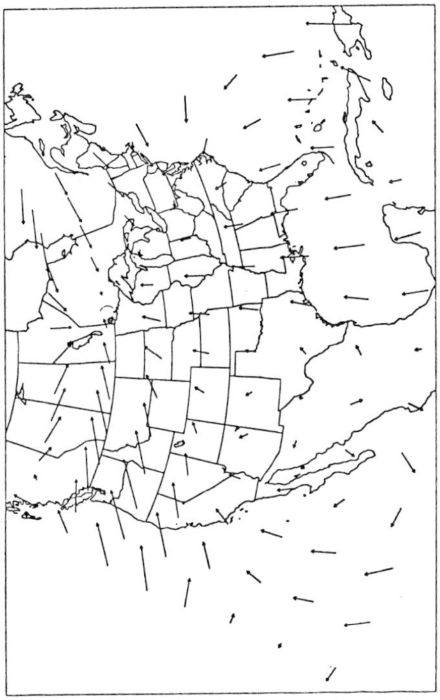

Figure 6: Estimated Wind Field.

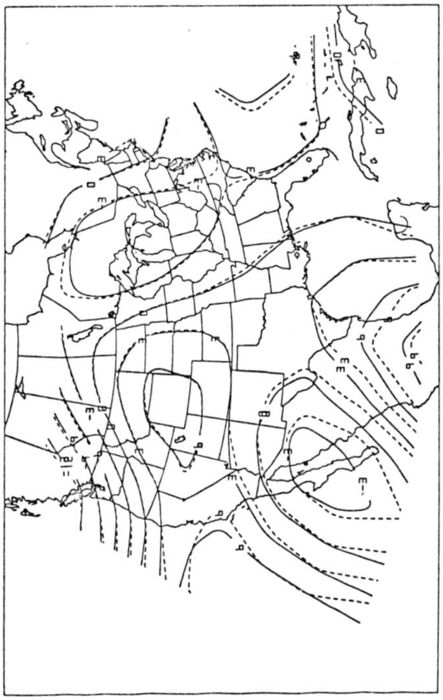

Figure 7: Model and Estimated Vorticity, X 10^{-5}/sec.

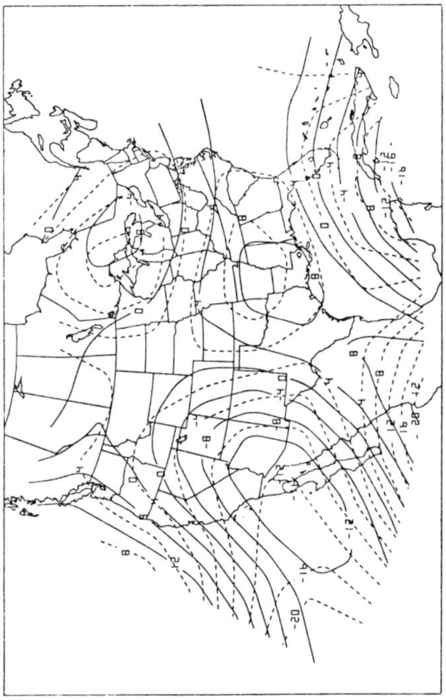

Figure 8: Model and Estimated Divergence, X 10^{-6}/sec.

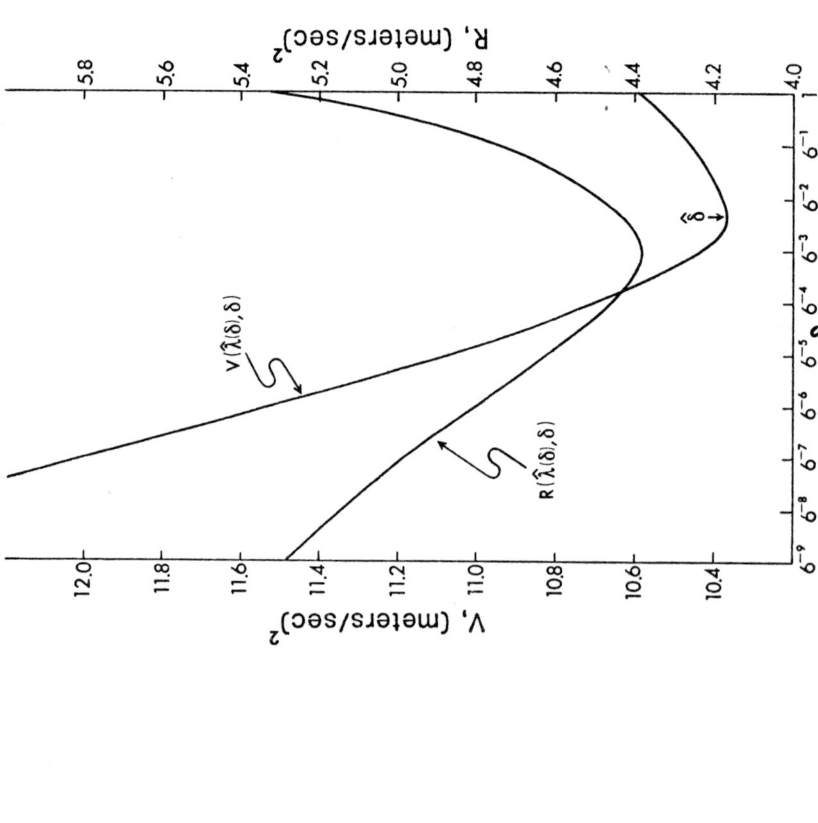

Figure 9: The cross validation function V and the predictive mean square error R of the wind field.

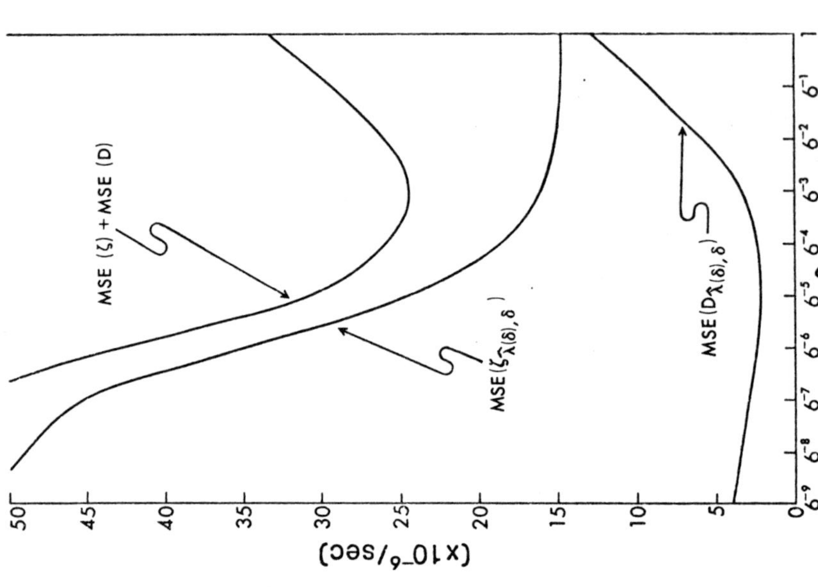

Figure 10: Mean square error in the estimated vorticity and divergence.

An estimate $\hat{\sigma}^2$ for the variance of the measurement error is available as

$$\hat{\sigma}^2(\hat{\lambda},\hat{\delta}) = \frac{||(I-A(\hat{\lambda},\hat{\delta})||^2}{Tr(I-A(\hat{\lambda},\hat{\delta}))}$$

since the numerator is the residual sum of squares and the denominator is the equivalent degrees of freedom for error. In this example $\hat{\sigma}$ was 2.58m/sec., very close to the "true" value of 2.5 meters/sec. In those occasional sticky cases encountered in practice where $V(\lambda,\delta)$ has multiple minima, if the order of magnitude of σ is known apriori, the examination of $\hat{\sigma}$ can usually be used to resolve ambiguity. See WAHBA (1981d), WENDELBERGER (1982). Bayesian confidence intervals are also available for these estimates, see Wahba (1981d).

We have concluded that this approach has much promise for applications.

6. References

Bates, D., and Wahba, G. (1982) in preparation.

Cramer, H., and Leadbetter (1967). Stationary and related stochastic processes, Chapter V. Wiley, New York.

Craven, P. and Wahba, G. (1979). Smoothing noisy data with spline functions: estimating the correct degree of smoothing by the method of generalized cross-validation. Numer. Math., 31, 377.

Freeden, W. (1981a). On spherical spline interpolation and approximation, Math. Meth. in The Appl. Sci. 3, 551-575.

Freeden, W. (1981b). On approximation by harmonic splines. Manuscripta Geodaetica, 6, 193-244.

Golub, G., Heath, M. and Wahba, G. (1979), Generalized cross-validation as a method for choosing a good ridge parameter, Technometrics 21, 215-223.

Julian, P.R., and Thiebaux, M. Jean (1975), On some properties of correlation functions used in optimum interpolation schemes, Monthly Weather Review, 103, 7, pp. 605-616.

Kassahara, A. (1976). Normal modes of ultra-long waves in the atmosphere. Monthly Weather Review 104, 6, 669-690.

Kimeldorf, G., and Wahba, G. (1970), A correspondence between Bayesian estimation of stochastic processes and smoothing by splines, Ann. Math. Statist., 41, 2.

Kimeldorf, G., and Wahba, G. (1971), Some results on Tchebycheffian spline functions, J. Math. Anal. and Applic., 33, 1.

Micchelli, C., and Wahba, G. (1981), Design problems for optimal surface interpolation in "Approximation Theory and Applications: Z. Ziegler, ed., Academic Press.

Nashed, M.Z. and Wahba, G. (1974), Generalized inverses in reproducing kernel spaces: an approach to regularization of linear operator equations, SIAM J. Math. Analysis, 5, 6.

Parzen, E. (1961), An approach to time series analysis. Ann. Math. Statist. 32, 951-989.

Schmidt, P.J., and Johnson, D.R. (1972), Use of approximating polynomials to estimate profiles of wind, divergence, and vertical motion. Monthly Weather Review, 100, 5, 249-353.

Shure, L., Parker, R.L., and Backus, G.E. (1981), Harmonic splines for geomagnetic modelling, to appear, PEPI.

Stanford, J. (1979), Latitudinal-Wavenumber power spectra of stratospheric temperature fluctuations, J. Atmospheric Sciences, 36, 5, pp. 921-931.

Utreras, F. (1981), Optimal smoothing of noisy data using spline functions, SIAM J. Sci. Stat. Comput. 2, 3, 349-362.

Wahba, G. (1977a) in invited discussion to Consistent nonparametric regression, C.J. Stone, Ann. Stat., 5, 4, 637-645.

Wahba, G. (1977b), Practical approximate solutions to linear operator equations when the data are noisy, SIAM J. Numerical Analysis, 14, 4.

Wahba, G. (1978), Improper priors, spline smoothing and the problem of guarding against model errors in regression, J. Roy. Stat. Soc. Ser. B., 40, 3.

Wahba, G. (1980), Spline bases, regularization, and generalized cross validation for solving approximation problems with large quantities of noisy data. Proceedings of the International Conference on Approximation Theory in Honor or George Lorenz, Jan. 8-11, 1980, Austin, Texas, Ward Cheney, ed. Academic Press (1980).

Wahba, G. (1981a), Spline interpolation and smoothing on the sphere, SIAM J. Scientific and Statistical Computing, 2, 1.

Wahba, G. (1981b), Some new techniques for variational objective analysis on the sphere using splines, Hough functions, and sample spectral data. Preprints of the Seventh Conference on Probability and Statistics in the Atmospheric Sciences, American Meteorological Society.

Wahba, G. (1981c), Data-based optimal smoothing of orthogonal series density estimates. Ann. Statist., 9, 1, 146-156.

Wahba, G. (1981d), Bayesian confidence intervals for the cross validated smoothing spline. University of Wisconsin-Madison Statistics Department Technical Report No. 645, Submitted.

Wahba, G., and Wendelberger, J. (1980), Some new mathematical methods for variational objective analysis using splines and cross-validation, Monthly Weather Review 108, 36-57.

Wendelberger, J. (1982), Ph.D. thesis, in preparation.

We thank Christopher Sheridan, who wrote the computer program, and Donald R. Johnson for many helpful discussions.

Professor Grace Wahba, Department of Statistics, University of Wisconsin-Madison, Madison, Wisconsin, 53706, USA.